TIM R.S.

—MRS.

GRID CONVERTERS FOR PHOTOVOLTAIC AND WIND POWER SYSTEMS

Dedicated to our families

GRID CONVERTERS FOR PHOTOVOLTAIC AND WIND POWER SYSTEMS

Remus Teodorescu
Aalborg University, Denmark

Marco Liserre
Politecnico di Bari, Italy

Pedro Rodríguez
Technical University of Catalonia, Spain

WILEY

A John Wiley and Sons, Ltd., Publication

This edition first published 2011
© 2011, John Wiley & Sons, Ltd

Registered office
John Wiley & Sons, Ltd, The Atrium, Southern Gate, Chichester, West Sussex, PO19 8SQ, United Kingdom

For details of our global editorial offices, for customer services and for information about how to apply for permission to reuse the copyright material in this book please see our website at www.wiley.com.

The right of the author to be identified as the author of this work has been asserted in accordance with the Copyright, Designs and Patents Act 1988.

All rights reserved. No part of this publication may be reproduced, stored in a retrieval system, or transmitted, in any form or by any means, electronic, mechanical, photocopying, recording or otherwise, except as permitted by the UK Copyright, Designs and Patents Act 1988, without the prior permission of the publisher.

Wiley also publishes its books in a variety of electronic formats. Some content that appears in print may not be available in electronic books.

Designations used by companies to distinguish their products are often claimed as trademarks. All brand names and product names used in this book are trade names, service marks, trademarks or registered trademarks of their respective owners. The publisher is not associated with any product or vendor mentioned in this book. This publication is designed to provide accurate and authoritative information in regard to the subject matter covered. It is sold on the understanding that the publisher is not engaged in rendering professional services. If professional advice or other expert assistance is required, the services of a competent professional should be sought.

MATLAB® is a trademark of The MathWorks, Inc., and is used with permission. The MathWorks does not warrant the accuracy of the text or exercises in this book. This book's use or discussion of MATLAB® software or related products does not constitute endorsement or sponsorship by The MathWorks of a particular pedagogical approach or particular use of MATLAB® software.

Library of Congress Cataloguing-in-Publication Data

Teodorescu, Remus.
　Grid converters for photovoltaic and wind power systems / Remus Teodorescu, Marco Liserre, Pedro Rodríguez.
　　p. cm.
　Includes bibliographical references and index.
　ISBN 978-0-470-05751-3 (hardback)
1. Electric current converters.　2. Photovoltaic power systems–Equipment and supplies.　3. Wind energy conversion systems–Equipment and supplies.　I. Liserre, Marco.　II. Rodríguez, Pedro.　III. Title.
　TK7872.C8T46 2011
　621.31′244–dc22

2010031106

A catalogue record for this book is available from the British Library.

Print ISBN: 9780470057513
E-PDF ISBN: 9780470667040
O-book ISBN: 9780470667057

Set in in 10/12pt Times by Aptara Inc., New Delhi, India
Printed and bound in Singapore by Markono Print Media Pte Ltd.

2 2011

Contents

About the Authors		xiii
Preface		xv
Acknowledgements		xvii
1	**Introduction**	1
1.1	Wind Power Development	1
1.2	Photovoltaic Power Development	3
1.3	The Grid Converter – The Key Element in Grid Integration of WT and PV Systems	4
	References	4
2	**Photovoltaic Inverter Structures**	5
2.1	Introduction	5
2.2	Inverter Structures Derived from H-Bridge Topology	6
	2.2.1 *Basic Full-Bridge Inverter*	7
	2.2.2 *H5 Inverter (SMA)*	11
	2.2.3 *HERIC Inverter (Sunways)*	13
	2.2.4 *REFU Inverter*	15
	2.2.5 *Full-Bridge Inverter with DC Bypass – FB-DCBP (Ingeteam)*	17
	2.2.6 *Full-Bridge Zero Voltage Rectifier – FB-ZVR*	19
	2.2.7 *Summary of H-Bridge Derived Topologies*	21
2.3	Inverter Structures Derived from NPC Topology	21
	2.3.1 *Neutral Point Clamped (NPC) Half-Bridge Inverter*	21
	2.3.2 *Conergy NPC Inverter*	23
	2.3.3 *Summary of NPC-Derived Inverter Topologies*	25
2.4	Typical PV Inverter Structures	25
	2.4.1 *H-Bridge Based Boosting PV Inverter with High-Frequency Transformer*	25
2.5	Three-Phase PV Inverters	26
2.6	Control Structures	27
2.7	Conclusions and Future Trends	28
	References	29

3	**Grid Requirements for PV**		**31**
3.1	Introduction		31
3.2	International Regulations		32
	3.2.1	IEEE 1547 Interconnection of Distributed Generation	32
	3.2.2	IEC 61727 Characteristics of Utility Interface	33
	3.2.3	VDE 0126-1-1 Safety	33
	3.2.4	IEC 61000 Electromagnetic Compatibility (EMC – low frequency)	34
	3.2.5	EN 50160 Public Distribution Voltage Quality	34
3.3	Response to Abnormal Grid Conditions		35
	3.3.1	Voltage Deviations	35
	3.3.2	Frequency Deviations	36
	3.3.3	Reconnection after Trip	36
3.4	Power Quality		37
	3.4.1	DC Current Injection	37
	3.4.2	Current Harmonics	37
	3.4.3	Average Power Factor	38
3.5	Anti-islanding Requirements		38
	3.5.1	AI Defined by IEEE 1547/UL 1741	39
	3.5.2	AI Defined by IEC 62116	40
	3.5.3	AI Defined by VDE 0126-1-1	40
3.6	Summary		41
	References		41
4	**Grid Synchronization in Single-Phase Power Converters**		**43**
4.1	Introduction		43
4.2	Grid Synchronization Techniques for Single-Phase Systems		44
	4.2.1	Grid Synchronization Using the Fourier Analysis	45
	4.2.2	Grid Synchronization Using a Phase-Locked Loop	51
4.3	Phase Detection Based on In-Quadrature Signals		58
4.4	Some PLLs Based on In-Quadrature Signal Generation		63
	4.4.1	PLL Based on a T/4 Transport Delay	63
	4.4.2	PLL Based on the Hilbert Transform	64
	4.4.3	PLL Based on the Inverse Park Transform	65
4.5	Some PLLs Based on Adaptive Filtering		68
	4.5.1	The Enhanced PLL	70
	4.5.2	Second-Order Adaptive Filter	72
	4.5.3	Second-Order Generalized Integrator	74
	4.5.4	The SOGI-PLL	78
4.6	The SOGI Frequency-Locked Loop		80
	4.6.1	Analysis of the SOGI-FLL	82
4.7	Summary		89
	References		89
5	**Islanding Detection**		**93**
5.1	Introduction		93
5.2	Nondetection Zone		94

5.3	Overview of Islanding Detection Methods		96
5.4	Passive Islanding Detection Methods		98
	5.4.1	*OUF–OUV Detection*	98
	5.4.2	*Phase Jump Detection (PJD)*	99
	5.4.3	*Harmonic Detection (HD)*	99
	5.4.4	*Passive Method Evaluation*	103
5.5	Active Islanding Detection Methods		104
	5.5.1	*Frequency Drift Methods*	104
	5.5.2	*Voltage Drift Methods*	110
	5.5.3	*Grid Impedance Estimation*	110
	5.5.4	*PLL-Based Islanding Detention*	114
	5.5.5	*Comparison of Active Islanding Detection Methods*	119
5.6	Summary		121
	References		121
6	**Grid Converter Structures for Wind Turbine Systems**		**123**
6.1	Introduction		123
6.2	WTS Power Configurations		124
6.3	Grid Power Converter Topologies		128
	6.3.1	*Single-Cell (VSC or CSC)*	128
	6.3.2	*Multicell (Interleaved or Cascaded)*	133
6.4	WTS Control		135
	6.4.1	*Generator-Side Control*	136
	6.4.2	*WTS Grid Control*	139
6.5	Summary		142
	References		142
7	**Grid Requirements for WT Systems**		**145**
7.1	Introduction		145
7.2	Grid Code Evolution		146
	7.2.1	*Denmark*	148
	7.2.2	*Germany*	148
	7.2.3	*Spain*	149
	7.2.4	*UK*	149
	7.2.5	*Ireland*	150
	7.2.6	*US*	150
	7.2.7	*China*	150
	7.2.8	*Summary*	151
7.3	Frequency and Voltage Deviation under Normal Operation		151
7.4	Active Power Control in Normal Operation		152
	7.4.1	*Power Curtailment*	153
	7.4.2	*Frequency Control*	154
7.5	Reactive Power Control in Normal Operation		155
	7.5.1	*Germany*	155
	7.5.2	*Spain*	157
	7.5.3	*Denmark*	157

	7.5.4	UK	157
	7.5.5	Ireland	158
	7.5.6	US	158
7.6	Behaviour under Grid Disturbances		158
	7.6.1	Germany	158
	7.6.2	Spain	160
	7.6.3	US-WECC	164
7.7	Discussion of Harmonization of Grid Codes		164
7.8	Future Trends		165
	7.8.1	Local Voltage Control	165
	7.8.2	Inertia Emulation (IE)	165
	7.8.3	Power Oscillation Dumping (POD)	166
7.9	Summary		166
	References		167

8	**Grid Synchronization in Three-Phase Power Converters**		**169**
8.1	Introduction		169
8.2	The Three-Phase Voltage Vector under Grid Faults		171
	8.2.1	Unbalanced Grid Voltages during a Grid Fault	175
	8.2.2	Transient Grid Faults, the Voltage Sags (Dips)	177
	8.2.3	Propagation of Voltage Sags	179
8.3	The Synchronous Reference Frame PLL under Unbalanced and Distorted Grid Conditions		182
8.4	The Decoupled Double Synchronous Reference Frame PLL (DDSRF-PLL)		186
	8.4.1	The Double Synchronous Reference Frame	186
	8.4.2	The Decoupling Network	187
	8.4.3	Analysis of the DDSRF	189
	8.4.4	Structure and Response of the DDSRF-PLL	192
8.5	The Double Second-Order Generalized Integrator FLL (DSOGI-FLL)		194
	8.5.1	Structure of the DSOGI	197
	8.5.2	Relationship between the DSOGI and the DDSRF	198
	8.5.3	The FLL for the DSOGI	200
	8.5.4	Response of the DSOGI-FLL	200
8.6	Summary		201
	References		203

9	**Grid Converter Control for WTS**		**205**
9.1	Introduction		205
9.2	Model of the Converter		206
	9.2.1	Mathematical Model of the L-Filter Inverter	207
	9.2.2	Mathematical Model of the LCL-Filter Inverter	209
9.3	AC Voltage and DC Voltage Control		210
	9.3.1	Management of the DC Link Voltage	211
	9.3.2	Cascaded Control of the DC Voltage through the AC Current	213

	9.3.3	Tuning Procedure of the PI Controller	216
	9.3.4	PI-Based Voltage Control Design Example	217
9.4	Voltage Oriented Control and Direct Power Control		219
	9.4.1	Synchronous Frame VOC: PQ Open-Loop Control	221
	9.4.2	Synchronous Frame VOC: PQ Closed-Loop Control	222
	9.4.3	Stationary Frame VOC: PQ Open-Loop Control	222
	9.4.4	Stationary Frame VOC: PQ Closed-Loop Control	224
	9.4.5	Virtual-Flux-Based Control	225
	9.4.6	Direct Power Control	226
9.5	Stand-alone, Micro-grid, Droop Control and Grid Supporting		228
	9.5.1	Grid-Connected/Stand-Alone Operation without Load Sharing	229
	9.5.2	Micro-Grid Operation with Controlled Storage	229
	9.5.3	Droop Control	231
9.6	Summary		234
	References		235
10	**Control of Grid Converters under Grid Faults**		**237**
10.1	Introduction		237
10.2	Overview of Control Techniques for Grid-Connected Converters under Unbalanced Grid Voltage Conditions		238
10.3	Control Structures for Unbalanced Current Injection		244
	10.3.1	Decoupled Double Synchronous Reference Frame Current Controllers for Unbalanced Current Injection	245
	10.3.2	Resonant Controllers for Unbalanced Current Injection	251
10.4	Power Control under Unbalanced Grid Conditions		256
	10.4.1	Instantaneous Active–Reactive Control (IARC)	258
	10.4.2	Positive- and Negative-Sequence Control (PNSC)	260
	10.4.3	Average Active–Reactive Control (AARC)	262
	10.4.4	Balanced Positive-Sequence Control (BPSC)	263
	10.4.5	Performance of the IARC, PNSC, AARC and BPSC Strategies	264
	10.4.6	Flexible Positive- and Negative-Sequence Control (FPNSC)	267
10.5	Flexible Power Control with Current Limitation		269
	10.5.1	Locus of the Current Vector under Unbalanced Grid Conditions	270
	10.5.2	Instantaneous Value of the Three-Phase Currents	272
	10.5.3	Estimation of the Maximum Current in Each Phase	274
	10.5.4	Estimation of the Maximum Active and Reactive Power Set-Point	277
	10.5.5	Performance of the FPNSC	279
10.6	Summary		285
	References		285
11	**Grid Filter Design**		**289**
11.1	Introduction		289
11.2	Filter Topologies		290
11.3	Design Considerations		291
11.4	Practical Examples of LCL Filters and Grid Interactions		296
11.5	Resonance Problem and Damping Solutions		300

		11.5.1 Instability of the Undamped Current Control Loop	300
		11.5.2 Passive Damping of the Current Control Loop	302
		11.5.3 Active Damping of the Current Control Loop	304
11.6		Nonlinear Behaviour of the Filter	306
11.7		Summary	311
		References	311
12		**Grid Current Control**	**313**
12.1		Introduction	313
12.2		Current Harmonic Requirements	313
12.3		Linear Current Control with Separated Modulation	315
		12.3.1 Use of Averaging	315
		12.3.2 PI-Based Control	317
		12.3.3 Deadbeat Control	320
		12.3.4 Resonant Control	326
		12.3.5 Harmonic Compensation	329
12.4		Modulation Techniques	335
		12.4.1 Single-Phase	338
		12.4.2 Three-Phase	340
		12.4.3 Multilevel Modulations	343
		12.4.4 Interleaved Modulation	347
12.5		Operating Limits of the Current-Controlled Converter	347
12.6		Practical Example	350
12.7		Summary	353
		References	353
Appendix A		**Space Vector Transformations of Three-Phase Systems**	**355**
A.1		Introduction	355
A.2		Symmetrical Components in the Frequency Domain	355
A.3		Symmetrical Components in the Time Domain	357
A.4		Components $\alpha\beta 0$ on the Stationary Reference Frame	359
A.5		Components $dq0$ on the Synchronous Reference Frame	361
		References	362
Appendix B		**Instantaneous Power Theories**	**363**
B.1		Introduction	363
B.2		Origin of Power Definitions at the Time Domain for Single-Phase Systems	365
B.3		Origin of Active Currents in Multiphase Systems	366
B.4		Instantaneous Calculation of Power Currents in Multiphase Systems	369
B.5		The p-q Theory	371
B.6		Generalization of the p-q Theory to Arbitrary Multiphase Systems	373
B.7		The Modified p-q Theory	374
B.8		Generalized Instantaneous Reactive Power Theory for Three-Phase Power Systems	376

B.9	Summary	377
	References	378

Appendix C Resonant Controller — **381**

C.1	Introduction	381
C.2	Internal Model Principle	381
C.3	Equivalence of the PI Controller in the dq Frame and the P+Resonant Controller in the $\alpha\beta$ Frame	382

Index — **385**

About the Authors

Remus Teodorescu received a Dipl.Ing. degree from Polytechnical University of Bucharest in 1989, and a PhD degree in power electronics from University of Galati in 1994. In 1998 he joined Aalborg University, Department of Energy, Power Electronics Section, where he currently works as a full professor.

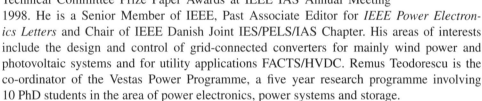

He has more than 150 papers published in IEEE conferences and transactions, one book and four patents (pending). He is the co-recipient of the Technical Committee Prize Paper Awards at IEEE IAS Annual Meeting 1998. He is a Senior Member of IEEE, Past Associate Editor for *IEEE Power Electronics Letters* and Chair of IEEE Danish Joint IES/PELS/IAS Chapter. His areas of interests include the design and control of grid-connected converters for mainly wind power and photovoltaic systems and for utility applications FACTS/HVDC. Remus Teodorescu is the co-ordinator of the Vestas Power Programme, a five year research programme involving 10 PhD students in the area of power electronics, power systems and storage.

Marco Liserre is engaged in teaching courses of power electronics, industrial electronics and electrical machines. His current research interests include industrial electronics applications to distributed power generation systems based on renewable energies.

He has been a visiting Professor at Aalborg University, Denmark, Alcala de Henares University, Spain, and at Christian-Albrechts University of Kiel, Germany. He has given lectures in different universities and tutorials for many conferences. Dr Liserre is an Associate Editor of the *IEEE Transactions on Industrial Electronics* and *IEEE Transactions on Sustainable Energy*. He is the Founder and the Editor-in-Chief of the *IEEE Industrial Electronics Magazine*, 2007–2009. He is the Founder and the Chairman of the Technical Committee on Renewable Energy Systems of the IEEE Industrial Electronics Society. He received the IES 2009 Early Career Award and is IEEE-IES Vice-President for publications. Dr Liserre was the Co-Chairman of the International Symposium on Industrial Electronics (ISIE 2010), held in Bari in 2010.

Pedro Rodríguez received his M.Sc. and Ph.D. degrees in electrical engineering from the Technical University of Catalonia (UPC-BarcelonaTech), Barcelona, Spain. In 1990 he joined the faculty of UPC as an Assistant Professor, where he is currently an Associate Professor in the Department of Electrical Engineering and the Head of the Research Center on Renewable Electrical Energy Systems (SEER). He was a Visiting Researcher at the Center for Power Electronics Systems, Virginia Polytechnic Institute and State University, (Blacksburg, USA). He stayed as a Postdoctoral Researcher at  the Department of Energy Technology (DET), Aalborg University (Denmark). Currently, he is a habitual visiting professor at the DET where he lectures Ph.D. courses and participates as a co-supervisor in the Vestas Power Program at the AAU.

Pedro Rodríguez has coauthored about 100 papers in technical journals and conference proceedings. He is the holder of four licensed patents on wind and photovoltaic systems. He has lectured courses on power converters applied to power systems in many universities around the world. His research activity lies in the field of electronic power processors applied to electrical distributed generation systems, being mainly focused on designing controllers for the grid interactive power processors, designing power electronics based power processors for green energy sources and proposing new technical solutions to improve stability and power quality in electrical networks. Dr. Rodriguez is a Senior Member of the IEEE, an Associate Editor of the *IEEE Transactions on Power Electronics*, and the Chair of the IEEE-IES Student and GOLD Members Activities Committee.

Preface

The penetration of wind and PV generated electrical energy into the grid system worldwide is increasing exponentially. A limiting factor is the more and more stringent grid requirements imposed by different grid operators aiming in maintain grid stability. Both wind power plants and PV power plants are connected to the grid through grid converters who besides transferring the generated dc power to the ac grid should now be able to exhibit advanced functions like: dynamic control of active and reactive power, stationary operation within a range of voltage and frequency, voltage ride-through, reactive current injection during faults, participation in grid balancing act like primary frequency control, etc. Therefore the aim of this book is to explain the topologies, modulation and control of grid converters for both PV and WP applications. In addition to the classical handbooks in power electronics this book is showing the PV or WT specific control functions according to the recent grid codes and is enhancing the classical synchronization and current control strategies with the general case when the grid is unbalanced.

The idea of this book originated in an biannual Industrial/PhD course "Power Electronics for renewable energy Systems" started in May 1995 in Aalborg University, Institute for Energy Technology and is successfully continued with over 250 PhDs or industry engineers attendees (by the end of 2010). The success of this course was due to the practical aspects involved as more than 40% of the time was spent in the lab for designing and testing control strategies on real grid converters. Thus, the initiative of writing this book together with Marco Liserre from Politecnico di Bari and Pedro Rodriguez from UPC Barcelona has been taken in order to ensure a unique reference for the course.

The book is intended as a textbook for graduating students with an electrical engineering background wanting to move in the electrical aspects of PV and WT power regenerations as well as for professionals in the PV or WT industry.

Chapter 1 is giving an overview of the latest developments in the PV and WT penetrations in the worldwide power systems as well as the forecast until 2014 which looks very promising despite the economical crisis 2008–2010. Chapter 2 discusses the various high efficiency topologies for PV inverters as well as some generic control structures. In Chapter 3 the grid requirements for PV installations are described. Chapter 4 gives a deep analysis of the basic PLL as the preferred tool for synchronization in single phase systems and discusses different quadrature signal generator methods. Chapter 5 introduces the reader to the problems related to undesired islands that may occur in power systems characterized by an high penetration of distributed power generation, and it present an overview of islading detection methods. Chapter 6 describes the most typical WT grid converters topologies together with generic control structures. The most recent grid requirements for WT grid connection, the so called Grid Codes are explained in Chapter 7. Chapter 8 extrapolate the knowledge of single-phase PLL structure for three phase systems. New robust synchronization structures are proposed in order to cope with the unbalance grid or frequency adaptation. In Chapter 9 the most used grid converter controls structures for WT are explained while Chapter 10 extrapolate the control issue for the case of grid faults where new control structures are proposed. In Chapter 11 the issue of designing grid interface filters is discussed along with methods to passively or actively damp the possible resonance. Finally Chapter 12 goes down to the modulation techniques for grid converters and to the methods for controlling the grid current and it briefly addresses the control of dc and ac voltages. The new resonant controllers are introduced and compared with the classical PI and dead-beat control is proposed as the fastest linear controller. Appendix 1 is familiarizing the reader with the issue of different coordinates transformation in three phase systems, Appendix 2 is giving the basic principle of Instantaneous Power Theory and Appendix 3 describes the concept of resonance controllers.

Acknowledgements

We would like to acknowledge the valuable support from the following persons:

- Frede Blaabjerg, Professor Aalborg University for his general support of this project.
- Philip C Kjær, chief specialist, Vestas Wind Systems A/S for his valuable advice and knowledge exchange in the generic field of wind turbines grid integration from a pure industrial perspective.
- Uffe Borup, technology manager at Danfoss Solar Inverter A/S for valuable support in the topic of photovoltaic inverters technology and control.
- Antonio Dell'Aquila, professor at Politecnico di Bari, for having continuously encouraged synergy between Politecnico di Bari and Aalborg University, which has given a fruitful basis to this project.
- Lars Helle, specialist, power electronics, Vestas Wind Systems A/S for his support in the Chapter 7.
- Tamas Kerekes, assistant professor Aalborg University for scrutinizing and commenting Chapter 2.
- Alvaro Luna, assistant professor at UPC Barcelona for his valuable contribution to Chapters 4, 8 and 10.
- Rosa Mastromauro, assistant professor at Politecnico di Bari, Italy for his valuable contribution on Chapter 11 and 12.
- Alberto Pigazo, assistant professor at Cantabria University, Spain, for his valuable contribution on Chapter 5.
- Francesco De Mango, GSE, for his valuable contribution on Chapter 5.
- Manuel Reyes Diaz, PhD student at University of Seville for his valuable contribution to Chapter 10.
- Ancuta G. Dragomir, MSc student at Aalborg University, for editing most of the figures.
- Nicky Skinner (post mortem), Project Editor, John Wiley & Sons Ltd for her patience and trust in this long project.

1

Introduction

In the last few years renewable energies have experienced one of the largest growth areas in percentage of over 30 % per year, compared with the growth of coal and lignite energy.

The goal of the European Community (the EU) is to reach 20 % in 2020, but the EU-27 energy is only 17 % of world energy. The US, with 22 % of energy share, has adopted similar goals under the pressure of public opinion concerned by environmental problems and in order to overcome the economical crisis. However, the policies of Asia and Pacific countries, with 35 % of energy share, will probably be more important in the future energy scenario. In fact, countries like China and India require continuously more energy (China energy share has increased 1 point every year from 2000).

The need for more energy of the emerging countries and the environmental concerns of the US and the EU increases the importance of renewable energy sources in the future energy scenario.

1.1 Wind Power Development

Grid-connected wind systems are being developed very quickly and the penetration of wind power (WP) is increasing.

The driving force in Europe was taken in March 2007, when EU Heads of State adopted a binding target of 20 % of energy generated from renewable sources by 2020. A similar plan for 25 % renewable energy sources until 2025 has been adopted in the US.

According to BTM Consult [1], the cumulative and annual installed wind power worldwide in 2009 is shown in Figure 1.1. Despite the economic crisis, 2009 was a very good year, with a worldwide installed wind power of 38.1 GW (35 % higher than 2008). The biggest markets in 2009 were China, with 36.1 %, and the US, with 26 %. The cumulative worldwide installed wind power by the end of 2009 was 160.1 GW. The average growth for the period 2004–2009 was 36.1 %, while the forecast average growth for the period 2010–2014 is 13.6 % (reduced due to the economic crisis of 2008–2010). The worldwide cumulative installed power forecasted by 2019 is close to 1 TW, leading to a global wind power penetration of 8.4 %.

Wind energy penetration (%) is defined as the total amount of wind energy produced annually (TWh) divided by the gross annual electricity demand (TWh). According to the

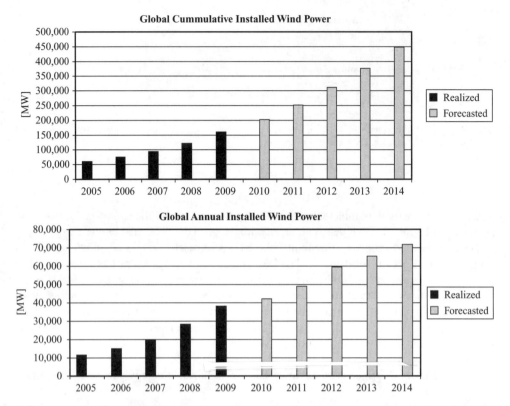

Figure 1.1 Wind power installed worldwide by 2009 and forecast until 2014: cumulative (left) and annual (right). Source: BTM Consult

EWEA (European Wind Energy Association) [2], the approximate wind energy penetration in Europe by 2008 was 3.8 %, with the highest penetration levels of 21 % in Denmark, 12 % in Spain and Portugal, 9 % in Ireland and 7 % in Germany. At the regional level much higher penetration levels were achieved, as, for example, 36 % in Schleswig-Holstein, Germany, and 70 % in Navarra, Spain.

According to the DOE (US Department of Energy) [3], the wind energy penetration level in the US reached 1.9 % by 2008, with highest state levels in Iowa 13.3 %, Minnesota 10.4 % and Texas 5.3 %. The worldwide wind energy penetration by 2008 was 1.5 %.

It is very difficult to define the maximum level of penetration as it is strongly dependent on the particularities of the grid in the considered area in terms of conventional generation, pricing, interconnection capacity, demand management and eventual storage capacity. Typically several regional or national grids are interconnected (as, for example, UCTE and NORDEL) and by agreement certain shared reserve generation and transmission capacity is provided in order to cope with $n - 1$ type contingency. The fluctuating wind power dispatch works also as a 'disturbance' in the system and this reserve capacity can also be successfully used for balancing purposes. Some studies have indicated that 20 % wind penetration may be achieved without major transmission or storage developments. Electrical utilities continue to study

Introduction

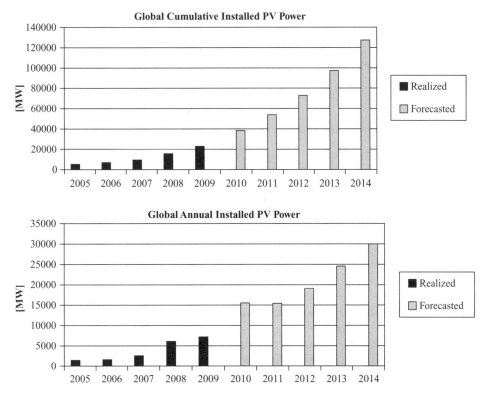

Figure 1.2 PV power installed worldwide by 2009 and forecast until 2014: cumulative (left) and annual (right). Source: EPIA

the effects of large-scale (20 % or more) penetration of wind generation on system stability and economics. Denmark has planned wind penetration of 50 % by 2025 [4]. The Danish grid is strongly interconnected to the European electrical grid through Norway, Germany and Sweden. Almost half of its wind power is exported to Norway, which can easily balance its almost entirely hydro-based power system.

In order to be able to increase the wind energy penetration, new grid interconnection requirements called grid codes have been developed by countries with high penetration.

1.2 Photovoltaic Power Development

The worldwide cumulative and annual photovoltaic (PV) power installed according to EPIA are shown in Figure 1.2 [5].

The year 2009 was also a good year for PV, as 6.4 GW was installed (equivalent to approx one-sixth of wind power installed). From an empirical point of view we can say that PV is growing at approximately the same rate as WP and is just approximately 6 years behind. The forecast for 2014 is 30 GW for PV, close to the 28.7 GW for WP forcast 6 years ago (for 2008). The worldwide cumulative PV power reached 22.8 GW by the end of 2009.

Today, there are several PV parks with installed power > 40 MW in Spain, Germany and Portugal. The PV penetration is quite low now but it is estimated by EPIA that it could be as high as 12 % in 2020.

Another important aspect is that the cost of PV panels have dropped during 2008 by around 40 % to levels under € 2/W for PV. The bulk penetration of the PV system is expected around 2015 when the cost of PV electricity is forecasted to became compatible with the cost of conventional energy.

1.3 The Grid Converter – The Key Element in Grid Integration of WT and PV Systems

Power converters is the technology that enables the efficient and flexible interconnection of different players (renewable energy generation, energy storage, flexible transmission and controllable loads) to the electric power system. Hence it is possible to foresee how the synchronous machine has a central role in the centralized power system and the grid converter, also denoted as the 'synchronous converter', will be a major player in a future power system based on smart grid technologies. While the electromagnetic field has a major role in the synchronous machine, the grid converter is based mainly on semiconductor technology and signal processing but its connection filter, where the inductor is dominant, still has a crucial role to play in transient behaviour.

The increase in the power that needs to be managed by the distributed generation systems leads to the use of more voltage levels, leading to more complex structures based on a single-cell converter (like neutral point clamped multilevel converters) or a multicell converter (like cascaded H-bridge or interleaved converters). In the design and control of the grid converter the challenges and opportunities are related to the need to use a lower switching frequency to manage a higher power level as well as to the availability of a more powerful computational device and of more distributed intelligence (e.g. in the sensors and in the PWM drivers).

The book analyses both basic and advanced issues related to synchronization with the grid, harmonic control and stability, and at the system level in order to detect and manage islanding conditions for PV power systems and control under grid faults for WT power systems. It is intended for both graduate students in electrical engineering as well as practising engineers in the WT and PV industry, with special focus on the design and control of grid converters.

References

[1] BTM Consult, 'World Market Update 2009 (Forecast 2010–2014)', March 2010. www.btm.dk.
[2] EWEA, 'Wind Energy – The Facts Executive Summary, March 2009'. http://www.ewea.org/fileadmin/ewea_documents/documents/publications/WETF/1565_ExSum_ENG.pdf.
[3] US Department of Energy, 'Wind Technologies Market Report 2008', Prepared by the National Renewable Energy Laboratory (NREL), DOE/GO-102009-2868, July 2009. http://www1.eere.energy.gov/windandhydro/pdfs/46026.pdf.
[4] PSO ForskEL, 'EcoGrid.dk, Phase I – Summary Report: Steps towards a Danish Power System with 50 % Wind Energy'. *R&D-Contract for Project 2007-1-7816, Funded by Energinet.dk*. http://www.e-pages.dk/energinet/137/fullpdf/full4aab3e1a6ad8.pdf.
[5] EPIA, 'Global Market Outlook for Photovoltaics until 2014', May 2010. http://www.epia.org/fileadmin/EPIA_docs/public/Global_Market_outlook_for_Photovoltaics_until_2014.pdf.

2

Photovoltaic Inverter Structures

2.1 Introduction

The PV inverter is the key element of grid-connected PV power systems. The main function is to convert the DC power generated by PV panels into grid-synchronized AC power.

Depending on the PV power plant configuration, the PV inverters can be categorized as:

- Module integrated inverters, typically in the 50–400 W range for very small PV plants (one panel).
- String inverters, typically in the 0.4–2 kW range for small roof-top plants with panels connected in one string.
- Multistring inverters, typically in the 1.5–6 kW range for medium large roof-top plants with panels configured in one to two strings.
- Mini central inverters, typically > 6 kW with three-phase topology and modular design for larger roof-tops or smaller power plants in the range of 100 kW and typical unit sizes of 6, 8, 10 and 15 kW.
- Central inverters, typically in the 100–1000 kW range with three-phase topology and modular design for large power plants ranging to tenths of a MW and typical unit sizes of 100, 150, 250, 500 and 1000 kW.

Historically the first grid-connected PV plants were introduced in the 1980s as thyristor-based central inverters. The first series-produced transistor-based PV inverter was PV-WR in 1990 by SMA [1]. Since the mid 1990s, IGBT and MOSFET technology has been extensively used for all types of PV inverters except module-integrated ones, where MOSFET technology is dominating.

Due to the high cost of solar energy, the PV inverter technology has been driven primarily by efficiency. Thus a very large diversity of PV inverter structures can be seen on the market.

In comparison with the motor drive inverters, the PV inverters are more complex in both hardware and functionality. Thus, the need to boost the input voltage, the grid connection filter, grid disconnection relay and DC switch are the most important aspects responsible for increased hardware complexity. Maximum power point tracking, anti-islanding, grid synchronization and data logger are typical functions required for the PV inverters.

Grid Converters for Photovoltaic and Wind Power Systems Remus Teodorescu, Marco Liserre, and Pedro Rodríguez
© 2011 John Wiley & Sons, Ltd

Actually, in contrast with electrical drive industry, which is 20 years older and driven by cost where the full-bridge topology is acknowledged worldwide, new innovative topologies have recently been developed for PV inverters with the main purpose of increasing the efficiency and reducing the manufacturing cost. As the lifetime of PV panels is typically longer than 20 years, efforts to increase the lifetime of PV inverters are also under way. Today, several manufacturers are offering extended service for 20 years.

The first method used to increase the efficiency was to eliminate the galvanic isolation typically provided by high-frequency transformers in the DC–DC boost converter or by a low-frequency transformer on the output. Thus a typical efficiency increase of 1–2 % can be obtained.

As the PV panels are typically built in a sandwich structure involving glass, silicon semiconductor and backplane framed by a grounded metallic frame, a capacitance to earth is appearing, creating a path for leakage current. This can compromise personal safety, which is typically based on a system that monitors the leakage current as an indication of faults, especially in residential applications. This capacitance can vary greatly, depending on construction or weather conditions, and in reference [2] typical values of 10 nF/kW for PV are measured using the full-bridge with unipolar modulation as a well-known source of common mode voltage resulting in leakage current.

Unfortunately, the transformerless structure requires more complex solutions, typically resulting in novel topologies in order to keep the leakage current and DC current injection under control in order to comply with the safety issues.

Another important design issue that is driving the development of new topologies is the ability to exhibit a high efficiency also at partial loads, i.e. during the periods with reduced irradiation levels. Actually a weighted efficiency called 'European efficiency' has been defined that takes into consideration the periods for different irradiation levels across Europe.

Today there are many PV inverter manufacturers in the market, such as SMA, Sunways, Conergy, Ingeteam, Danfoss Solar, Refu, etc., offering a wide range of transformerless PV inverters with very high European efficiency (>97 %) and maximum efficiency of up to 98 %.

The topology development for the transformerless PV inverters has been taking the starting point in two 'well-proven' converter families:

- H-bridge.
- Neutral point clamped (NPC).

The aim of this chapter is to explain some of the most relevant actual transformerless PV inverter structures as derivatives of these main families. The level of diversity is high as some structures require a boost DC–DC converter with or without isolation. These boost converters are well known and will not be described in detail. Some typical combined boosted inverter structures are presented at the end of this chapter.

For the module-integrated inverters, due to the low power level a very large diversity of new topologies is reported, but due to the very low actual market share of this type of inverter they are not explored in this chapter.

2.2 Inverter Structures Derived from H-Bridge Topology

The H-bridge or full-bridge (FB) converter family, first developed by W. Mcmurray in 1965 [3], has been an important reference in the power electronic converter technology development.

Photovoltaic Inverter Structures

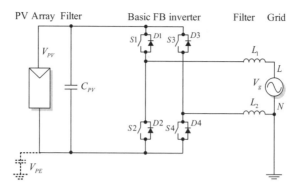

Figure 2.1 Basic FB inverter

It was the first structure able to take advantage of the first available force-commuted semiconductor devices (thyristors). The H-bridge topology is very versatile, being able to be used for both DC–DC and DC–AC conversion and can also be implemented in FB form (with two switching legs) or in half-bridge form (with one switching leg).

2.2.1 Basic Full-Bridge Inverter

The practical PV inverter topology based on the full-bridge (FB) inverter is shown in Figure 2.1. Three main modulation strategies can be used:

- Bipolar (BP) modulation.
- Unipolar (UP) modulation.
- Hybrid modulation.

In the case of *bipolar (BP) modulation* the switches are switched in diagonal, i.e. $S1$ synchronous with $S4$ and $S3$ with $S2$. Thus AC voltage can be generated as shown in Figure 2.2(a) and (b) for the positive and negative output currents respectively.

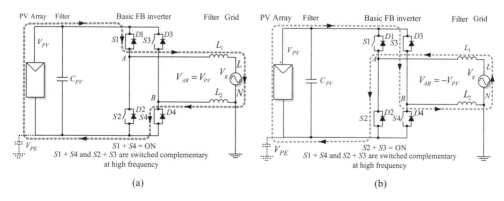

Figure 2.2 FB with BP modulation strategy in the case of: (a) positive output current and (b) negative output current

The main features of this converter are:

- Leg A and leg B are switched synchronously in the diagonal ($S1 = S3$ and $S2 = S4$) with high frequency and the same sinusoidal reference.
- No zero output voltage state is possible.

Advantages:

- V_{PE} has only a grid frequency component and no switching frequency components, yielding a very low leakage current and EMI.

Drawbacks:

- The switching ripple in the current equals $1 \times$ switching frequency, yielding higher filtering requirements (no artificial frequency increase in the output!).
- The voltage variation across the filter is bipolar ($+V_{PV} \rightarrow -V_{PV} \rightarrow +V_{PV}$), yielding high core losses.
- Lower efficiency of up to 96.5 % is due to reactive power exchange between $L_{1(2)}$ and C_{PV} during freewheeling and high core losses in the output filter, due to the fact that two switches are simultaneously switched every switching period.

Remark:

Despite its low leakage current the FB with BP modulation is not suitable for use in transformerless PV applications due to the reduced efficiency.

In the case of *unipolar modulation*, each leg is switched according to its own reference. Thus AC current can be generated as shown in Figure 2.3.

The main features of this converter are:

- Leg A and leg B are switched with high frequency with mirrored sinusoidal reference.
- Two zero output voltage states are possible: $S1, S3 =$ ON and $S2, S4 =$ ON.

Advantages:

- The switching ripple in the current equals $2 \times$ switching frequency, yielding lower filtering requirements.
- Voltage across the filter is unipolar ($0 \rightarrow +V_{PV} \rightarrow 0 \rightarrow -V_{PV} \rightarrow 0$), yielding lower core losses.
- High efficiency of up to 98 % is due to reduced losses during zero voltage states.

Drawbacks:

- V_{PE} has switching frequency components, yielding high leakage current and EMI.

Photovoltaic Inverter Structures

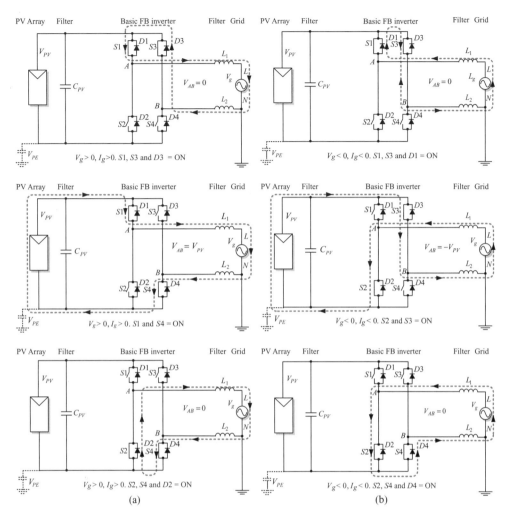

Figure 2.3 The switching states of FB with UP modulation in the case of generating: (a) positive current and (b) negative current

Remark:

Despite its high efficiency and low filtering requirements FB with UP modulation is not suitable for use in transformerless PV applications due to the high-frequency content of the V_{PE}.

In the case of *hybrid modulation* [4], one leg is switched at grid frequency and one leg at high frequency. Thus AC current can be generated as shown in Figure 2.4(a) and (b).

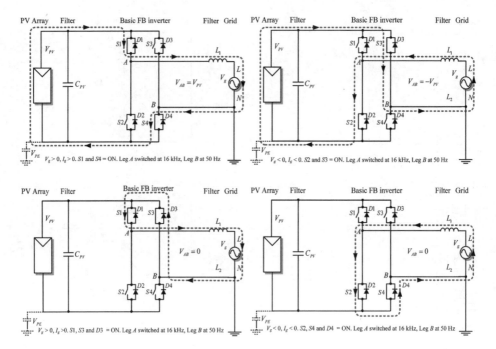

Figure 2.4 The switching states of FB with hybrid modulation in the case of generating: (a) positive current and (b) negative current

The main features of this converter are:

- Leg A is switched with grid low frequency and leg B is switched with high PWM frequency.
- Two zero output voltage states are possible: $S1$, $S2 =$ ON and $S3$, $S4 =$ ON.

Advantages:

- Voltage across the filter is unipolar ($0 \rightarrow +V_{PV} \rightarrow 0 \rightarrow -V_{PV} \rightarrow 0$), yielding lower core losses.
- Higher efficiency of up to 98 % is due to no reactive power exchange between $L_{1(2)}$ and C_{PV} during zero voltage and to lower frequency switching in one leg.

Drawbacks: also a drawback is the fact that this modulation only works for a two quadrant operation.

- The switching ripple in the current equals $1 \times$ switching frequency, yielding higher filtering requirements (no artificial frequency increase in the output!).
- V_{PE} has square wave variation at grid frequency, leading to high leakage current peaks and large EMI filtering requirements.

Remark:

Despite its high efficiency FB with hybrid modulation is not suitable for use in transformerless PV applications due to the square-wave variation of the V_{PE}.

2.2.2 H5 Inverter (SMA)

In 2005 SMA patented a new inverter topology called H5 [5]. This topology is depicted in the Figure 2.5 and, as its name indicates, it is a classical H-bridge with an extra fifth switch in the positive bus of the DC link which provides two vital functions:

- Prevents the reactive power exchange between $L_{1(2)}$ and C_{PV} during the zero voltage state, thus increasing efficiency.
- Isolates the PV module from the grid during the zero voltage state, thus eliminating the high-frequency content of V_{PE}.

The switching states for positive and negative generated AC currents are depicted in Figure 2.6.
The main features of this converter are:

- S5 and S4 (S2) are switched at high frequency and S1 (S3) at grid frequency.
- Two zero output voltage states are possible: $S5 =$ OFF and $S1$ ($S3$) $=$ ON.

Advantages:

- Voltage across the filter is unipolar ($0 \rightarrow +V_{PV} \rightarrow 0 \rightarrow -V_{PV} \rightarrow 0$), yielding lower core losses.

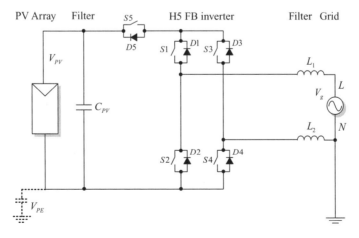

Figure 2.5 H5 inverter topology (SMA)

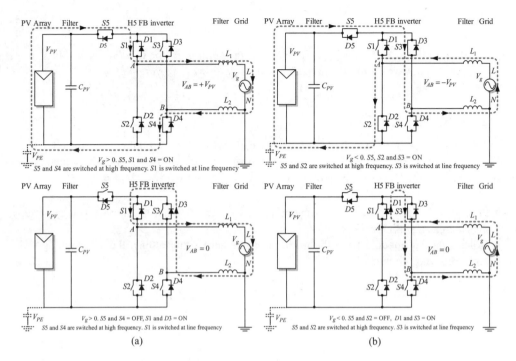

Figure 2.6 The switching states of the H5 inverter in the case of generating: (a) positive current and (b) negative current

- Higher efficiency of up to 98 % is due to no reactive power exchange between $L_{1(2)}$ and C_{PV} during zero voltage and to lower frequency switching in one leg.
- V_{PE} has only a grid frequency component and no switching frequency components, yielding a very low leakage current and EMI.

Drawbacks:

- One extra switch.
- Three switches are conducting during the active vector, leading to higher conduction losses but without affecting the overall high efficiency.

Remark:

> The H5 features all the advantages of FB with hybrid modulation and eliminates the high-frequency content of V_{PE} by isolating the PV panels from the grid during zero voltage state using the extra switch. This topology is thus very suitable for use in transformerless PV applications due to high efficiency and low leakage current and EMI. It is currently commercialized by SMA in the series called SunnyBoy 4000/5000 TL with European efficiency higher than 97.7 % and maximum efficiency of 98 % (Photon International, October 2007).

2.2.3 HERIC Inverter (Sunways)

In 2006, Sunways patented a new topology also derived from the classical H-bridge called HERIC (highly efficient and reliable inverter concept) by adding a bypass leg in the AC side using two back-to-back IGBTs (insulated gate bipolar transistors), as shown in Figure 2.7 [6].

The AC bypass provides the same two vital functions as the fifth switch in case of the H5 topology:

- Prevents the reactive power exchange between $L_{1(2)}$ and C_{PV} during the zero voltage state, thus increasing efficiency.
- Isolates the PV module from the grid during the zero voltage state, thus eliminating the high-frequency content of V_{PE}.

The switching states for positive and negative generated AC currents are depicted in Figure 2.8.

The main features of this converter are:

- S1–S4 and S2–S3 are switched at high frequency and S+ (S−) at grid frequency.
- Two zero output voltage states are possible: S+ = on and S− = on (providing the bridge is switched off).

Advantages:

- Voltage across the filter is unipolar ($0 \to +V_{PV} \to 0 \to -V_{PV} \to 0$), yielding lower core losses.
- Higher efficiency of up to 97 % is due to no reactive power exchange between $L_{1(2)}$ and C_{PV} during zero voltage and to lower frequency switching in one leg.
- V_{PE} has only a grid frequency component and no switching frequency components, yielding very low leakage current and EMI.

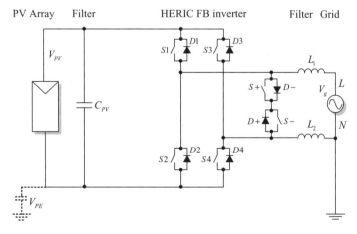

Figure 2.7 HERIC topology (Sunways)

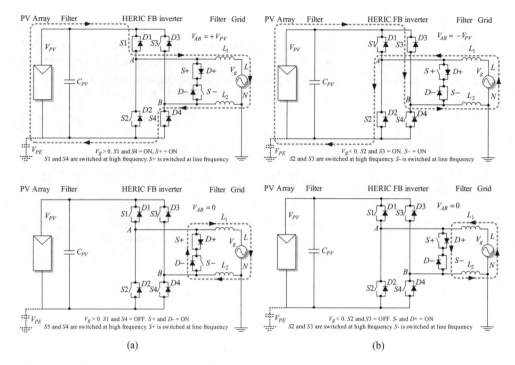

Figure 2.8 The switching states of the HERIC inverter in the case of generating: (a) positive current and (b) negative current

Drawbacks:

- Two extra switches.

Remark:

The HERIC improves the performance of FB with BP modulation by adding the zero voltage obtained with the AC bypass, thus increasing the efficiency. This topology is therefore very suitable for use in transformerless PV applications due to high efficiency and low leakage current and EMI. It is currently commercialized by Sunways in the AT series (2.7–5 kW) with reported European efficiency of 95 % and maximum efficiency of 95.6 % (Photon International, July 2008).

The behavior of HERIC and H5 are quite similar as both realize the decoupling of the PV generator from the grid during the zero voltage state on the AC side and DC side respectively. They both use two switches switched at high frequency and one switched at grid frequency, and H5 has three switches conducting at the same time, while HERIC has only two.

2.2.4 REFU Inverter

In 2007, Refu Solar patented a new topology also derived from the classical H-bridge. The topology actually uses a half-bridge within the AC side bypass and a bypassable DC–DC converter as shown in Figure 2.9 [7].

The AC bypass provides the same two vital functions as in the case of HERIC:

- Prevents the reactive power exchange between L and C_{PV} during the zero voltage state, thus increasing efficiency.
- Isolates the PV module from the grid during the zero voltage state, thus eliminating the high-frequency content of V_{PE}.

The AC bypass is implemented in a different way by comparison with HERIC, i.e. by using unidirectional switches composed of standard IGBT modules with a diode in series to cancel the free-wheeling path. Another specific characteristic of this topology is the use of a boost converter, which is activated only when the input DC voltage is lower than the grid voltage. The switching states for positive and negative generated AC currents are depicted in Figure 2.10.

The main features of this converter are:

- $S1$ ($S2$) are switched at high frequency when there is no need for boost: $V_{PV} > |V_g|$.
- $S3$ ($S4$) are switched at high frequency when the boost is enabled: $V_{PV} < |V_g|$.
- $S+(S-)$ are switched at grid frequency depending on voltage polarity.

Advantages:

- Voltage across the filter is unipolar ($0 \rightarrow +V_{PV} \rightarrow 0 \rightarrow -V_{PV} \rightarrow 0$), yielding lower core losses.

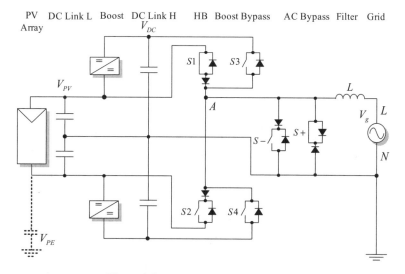

Figure 2.9 The REFU inverter topology

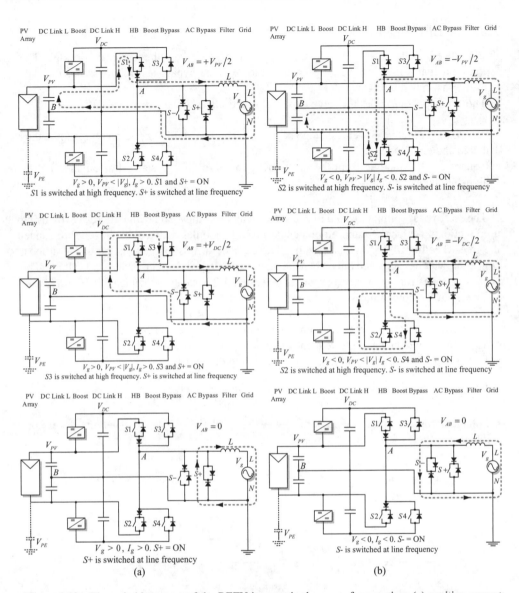

Figure 2.10 The switching states of the REFU inverter in the case of generating: (a) positive current and (b) negative current

- Higher efficiency of up to 98 % is due to no reactive power exchange between L and C_{PV} during zero voltage, boost only when necessary and to lower frequency switching in one leg.
- V_{PE} has only a grid frequency component and no switching frequency components, yielding very low leakage current and EMI.

Drawbacks:

- Needs double DC voltage.
- Two extra switches, but switched at low frequency.

Remark:

The REFU topology is an improvement on the half-bridge topology by adding the AC bypass to create zero voltage with minimum losses. This topology is very suitable for use in transformerless PV applications due to high efficiency and low leakage current and EMI. It is currently commercialized by Refu in the three-phase series RefuSol® (11/15 kW) with reported European efficiency of 97.5 % and maximum efficiency of 98 % (Photon International, September 2008).

2.2.5 *Full-Bridge Inverter with DC Bypass – FB-DCBP (Ingeteam)*

Another 'modified' FB topology is the full-bridge with DC bypass as patented (pending) by Ingeteam [8] and published in reference [9]. This topology is depicted in Figure 2.11 and is a classical H-bridge with two extra switches in the DC link and also two extra diodes clamping the output to the grounded middle point of the DC bus. The DC switches provide the separation of the PV panels from the grid during the zero voltage states and the clamping diodes ensure that the zero voltage is grounded, in opposition to HERIC or H5 where the zero voltage is floating. Essentially both solutions ensure 'jump-free' V_{PE}, leading to low leakage current and high efficiency due to prevention of reactive power exchange between $L_{1(2)}$ and $C_{PVI(2)}$ during zero voltage.

The switching states for positive and negative generated AC currents are depicted in Figure 2.12.

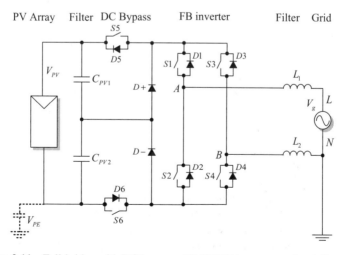

Figure 2.11 Full-bridge with DC bypass – FB-DCBP inverter topology (Ingeteam)

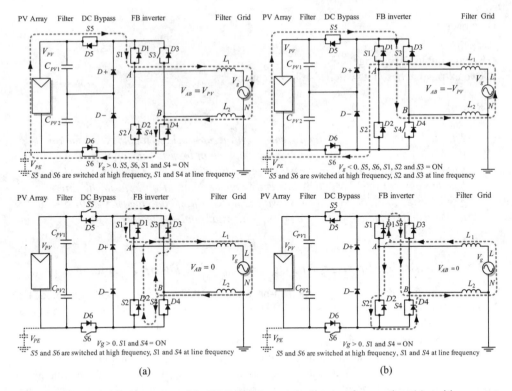

Figure 2.12 The switching states of the FB-DCBP inverter in the case of generating: (a) positive current and (b) negative current

The main features of this converter are:

- $S5$ and $S6$ are switched at high frequency and $S1$ ($S2$) and $S4$($S3$) at grid frequency.
- Zero output voltage is achieved by turning the DC bypass switches $S5$ and $S6$ OFF. When $S5$ and $S6$ are turned OFF and $S2$ and $S3$ are turned ON, the current splits into two paths: $S1$ and the freewheeling diode of $S3$ ($D3$), and $S4$ and the freewheeling diode of $S2$ ($D2$). Thus, $S2$ and $S3$ are turned ON with no current and therefore no switching losses appear. The path of the current during zero voltage state will be $S4$-$D2$ or $S1$-$D3$ for positive grid current, while the negative grid current will flow through $S2$-$D4$ or $S3$-$D1$. D+ and D− are just used for clamping the bypass switches to the half of the DC-link voltage [9].

Advantages:

- Voltage across the filter is unipolar ($0 \rightarrow +V_{PV} \rightarrow 0 \rightarrow -V_{PV} \rightarrow 0$), yielding lower core losses.
- The rating of the DC bypass switches is half of the DC voltage.
- Higher efficiency is due to no reactive power exchange between $L_{1(2)}$ and $C_{PVI(2)}$ during zero voltage and to a lower switching frequency in the FB and low voltage rating of $S5$ and $S6$.
- V_{PE} has only a grid frequency component and no switching frequency components, yielding a very low leakage current and EMI.

Drawbacks:

- Two extra switches and two extra diodes.
- Four switches are conducting during the active vector, leading to higher conduction losses but without affecting the overall high efficiency.

Remark:

> *The FB-DCBP topology is very suitable for use in transformerless PV applications due to high efficiency and low leakage current and EMI. This topology is currently commercialized by Ingeteam in the Ingecon® Sun TL series (2.5/3.3/6 kW) with reported European efficiency of 95.1 % and maximum efficiency of 96.5 % (Photon International, August 2007).*

2.2.6 Full-Bridge Zero Voltage Rectifier – FB-ZVR

Another 'modified' FB topology is the full-bridge zero voltage rectifier [10], depicted in Figure 2.13. This topology is derived from HERIC, where the bidirectional grid short-circuiting switch is implemented using a diode bridge and one switch (S5) and a diode clamp to the DC midpoint. Zero voltage is achieved by turning the FB off and turning S5 on.

The switching states for positive, negative and zero output voltage states are shown in Figure 2.14.

The main features of this converter are:

- The switches within the FB are switched diagonally like in bipolar modulation. The zero state is introduced after each switching by turning all switches of the bridge off and turning S5 on.

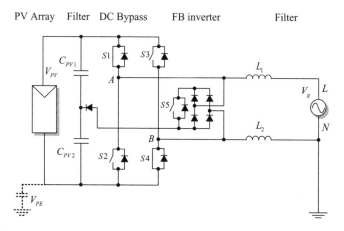

Figure 2.13 Full-bridge zero voltage rectifier – FB-ZVR inverter topology

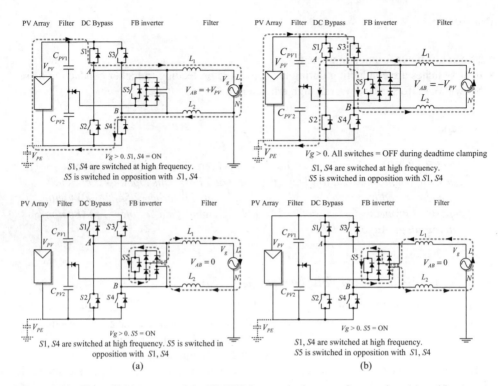

Figure 2.14 The switching states of the FB-ZVR inverter in the case of generating: (a) positive current and (b) negative current

Advantages:

- Voltage across the filter is unipolar ($0 \rightarrow +V_{PV} \rightarrow 0 \rightarrow -V_{PV} \rightarrow 0$), yielding lower core losses.
- High efficiency of up to 96 % is due to no reactive power exchange between $L_{1(2)}$ and C_{PV} during zero voltage and to lower frequency switching in one leg.
- V_{PE} has only a grid frequency component and no switching frequency components, yielding very low leakage current and EMI.

Drawbacks:

- One extra switch and four diodes.
- During deadtime clamping, bipolar output voltage is obtained, leading to increased losses across the filter.

Remark:

The FB-ZVR inherits the advantages of the HERIC in terms of high efficiency and low leakage. Due to the high switching frequency of S5, the efficiency is lower than at HERIC, but it provides the advantage that can work at any power factor.

2.2.7 Summary of H-Bridge Derived Topologies

Actually, HERIC, H5, REFU and FB-DCBP topologies convert the two-level FB (or HB) inverter into a three-level one. This increases the efficiency as both the switches and the output inductor are subject to half of the input voltage stress. The zero voltage state is achieved by shorting the grid using the higher switches of the bridge (H5) or by using an additional AC bypass (HERIC or REFU) or DC bypass (FB-DCBP). H5 and HERIC isolate the PV panels from the grid during zero voltage while REFU and FB-DCBP clamp the neutral to the midpoint of the DC link. Both REFU and HERIC use the AC by-pass but REFU uses two switches in antiparallel and HERIC uses two switches in series (back to back). Thus the conduction losses in the AC bypass are lower for the REFU topology. REFU and H5 have slightly higher efficiencies as they have only one switch that switches with high-frequency while HERIC and FB-DCBP have two.

FB-ZVR is derived from HERIC but uses a different implementation of the bidirectional switch, using a diode bridge and one switch. Constant V_{PE} but moderate high efficiency (lower than HERIC but higher than FB-BP) are obtained and can also work with nonunitary PF.

In the following section another family of converters, called neutral point clamped (NPC), achieving more or less the same performance but at the expense of more switches is explored.

2.3 Inverter Structures Derived from NPC Topology

The NPC topology has been introduced by Nabae, Magi and Takahashi in 1981 [11] showing great improvements in terms of lower dV/dt and switch stress in comparison with the classical two-level full-bridge inverter. The NPC topology is also very versatile and can be used in both single-phase (full-bridge or half-bridge) and three-phase inverters.

2.3.1 Neutral Point Clamped (NPC) Half-Bridge Inverter

The main concept is that zero voltage can be achieved by 'clamping' the output to the grounded 'middle point' of the DC bus using $D+$ or $D-$ depending on the sign of the current (Figure 2.15). The switching states for generating positive and negative current are depicted in Figure 2.16.

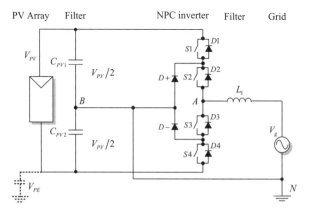

Figure 2.15 Neutral clamped half-bridge

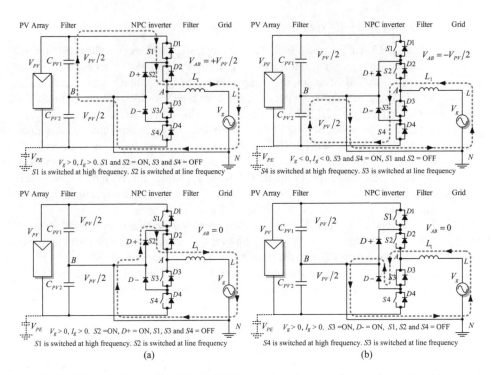

Figure 2.16 The switching states of the NPC-HB inverter in the case of generating: (a) positive current and (b) negative current

The main features of this converter are:

- $S1$ ($S4$) are switched at high frequency and $S2$ ($S3$) at grid frequency.
- Two zero voltage states are possible: $S2, D+ =$ ON and $S3, D- =$ ON. For operation out of unitary power factors $S1$ and $S3$ switch in opposition for $V_g > 0$, $I_g < 0$, and $S2$ and $S4$ for $V_g < 0$, $I_g > 0$.

Advantages:

- Voltage across the filter is unipolar ($0 \rightarrow +V_{PV} \rightarrow 0 \rightarrow -V_{PV} \rightarrow 0$), yielding lower core losses.
- Higher efficiency of up to 98 % is due to no reactive power exchange between $L_{1(2)}$ and C_{PV} during zero voltage and to lower switching frequency in one leg.
- The voltage rating of outer switches can be reduced to $V_{PV}/4$, resulting in reduced switching losses.
- V_{PE} is constant and is equal to $-V_{PV}/2$ without switching frequency components, yielding very low leakage current and EMI.

Drawbacks:

- Two extra diodes.
- Requires double voltage input in comparison with FB.

- Unbalanced switch losses: higher on the higher/lower switches and lower on the middle switches.
- Any inductance introduced in the neutral connection by, for example, EMI filters generates high-frequency common-mode voltage, which will lead to leakage current.

Remark:

The NPC has very similar performances in comparison with H5, HERIC or REFU, being very suitable for use in transformerless PV applications due to high efficiency and low leakage current and EMI. It is currently used by Danfoss Solar inverters in the TripleLynx series (three-phase 10/12.5/15 kW) with reported European efficiency of 97 % and maximum efficiency of 98 % (Photon Magazine, July 2010).

2.3.2 Conergy NPC Inverter

A 'variant' of the classical NPC is a half-bridge with the output clamped to the neutral using a bidirectional switch realized with two series back-to-back IGBTs as patented by Conergy [12] (see Figure 2.17). An alternative realization of the same concept is presented in reference [13], where the unidirectional clamping switches are connected in parallel rather than in series and a full-bridge is used instead of a half-bridge.

The main concept of the Conergy NPC inverter is that zero voltage can be achieved by 'clamping' the output to the grounded 'middle point' of the DC bus using $S+$ or $S-$ depending on the sign of the current. The switching states for generating positive and negative currents are depicted in Figure 2.18.

The main features of this converter are:

- $S1$ ($S2$) and $S+(S-)$ are switched at high frequency.
- Two zero voltage states are possible: $S+, D+ = $ ON ($S-, D- = $ ON).

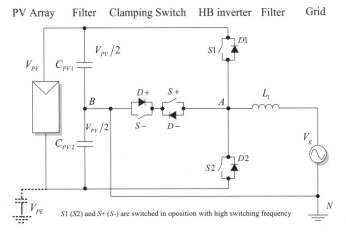

Figure 2.17 Conergy neutral point clamped inverter

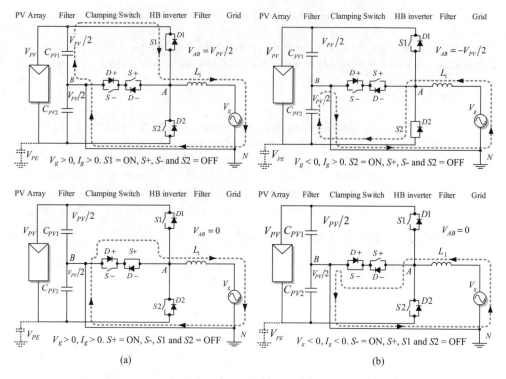

Figure 2.18 The switching states of the Conergy NPC inverter in the case of generating: (a) positive current and (b) negative current

Advantages:

- Voltage across the filter is unipolar ($0 \rightarrow +V_{PV} \rightarrow 0 \rightarrow -V_{PV} \rightarrow 0$), yielding lower core losses.
- Higher efficiency of up to 98 % is due to no reactive power exchange between $L_{1(2)}$ and C_{PV} during zero voltage and to a reduced voltage drop as only one switch is conducting during active states of the Conergy NPC inverter.
- V_{PE} is constant and is equal to $-V_{PV}/2$ without switching frequency components, yielding a very low leakage current and EMI.
- Balanced switching losses in contrast with the classical NPC.

Drawbacks:

- The voltage rating of $S1$ and $S2$ is double in comparison with the outer switches in the NPC.
- Requires double voltage input in comparison with FB.
- Any inductance introduced in the neutral connection by, for example, EMI filters generates high-frequency common-mode voltage, which will lead to leakage current.

Remark:

The Conergy NPC has slightly higher efficiency in comparison with NPC as during the active state only one switch is conducting, which is very suitable for use in transformerless PV applications due to high efficiency and low leakage current and EMI. It is currently used in the market by Conergy in the IPG series (2–5 kW) string inverter series with reported European efficiency of 95.1 % and maximum efficiency of 96.1 % (Photon International, July 2007).

2.3.3 Summary of NPC-Derived Inverter Topologies

The classical NPC and its 'variant' Conergy NPC are both three-level topologies featuring the advantages of unipolar voltage across the filter, high efficiency due to clamping of PV panels during the zero voltage state and practically no leakage due to the grounded DC link midpoint. Due to higher complexity in comparison with FB-derived topologies, these structures are typically used in three-phase PV inverters with ratings over 10 kW (mini-central). These topologies are also very attractive for high power in the range of hundreds of kW (central inverters) where the advantages of multilevel inverters are even more important.

2.4 Typical PV Inverter Structures

In this chapter different innovative topologies for the transformerless PV inverters are presented. However, most of them will require boosting so the final structure will be different. In the following, some typical complete PV structures are described.

2.4.1 H-Bridge Based Boosting PV Inverter with High-Frequency Transformer

A typical structure of an H-bridge based boosting PV inverter is shown in Figure 2.19. The FB DC–DC converter boost factor is controlled by shifting the switching phase between the two legs [14]. The FB inverter can be easily replaced with higher-efficiency versions (H5 or HERIC).

2.4.1.1 Boosting Inverter with Low-Frequency Transformer

A typical structure using a classical boost DC–DC converter is shown in Figure 2.20. The transformer is placed on the low-frequency side.

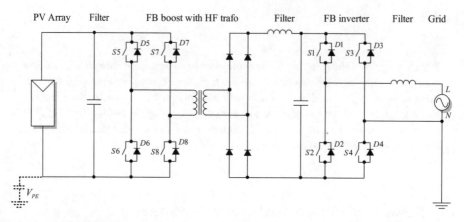

Figure 2.19 Boosting inverter with a HF transformer based on the H-bridge

2.5 Three-Phase PV Inverters

Most of the three-phase PV inverters are not typically true three-phase three-wire inverters but rather three-phase four-wire ones. Actually they work as three independent single-phase inverters.

This solution has two advantages:

- It allows using the existing single-phase inverters.
- It allows using the 'mild' anti-islanding requirement from the German standard VDE-0126-1-1 (2006), which states that the impedance monitoring can be replaced by line-to-line voltage monitoring if the control of each phase current is done independently.

Companies like SMA are promoting the concept that three-phase systems can be built up by using a single-phase building block, the so-called minicentral inverters (like, for example, Sunny Mini Central 8000TL).

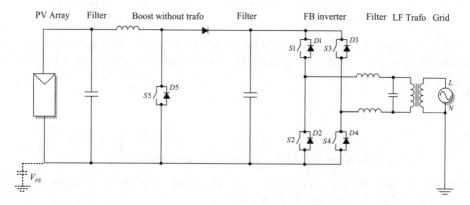

Figure 2.20 Boosting inverter with a LF transformer based on the boost converter

Other companies like Conergy, Refusol and Danfoss Solar are promoting three-phase inverters in the range of 10–15 kW based on the same concept but built like a three-phase unit. A recent comparative study among three-phase transformerless topologies [15] revealed that three-phase NPC exhibits the best performances in comparison with full-bridge with a split DC link in terms of low leakage, efficiency and performance. The real problem of using a true three-phase three-wire topology is that the DC voltage needs to be relatively high, at least around 600 V for a 400 V three-phase grid and is limited to 1000 V due to safety requirements (maximum installation voltage). The variation range can be too narrow in comparison with the variations required by the MPPT due to temperature changes and grid voltage allowed variations. On the contrary, single-phase inverters need at least around 400 V DC voltage having a larger variation range yielding more flexibility.

2.6 Control Structures

Due to the very large variety of transformerless PV inverter topologies, the control structures are also very different. The modulation algorithm has to be specific for each topology. In the following a generic, topology invariant control structure will be presented for a typical transformerless topology with boost stage, as shown in Figure 2.21.

As can be seen, three different classes of control functions can be defined:

1. **Basic functions – common for all grid-connected inverters**
    - Grid current control
        - THD limits imposed by standards
        - Stability in the case of large grid impedance variations
        - Ride-through grid voltage disturbances

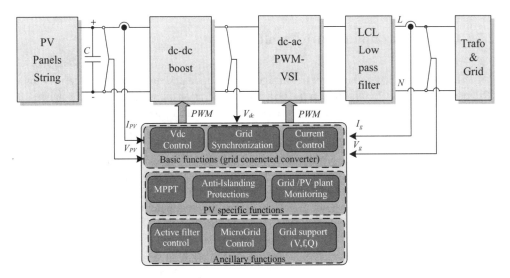

Figure 2.21 Generic control structure for a PV inverter with boost stage

- DC voltage control
  - Adaptation to grid voltage variations
  - Ride-through grid voltage disturbances
- Grid synchronization
  - Operation at the unity power factor as required by standards
  - Ride-through grid voltage disturbances
2. **PV specific functions – common for all PV inverters**
- Maximum power point tracking (MPPT)
  - Very high MPPT efficiency during steady state (typically > 99 %)
  - Fast tracking during rapid irradiation changes (dynamical MPPT efficiency)
  - Stable operation at very low irradiation levels
- Anti-islanding (AI), as required by standards (VDE 0126, IEEE 1574, etc.)
- Grid monitoring
  - Synchronization
  - Fast voltage/frequency detection for passive AI
- Plant monitoring
  - Diagnostic of PV panel array
  - Partial shading detection
3. **Ancillary functions**
- Grid support
  - Local voltage control
  - Q compensation
  - Harmonic compensation
  - Fault ride-through

In this book, the common basic functions (grid current control, DC voltage control and synchronization) as well as anti-islanding methods are covered.

2.7 Conclusions and Future Trends

PV inverter structures are evolving at a high pace. A high number of new patented transformerless topologies based on either H-bridge or NPC appeared on the market with very high efficiency, up to 98 %. In this chapter the principle of operation along with the performances of these topologies are presented as this represents a really high value for the power electronics community.

The obvious trend is more silicon for lower losses, as the number of switches has increased. The PV inverter market is driven by efficiency rather than cost, mainly due to the still very high price for PV energy. To increase even more the efficiency could be quite difficult using the current technology, but new research shows a real good potential in replacing the silicon switches by silicon-carbide ones. In reference [16] an efficiency increase of roughly 1 % was demonstrated on an HERIC topology by simply replacing the IGBT with a SiC MosFet. It is thus expected that in a few years the SiC MosFets will become commercially available, along with the SiC diodes, which are currently used today in very high efficiency boost converters.

Another trend in the design of PV inverters will be influenced by the grid requirements. At the moment, in many countries it is required that eventual islanding should be quickly detected and the inverter should be disconnected from the grid immediately in order to avoid any personal safety issues, especially for residential PV systems. However, as the PV weight in the grid integration is expected to grow very fast, it is possible that grid requirements will change and will require fault ride-through capability in order to stabilize the power system. Just like the case for wind power systems, this requirement has been introduced after a long period when its share in the power generation became important. This will most probably apply for large PV plants connected to distribution systems.

Finally, integration of power components is an important factor as it is known from the electric drives sector that this will reduce the costs in the long term. The problem with PV inverters is that there are so many topologies and it is actually very difficult to find standard modules for implementation. A good example is SMA, which managed to produce customized power modules for the H5 topology. Semikron and Vincotech (previously a Tyco division) are now offering power modules for NPC topologies, Mitsubishi is offering intelligent power modules (IPM) with one or two boost converters plus an H-bridge inverter specially designed for PV applications, and this trend is expected to continue also with other major device manufacturers as the PV inverter market is growing very fast.

References

[1] Meinhardt, M., Cramer, G., Burger, B. and Zacharias, P. 'Multi-String-Converter with Reduced Specific Costs and Enhanced Functionality'. *Solar Energy*, **69**(Suppl. 6), July–December 2001, 217–227.
[2] Lopez, O., Teodorescu, R., Freijedo, F. and DovalGandoy, J., 'Leakage Current Evaluation of a Single-Phase Transformerless PV Inverter Connected to the Grid'. In *Applied Power Electronics Conference, APEC 2007 – Twenty Second Annual IEEE*, 25 February–1 March 2007, pp. 907–912.
[3] Mcmurray, W., 'Inverter Circuits'. US Patent 3207974, September 1965.
[4] Lai, R.-S. and Ngo, K. D. T., 'A PWM Method for Reduction of Switching Loss in a Full-Bridge Inverter', *IEEE Transactions on Power Electronics*, **10**(3), May 1995, 326–332 .
[5] Victor, M. *et al.*, US Patent Application, Publication Number US 2005/0286281 A1, 29 December 2005.
[6] Schmid, H. *et al.* US Patent 7046534, issued 16 May 2006.
[7] Hantschel, J., German Patent Application, Publication Number DE102006010694 A11, 20 September 2007.
[8] Gonzalez, S. R. *et al.* International Patent Application, Publication Number WO2008015298, 2 July 2007.
[9] Gonzalez, R., Lopez, J., Sanchis, P. and Marroyo, L., 'Transformerless Inverter for Single-Phase Photovoltaic Systems'. *IEEE Transactions on Power Electronics*, **22**(2), March 2007, 693–697.
[10] Kerekes, T., Teodorescu, R., Rodriguez, P., Vazquez, G. and Aldabas, E., 'A New High-Efficiency Single-Phase Transformerless PV Inverter Topology', *IEEE Transactions on Industrial Electronics*, 2010.
[11] Nabae, A., Magi, H. and Takahashi, I., 'A New Neutral-Point-Clamped PWM Inverter'. *IEEE Transactions on Industry Applications*, **IA-17**(5), September/October 1981, 518–523.
[12] Knaup, P., International Patent Application, Publication Number WO 2007/048420 A1, 3 May 2007.
[13] Calais, M., Agelidis, V.G. and Meinhardt, M., 'Multilevel Converters for Single-Phase Grid Connected Photovoltaic Systems: An Overview'. *Solar Energy*, **66**(5), August 1999, 325–335.
[14] Mohan, N., Undeland, T. and Robbins, P. W., *Power Electronics. Converters, Applications and Design*, John Wiley & Sons, Ltd, Chichester, 2003, ISBN 0471226939.
[15] Kerekes, T., Teodorescu, R., Klumpner, C., Sumner, M., Floricau, D. and Rodriguez, P., 'Evaluation of Three-Phase Transformerless Photovoltaic Inverter Topologies'. In *Power Electronics and Applications, 2007 European Conference*, 2–5 September 2007, pp. 1–10.
[16] Burger, B. and Schmidt, H., '25 Years Transformerless Inverters'. In *Proceedings of PVSEC*, 2007.

3

Grid Requirements for PV

3.1 Introduction

Grid-connected PV systems are being developed very fast and systems from a few kW to tenths of a MW are now in operation. As an important source of distributed generation (DS) the PV systems need to comply with a series of standard requirements in order to ensure the safety and the seamless transfer of the electrical energy to the grid.

Typically, local regulations imposed by the grid operators apply in most countries but large efforts are made worldwide to impose some standard grid requirements that can be adopted by various countries. The most relevant international bodies that are developing worldwide standards for grid requirements are: IEEE (Institute of Electrical and Electronic Engineers) in the US, IEC (International Electrotechnical Commission) in Switzerland and DKE (German Commission for Electrical, Electronic and Information Technologies of DIN and VDE) in Germany, the dominant PV market.

The grid requirements are a very important specification that is having a big impact on the design and performances of the PV inverter. For example, in the US market the transformerless PV inverters are not yet popular as the utilities are still reticent in allowing them due to some older local regulations (NEC). Another example is the recent three-phase mini-central PV inverter in the range of 8–15 kW tailored to the new grid requirements in some European countries, which are limiting the feed-in tariff to about these levels. Therefore the PV industry is very sensitive to the grid regulations and is tailoring the products in accordance with them.

The purpose on this chapter is to introduce the most relevant standard requirements from the main market areas with the focus on the most recent developments and to describe the technical challenges that should be addressed by the PV inverter technology.

First in Section 3.2 the most important and recent standard regulations are introduced, focusing on Europe and US markets. Then the most relevant requirements: response to abnormal grid conditions, power quality and anti-islanding are described in Sections 3.3, 3.4 and 3.5 respectively, with a comparison made among the different standards.

Grid Converters for Photovoltaic and Wind Power Systems Remus Teodorescu, Marco Liserre, and Pedro Rodríguez
© 2011 John Wiley & Sons, Ltd

3.2 International Regulations

3.2.1 IEEE 1547 Interconnection of Distributed Generation

In the US, several standards for interconnection have come and gone since the enactment of the Public Utilities Regulatory Policy Act in 1978 [1]. For PV and other inverter-based technologies, IEEE 929-2000, *Recommended Practice for Utility Interface of Photovoltaic (PV) Systems* [2], has the longest history and probably is the most used standard. It applies to residential and other small-scale interconnected PV, but many aspects have been applied to other generating technologies that use inverters. It dates back to 1981 and originated from the IEEE Standards Coordinating Committee 21 (SCC21) for PV systems, which has been expanded to include fuel cells, PVs, dispersed generation and energy storage. IEEE 929 has since been effectively incorporated into many state rules and in UL 1741 [3].

Largely derived from IEEE 929, UL 1741, *Standard for Inverters, Converters, and Controllers for Use in Independent Power Systems*, elaborated by Underwriters Laboratories Inc., an important standardization body in the US, became an important safety listing in 1999 and is applied to small grid-tied inverters. It has taken on a greater role since 2002 when the first large-scale three-phase inverters were included under this UL listing. Like other UL standards, 1741 addresses issues related to construction, electrical safety and principles derived from the National Electric Code (NEC) and related UL standards. UL 1741 is unique in its incorporation of grid performance requirements. Most other UL standards are limited to electrical safety and do not address performance. The practical implementation of UL 1741 has had a huge impact on the viability of projects incorporating PV and other inverter-based technologies. Where it has been adopted in state and utility rules, it has greatly simplified the interconnection process for developers and utilities alike.

The review process in many cases has evolved from detailed interconnection studies to simple checklists. Today the single most influential standard for interconnection of all forms of DR is IEEE 1547-2003 [4], *Standard for Interconnecting Distributed Resources with Electric Power Systems*. IEEE 1547 is the result of a recent effort by SCC21 to develop a single interconnection standard that applies to all technologies. The IEEE 1547 standard has benefited greatly from earlier utility industry work documented in IEEE and IEC standards (e.g. IEEE 929, 519, 1453; IEC EMC series 61000; etc.) and ANSI C37 series of protective relaying standards. IEEE 1547 addresses all types of interconnected generation up to 10 MW and establishes mandatory requirements. This sets it apart from several previous IEEE guides or recommended practices on DG, which convey only suggestions and recommendations.

The IEEE 1547 standard focuses on the technical specifications for, and the testing of, the interconnection itself. It includes general requirements, response to abnormal conditions, power quality, islanding and test specifications as well as requirements for design, production, installation evaluation, commissioning and periodic tests. The requirements are applicable for interconnection with EPSs at typical distribution levels (medium voltage) but also low-voltage distribution networks are considered.

IEEE 1547.1-2005 [5], *Standard for Conformance Test Procedures for Equipment Interconnecting Distributed Resources with Electric Power Systems*, is a new standard deriving from IEEE 1547, specifying the type, production and commissioning tests that shall be performed to demonstrate that the interconnection functions and equipment of the distributed resources (DR) conform to IEEE Std 1547.

In May 2007 the following paragraph has been added to UL1741: 'For utility-interactive equipment, these requirements are intended to supplement and be used in conjunction with the *Standard for Interconnecting Distributed Resources with Electric Power Systems*, IEEE 1547, and the *Standard for Conformance Test Procedures for Equipment Interconnecting Distributed Resources with Electric Power Systems*, IEEE 1547.1.' Thus in terms of grid requirements UL 1741 acknowledges IEEE 1547.

3.2.2 IEC 61727 Characteristics of Utility Interface

A huge effort in harmonizing the grid requirements has been done by IEC that promotes international cooperation within standardizations for electrical and electronic issues. The TC-82 Committee on Solar Photovoltaic Energy Systems is developing a large range of standards for the PV industry. In the grid interconnection requirements, the TC-82 has developed the standard IEC 61727 [6], *Photovoltaic (PV) Systems – Characteristics of the Utility Interface*, published in December 2004. The standard applies to utility-interconnected PV power systems operating in parallel with the utility and utilizing static (solid-state) nonislanding inverters for the conversion of DC to AC and lays down requirements for interconnection of PV systems to the utility distribution system.

Another related standard of 2005, IEC 62116 Ed. 1 [7], *Testing Procedure of Islanding Prevention Measures for Utility Interactive Photovoltaic Inverters*, describing the testing procedure for the requirements stated in IEC 61727, has been recently approved and is expected to be published by the end of 2007. Although there are slight differences, the requirements from IEC 61727 are harmonizing very well with the ones from IEEE 1574, especially for anti-islanding detection, which is the critical issue.

3.2.3 VDE 0126-1-1 Safety

Germany is the dominant PV market and therefore the German regulations issued by the VDE Testing and Certification Institute is very important. During the 1990s the so-called ENS safety device, *Die selbsttätig wirkende Freischaltstelle besteht aus zwei voneinander unabhängigen Einrichtungen zur Netzüberwachung mit zugeordnete allpoligen Schaltern in Reihe* or *Automatic Disconnection Device between a Generator and the Public Low-Voltage Grid*, was introduced, first as an external hardware device, but later in software. According to the former VDE 0126-1999, the ENS device should be able to detect a jump of 0.5 Ω in the grid impedance in a power-balanced situation. This was only possible by using active methods based on distorting the grid and measuring the response. After some years of experience in the field it was agreed that the requirements are too tight and frequently led to nuisance trips affecting the yield and also power quality degradation. Especially cases where many inverters operating close to each other increased malfunctioning have been reported.

Eventually, the newly revised standard VDE 0126-1-1-2006 [8] has relaxed the tight thresholds for disconnection in the case of abrupt grid impedance changes (from 0.5 to 1.0 Ω) and even allow as an alternative method an anti-islanding requirement very similar to the one in IEEE 1547.1 based on a resonant RLC load. These changes are expected to contribute to increased grid stability, without altering safety.

Apart from the ENS, the VDE 0126-1-1 also includes over/undervoltage and frequency detection and describes the test procedures for a fail-safe protective interface that has to disconnect automatically the PV inverter from the grid in cases of DC injection, fault current and low isolation to earth. In order to accommodate the transformerless PV inverters, a leakage current limit (300 mA) is imposed and active monitoring of the fault current with sensitivity down to 30 mA and active monitoring of isolation (>1 kΩ/V) is required. Also, for the fail-safe disconnection circuit, a redundant circuit is required. This means that additional hardware is needed in order to achieve these safety functions, thus increasing the complexity and cost. In Germany it is not allowed to apply ENS in an installation with an AC output power $\geq$ 30 kW.

3.2.4 IEC 61000 Electromagnetic Compatibility (EMC – low frequency)

IEC 61000-3-2 [9] deals with the limitation of harmonic currents injected into the public supply system. It specifies limits of harmonic components of the input current, which may be produced by equipment tested under specified conditions. This part of IEC 61000 is applicable to electrical and electronic equipment having an input current up to and including 16 A per phase, and is intended to be connected to public low voltage. For equipment with current higher than 16 A but lower than 75 A the corresponding standard IEC 61000-3-12 [10] applies.

IEC 61000-3-3 [11] is concerned with the limitation of voltage fluctuations and flicker impressed on the public low-voltage system. It specifies limits of voltage changes that may be produced by an equipment tested under specified conditions and gives guidance on methods of assessment. This part of IEC 61000 is applicable to electrical and electronic equipment having an input current equal to or less than 16 A per phase, intended to be connected to public low-voltage distribution systems of between 220 and 250 V line to neutral at 50 Hz, and not subject to conditional connection. For equipment with current higher than 16 A but lower than 75 A the corresponding standard IEC 61000-3-11 [12] applies.

3.2.5 EN 50160 Public Distribution Voltage Quality

The voltage quality in the public distribution system is regulated in Europe by EN 50160 [13], which gives the main voltage parameters and their permissible deviation ranges at the customer's point of common coupling in public low-voltage (LV) and medium-voltage (MV) electricity distribution systems, under normal operating conditions. The parameters of the supply voltage shall be within the specified range during 95 % of the test period, while the permitted deviations in the remaining 5 % of the period are much greater. EN 50160 is principally informative and accepts no responsibility when the limits are exceeded. The following parameters are of interest for designing the control of PV inverters:

- Voltage harmonic levels (see Table 3.1). Maximum voltage THD is 8 %.
- Voltage unbalance for three-phase inverters. Maximum unbalance is 3 %.
- Voltage amplitude variations: maximum ± 10 %.
- Frequency variations: maximum ± 1 %.
- Voltage dips: duration < 1 sec. deep < 60 %.

For the PV inverter, point-of-view compatibility with this voltage power quality standard is important as it can demonstrate that the inverter is able to operate with the whole range

Table 3.1 Public distribution grid voltage harmonics limits – EN 50160

Odd harmonics				Even harmonics	
Not multiple of 3		Multiple of 3			
Order h	Relative voltage (%)	Order h	Relative voltage (%)	Order h	Relative voltage (%)
5	6	3	5	2	2
7	5	9	1.5	4	1
11	3.5	15	0.5	6 to 24	0.5
13	3	21	0.5		
17	2				
19	1.5				
23	1.5				
25	1.5				

of disturbances. The voltage and frequency variations are surpassed by other PV specific standards, as shown in Section 3.3. For the permitted disturbances in terms of dips, there is not yet any ride-through requirement, but in the future as PV increases penetration of the market such a requirement (similar to the wind power grid codes) is expected.

In the following an analytical and comparative analysis of the main grid requirements from the three main standard groups IEEE 1574/UL 1741, IEC 61727 and VDE 0126-1-1 is given.

3.3 Response to Abnormal Grid Conditions

The PV inverters need to disconnect from the grid in case of abnormal grid conditions in terms of voltage and frequency. This response is to ensure the safety of utility maintenance personnel and the general public, as well as to avoid damage to connected equipment, including the photovoltaic system.

3.3.1 Voltage Deviations (see Table 3.2)

All discussions regarding system voltage refer to the local nominal voltage. The voltages in RMS (root mean square) are measured at the point of utility connection. The disconnection

Table 3.2 Disconnection time for voltage variations

IEEE 1547		IEC 61727		VDE 0126-1-1	
Voltage range (%)	Disconnection time (sec.)	Voltage range (%)	Disconnection time (sec.)	Voltage range (%)	Disconnection time (sec.)
$V < 50$	0.16	$V < 50$	0.10	$110 \leq V < 85$	0.2
$50 \leq V < 88$	2.00	$50 \leq V < 85$	2.00		
$110 < V < 120$	1.00	$110 < V < 135$	2.00		
$V \geq 120$	0.16	$V \geq 135$	0.05		

Table 3.3 Disconnection time for frequency variations

IEEE 1547		IEC 61727		VDE 0126-1-1	
Frequency range (Hz)	Disconnection time (sec.)	Frequency range (Hz)	Disconnection time (sec.)	Frequency range (Hz)	Disconnection time (sec.)
$59.3 < f < 60.5$[a]	0.16	$f_n - 1 < f < f_n + 1$	0.2	$47.5 < f < 50.2$	0.2

[a]For systems with power < 30 kW the lower limit can be adjusted in order to allow participation in the frequency control.

time refers to the time between the abnormal condition occurring and the inverter ceasing to energize the utility line. The inverter controls shall actually remain connected to the utility to allow sensing of utility electrical conditions for use by the 'reconnect' feature. The purpose of the allowed time delay is to ride through short-term disturbances to avoid excessive nuisance tripping.

Observation: The required disconnection time for VDE 0126-1-1 is much shorter (0.2 sec.) and fast voltage monitoring is thus required.

3.3.2 Frequency Deviations (see Table 3.3)

The purpose of the allowed range and time delay is to ride through short-term disturbances to avoid excessive nuisance tripping in weak-grid situations.

Observation: The VDE 0126-1-1 allows a much lower frequency limit and thus frequency adaptive synchronization is required.

3.3.3 Reconnection after Trip

After a disconnection caused by abnormal voltage or frequency conditions the inverter can be reconnected only in the conditions given in Table 3.4.

Observation: The time delay in IEC 61727 is an extra measure to ensure resynchronization before reconnection in order to avoid possible damage.

Table 3.4 Conditions for reconnection after trip

IEEE 1547	IEC 61727	VDE 0126-1-1
$88 < V < 110$ (%)	$85 < V < 110$ (%) AND	
AND	$f_n - 1 < f < f_n + 1$ (Hz) AND	
$59.3 < f < 60.5$ (Hz)	Minimum delay of 3 min.	

Grid Requirements for PV

Table 3.5 DC current injection limitation

IEEE 1574	IEC 61727	VDE 0126-1-1
$I_{DC} < 0.5\ (\%)$ of the rated RMS current	$I_{DC} < 1\ (\%)$ of the rated RMS current	$I_{DC} < 1$ A Maximum trip time 0.2 sec.

3.4 Power Quality

The quality of the power provided by the photovoltaic system for the local AC loads and for the power delivered to the utility is governed by practices and standards on voltage, flicker, frequency, harmonics and power factor. Deviation from these standards represents out-of-bounds conditions and may require disconnection of the photovoltaic system from the utility.

3.4.1 DC Current Injection

DC current injection in the utility can saturate the distribution transformers, leading to overheating and trips. For the conventional PV systems with galvanic isolation, this problem is minimized, but with the new generation of transformerless PV inverters increased attention is required in this matter. Thus the limits of injected DC current given in Table 3.5 are accepted.

Observation: For IEEE 1574 and IEC 61727 the DC component of the current should be measured by using harmonic analysis (fast Fourier transform, or FFT) and there is no maximum trip time condition. During the test the measured DC component should be below the limits for different loading conditions (1/3, 2/3 and 3/3 of the nominal load). For VDE 0126-1-1 this condition requires a special designed current sensor that can detect this threshold and disconnect within the required trip time.

3.4.2 Current Harmonics

The PV system output should have low current-distortion levels to insure that no adverse effects are caused to other equipment connected to the utility system. The levels given in Table 3.6 are accepted.

Table 3.6 Maximum current harmonics

	IEEE 1547 and IEC 61727					
Individual harmonic order (odd)[a] (%)	$h < 11$	$11 \leq h < 17$	$17 \leq h < 23$	$23 \leq h < 35$	$35 \leq h$	Total harmonic distortion THD (%)
	4.0	2.0	1.5	0.6	0.3	5.0

[a] Even harmonics are limited to 25 % of the odd harmonic limits above.

Table 3.7 Current harmonic limits set by IEC 61000-3-2 (class A)

Odd harmonics		Even harmonics	
Order h	Current (A)	Order h	Current (A)
3	2.30	2	1.08
5	1.14	4	0.43
7	0.77	6	0.30
9	0.40	$8 \leq h \leq 40$	$0.23 \times 8/h$
11	0.33		
13	0.21		
$13 \leq h \leq 39$	$0.15 \times 15/h$		

Observation: The test voltage for IEEE 1574/IEC 61727 should be produced by an electronic power source with a voltage THD (thermohydrodynamic) < 2.5 % and individual voltage harmonics lower than 50 % of the current harmonic limits. The practice is to use an ideal sinusoidal power source so as not to influence the results by the background distortion.

As in Europe IEC 61727 is not yet approved, the practice is that the harmonic limits are set by the IEC 61000-3-2 for class A equipments (see Table 3.7).

Observation: The current limits in IEC 61000-3-2 are given in amperes and are in general higher than the ones in IEC 61727. For equipments with a higher current than 16 A but lower than 75 A, another similar standard IEEE 61000-3-12 [10] applies.

3.4.3 Average Power Factor

Only in IEC 61727 is it stated that the PV inverter shall have an average lagging power factor greater than 0.9 when the output is greater than 50 %. Most PV inverters designed for utility-interconnected service operate close to the unity power factor.

In IEEE 1574 there is no requirement for the power factor as this is a general standard that should also allow distributed generation of reactive power. No power factor requirements are mentioned in VDE 0126-1-1.

Observation: Usually the power factor requirement for PV inverters should now be interpreted as a requirement to operate at a quasi-unity power factor without the possibility of regulating the voltage by exchanging reactive power with the grid. For high-power PV installations connected directly to the distribution level local grid requirements apply as they may participate in the grid control. For low-power installations it is also expected that in the near future the utilities will allow them to exchange reactive power, but new regulations are still expected.

3.5 Anti-islanding Requirements

Definitely the most technical challenging requirement is the so-called anti-islanding. Islanding for grid-connected PV systems takes place when the PV inverter does not disconnect for a very short time after the grid is tripped, i.e. it is continuing to operate with local load. In the typical

Grid Requirements for PV

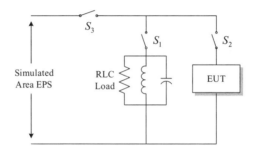

NOTES
1 – Switch S_1 may be replaced with individual switches on each of the RLC load components
2 – Unless the EUT has a unity output p.f., the receiver power component of the EUT is considered to be a part of the islanding load circuit in the figure.

Figure 3.1 Test setup for the anti-islanding requirement in IEEE 1547.1

case of a residential electrical system co-supplied by a roof-top PV system, grid disconnection can occur as a result of a local equipment failure detected by ground fault protection or of an intentional disconnection of the line for servicing. In both situations, if the PV inverter does not disconnect the following consequences can occur:

- Retripping the line or connected equipment can cause damage due to an out-of-phase closure.
- Safety hazard for utility line workers who assume de-energized lines during islanding.

In order to avoid these serious consequences safety measures called anti-islanding (AI) requirements have been issued and are embodied in standards.

3.5.1 AI Defined by IEEE 1547/UL 1741

In IEEE 1574 the requirement is that after an unintentional islanding where the distributed resources (DR) continue to energize a portion of the power system (island) through the PCC, the DR shall detect the islanding and cease to energize the area within 2 sec. In IEEE 1547.1 the test setup is described as shown in Figure 3.1, where EUT represents the equipment under test, i.e. the PV inverter.

The test conditions require that an adjustable RLC load should be connected in parallel between the PV inverter and the grid. The resonant LC circuit should be adjusted to resonate at the rated grid frequency f and to have a quality factor of $Q_f = 1$, or, in other words, the reactive power generated by C [VAR] should equal the reactive power absorbed by L [VAR] and should equal the power dissipated in R [W] at the nominal power P and rated grid voltage V. Thus the values for the local RLC load can be calculated as

$$\begin{cases} R = \dfrac{V^2}{P} \\ L = \dfrac{V^2}{2\pi f P Q_f} \\ C = \dfrac{P Q_f}{2\pi f V^2} \end{cases} \quad (3.1)$$

The parameters of the RLC load should be fine-tuned until the grid current through S_3 should be lower than 2 % of the rated value on a steady-state base. In this balanced condition, S_3 should be open and the time before disconnection should be measured and should be lower than 2 sec.

For three-phase PV inverters each phase should be tested with respect to the neutral individually. For three-wire three-phase PV inverters the local RLC load should be connected between phases.

The UL 1741 standard in theUS has been harmonized with the anti-islanding requirements stated in IEEE 1547.

Observation: The main difference with respect to the previous IEEE 929-2000 standard is that the requirement for the quality factor of the local RLC load has been reduced from 2.5 to 1.0, thus making compliance easier to achieve technically.

3.5.2 AI Defined by IEC 62116

In the draft version of IEC 62116-2006 similar AI requirements as those for IEEE 1547 are proposed. The test can also be utilized by other inverter interconnected DER. In the normative reference IEC 61727-2004 the ratings of the system valid in this standard have a rating of 10 kVA or less; however, the standard is subject to revision. The test circuit is the same as in the IEEE 1547.1 test (Figure 3.1) and a power balance is required before the island detection test. The requirement for passing the test contains more test cases but the conditions for confirming island detection do not have a significant deviation compared to the IEEE 1547.1 test.

The inverter is tested at three levels of output power (A: 100–105 %, B: 50–66 % and C: 25–33 % of inverters output power). Case A is tested under maximum allowable inverter input power and case C at minimum allowable inverter output power if > 33 %. The voltage at the input of the inverter also has specific conditions (see reference [8]). All conditions are to be tested at no deviation in real and reactive load power consumption than for condition A in a step of 5 % both real and reactive power iterated deviation from -10 to 10 % from the operating output power of the inverter. Conditions B and C are evaluated by deviating the reactive load in an interval of ± 5 % in a step of 1 % of the inverter output power.

The maximum trip time is the same as in IEEE 1547.1 standards 2 sec. In IEC 61727, there is no specific description of the anti-islanding requirements. Instead reference to IEC 62116 is made.

3.5.3 AI Defined by VDE 0126-1-1

The VDE 0126-1-1 allows the compliance with one of the following anti-islanding methods:

1. *Impedance measurement.* The test circuit is depicted in Figure 3.2.
    The procedure is based on local balancing of the active and reactive power using the variable RLC circuit and the switch S is opened in order to increase the grid impedance by 1.0 Ω. The inverter should disconnect within the required time, which is 5 sec. The test should be repeated for different values of the simulated grid impedance (R_2, L_2) in the range of 1 Ω (maximum of 0.5 Ω inductive reactance).

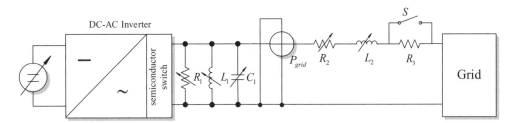

Figure 3.2 Test setup for anti-islanding requirements in VDE 0126-1-1

2. *Disconnection detection with RLC resonant load*. The test circuit is the same as the one from IEEE 1547.1, depicted in Figure 3.1, and the test conditions are that the RLC resonant circuit parameters should be calculated for a quality factor of $Q_f > 2$ using (3.1). With balanced power the inverter should disconnect after the disconnection of S_2 in a maximum of 5 sec. for the following power levels: 25 %, 50 % and 100 %.

For three-phase PV inverters a passive anti-islanding method is accepted by monitoring all three phases of voltage with respect to the neutral. This method is conditioned by having individual current control in each of the three phases.

Finding a software-based anti-islanding method has been a very challenging task, resulting in a large number of research works and publications. In Chapter 5 the most relevant AI methods are explained.

3.6 Summary

In this chapter an overview of the most relevant standards related to the grid connection requirements of PV inverters has been given. Great effort has been made by the international standard bodies in order to 'harmonize' the grid requirements for PV inverters worldwide. Recently the IEEE 1574 standard has made a big step in the direction of issuing a standard that includes grid requirements not only for PV inverters but for all distributed resources under 10 MVA. Underwriters Laboratories in the US have revised UL 1471 this year by accepting the grid requirements of IEEE 1574 and also IEC 62116 was revised to harmonize with the requirements of IEEE 1574 in anti-islanding requirements. Even the very specific German standard VDE 0126-1-1 was revised in 2006 where the grid impedance measurement has become optional and an alternative requirement very similar to IEEE 1574 was included. All these positive actions need to be adopted in different countries that still use their own local regulations.

The most relevant conditions from these standards are highlighted in order to envisage the impact on the control strategies. For designing purposes, readers are strongly recommended to access the complete texts of the standards and deal with all the related details.

References

[1] Dugan, R. C., Key, T. S. and Ball, G. J., 'Distributed Resources Standards'. *IEEE Industry Applications Magazine*, **12** (1), January–February 2006, 27–34.

[2] IEEE Std 929-2000, *IEEE Recommended Practice for Utility Interface of Photovoltaic (PV) Systems*, April 2000. ISBN 0-7381-1934-2 SH94811.
[3] UL Std 1741, *Inverters, Converters, and Controllers for Use in Independent Power Systems*, Underwriters Laboratories Inc. US, 2001.
[4] IEEE Std 1547-2003, *Standard for Interconnecting Distributed Resources with Electric Power Systems*, IEEE, June 2003. ISBN 0-7381-3720-0 SH95144.
[5] IEEE Std 1547.1-2005, *Standard Conformance Test Procedures for Equipment Interconnecting Distributed Resources with Electric Power Systems*, IEEE, July 2005. ISBN 0-7381-4736-2 SH95346.
[6] IEC 61727 Ed. 2, *Photovoltaic (PV) Systems – Characteristics of the Utility Interface*, December 2004.
[7] IEC 62116 CDV Ed. 1, *Test Procedure of Islanding Prevention Measures for Utility-Interconnected Photovoltaic Inverters*, IEC 82/402/CD, 2005.
[8] VDE V 0126-1-1, *Automatic Disconnection Device between a Generator and the Public Low-Voltage Grid*, Document 0126003, VDE Verlag, 2006.
[9] IEC 61000-3-2 Ed. 3.0, *Electromagnetic Compatibility (EMC) – Part 3-2: Limits – Limits for Harmonic Current Emissions (Equipment Input Current ≤ 16 A per Phase)*, November 2005. ISBN 2-8318-8353-9.
[10] IEC 61000-3-12, Ed. 1, *Electromagnetic Compatibility (EMC) – Part 3-12:Limits – Limits for Harmonic Currents Produced by Equipment Connected to Public Low-Voltage Systems with Input Current >16 A and ≤ 75 A per Phase*, November 2004.
[11] EN 61000-3-3, Ed. 1.2 *Electromagnetic Compatibility (EMC) – Part 3-3: Limits – Limitation of Voltage Changes, Voltage Fluctuations and Flicker in Public Low-Voltage Supply Systems, for Equipment with Rated Current ≤ 16 A per Phase and Not Subject to Conditional Connection*, November 2005. ISBN 2-8318-8209-5.
[12] IEC 61000-3-11, Ed. 1, *Electromagnetic Compatibility (EMC) – Part 3-11: Limits – Limitation of Voltage Changes, Voltage Fluctuations and Flicker in Public Low-Voltage Supply Systems – Equipment with Rated Current ≤ 75 A and Subject to Conditional Connection*, August 2000.
[13] Standard EN 50160, Voltage Characteristics of Public Distribution System, *CENELEC*: European Committee for Electrotechnical Standardization, Brussels, Belgium, November 1999.

4

Grid Synchronization in Single-Phase Power Converters

4.1 Introduction

Electrical grids are complex and dynamic systems affected by multiple eventualities such as continuous connection and disconnection of loads, disturbances and resonances resulting from the harmonic currents flowing through the lines, faults due to lightning strikes and mistakes in the operation of electrical equipment. Consequently, grid variables cannot be considered as constant magnitudes when a power converter is connected to the grid, but they should be continuously monitored in order to ensure that the grid state is suitable for the correct operation of the power converter. Moreover, when the power managed by this power converter cannot be neglected with respect to the rated power of the grid at the point of connection, the grid variables can be significantly affected by the action of such a power converter. Therefore, power converters cannot be considered as simple grid-connected equipment since they keep an interactive relationship with the grid and can actively participate in supporting the grid frequency and voltage, mainly when high levels of power are considered for the power converter. This implies, however, that the grid stability and safety conditions can be seriously affected in networks with extended usage of power converters, as is the case of distributed energy systems based on renewable energies. For this reason, many international grid codes have been in force during the last few years in order to regulate the behaviour of photovoltaic and wind energy systems in both regular steady-state and abnormal transient conditions, e.g. in the presence of grid faults.

Monitoring of the grid variables is a necessary task to be implemented in the power converter interfacing renewable energy sources to the grid. The grid codes state the voltage and frequency boundaries within which the photovoltaic and wind generators should remain connected to the grid while ensuring stable operation. Hence, the power converter of these renewable energy systems should accurately screen the grid variables at the point of common coupling in order to trip the disconnection procedure when they go beyond the limits set by the grid codes. These grid codes also specify certain dynamic requirements in the connection to the grid of distributed generators; e.g. the instantaneous response of wind turbines in the presence of

---

Grid Converters for Photovoltaic and Wind Power Systems   Remus Teodorescu, Marco Liserre, and Pedro Rodríguez
© 2011 John Wiley & Sons, Ltd

transient grid faults is clearly delimited in recent grid codes. Consequently, grid monitoring algorithms to be implemented in grid-connected converters should detect the grid state in a fast and precise way in order to fulfil both precision and time response requirements demanded by the grid codes.

Grid monitoring and grid synchronization are two closely linked concepts. Actually, grid synchronization of a power converter is not nothing but an instantaneous monitoring of the state of the grid to which the power converter is connected. Grid synchronization is an adaptive process by means of which an internal reference signal generated by the control algorithm of a grid-connected power converter is brought into line with a particular grid variable, usually the fundamental component of the grid voltage.

Grid synchronization is a fundamental issue in the connection of power converters to the grid since it allows the grid and the synchronized power converter to work in unison. Information generated by the grid synchronization algorithm is used at different levels of the control system of a grid-connected converter. As explained in Appendix A, information about the phase-angle of the grid voltage is necessary to transform the grid variables from the natural reference frame to the synchronous reference frame, which makes it possible to deal with DC variables in the regulation of AC currents or voltages supplied to the grid by the power converter. Likewise, knowing the magnitude and phase-angle of the grid voltage allows regulation of the active and reactive power delivered to the grid.

The purpose of this chapter is to introduce the basis of the synchronization problem in single-phase systems and to present some of the most relevant structures of synchronization systems used in single-phase networks.

4.2 Grid Synchronization Techniques for Single-Phase Systems

Grid synchronization of single-phase grid-connected converters lies in the accurate detection of the attributes of the grid voltage in order to tune an internal oscillator of the power converter controller to the oscillatory dynamics imposed by the grid. Usually, the main attributes of interest for interfacing renewable energies to the grid by using power converters are the amplitude and the phase-angle of the fundamental frequency component of the grid voltage. However, the detection of other harmonic components can also be interesting for implementing extra functionalities in the grid-connected power converter of distributed generators, such as power conditioning, resonance damping or grid impedance detection. Therefore, grid synchronization techniques bear a certain similarity to the harmonic detection methods used in power systems and can be classified into two main groups, namely the frequency-domain and the time-domain detection methods.

The frequency-domain detection methods are usually based on some discrete implementation of the Fourier analysis. The Fourier series, the discrete Fourier transform (DFT) and the recursive discrete Fourier transform (RDFT) will be briefly presented in the following as possible grid synchronization techniques in single-phase systems. By definition, the frequency analysis assumes that the fundamental frequency of the processed signal is a well-known and constant magnitude. The sample frequency of the signal processor should be an integer multiple of the fundamental grid frequency.

The time-domain detection methods are based on some kind of adaptive loop that enables an internal oscillator to track the component of interest of the input signal. The most extended synchronization method in engineering applications, the phase-locked loop (PLL), will be presented and discussed in this chapter. The application of a simple PLL structure for synchronizing with the low-frequency voltage of a conventional grid (50/60 Hz) will provide evidence of the need to improve its structure by using some kind of quadrature signal generator (QSG). Finally, the frequency-locked loop (FLL) will be introduced as a very effective synchronization technique to be implemented in a grid-connected power converter, mainly when the grid is affected by transient disturbances due to grid faults.

4.2.1 Grid Synchronization Using the Fourier Analysis

Fourier analysis is a mathematical tool allowing a given function to be transformed from the time domain to the frequency domain and vice versa. This duality in the conception of mathematical functions entails significant operating advantages; e.g. linear differential equations with constant coefficients in the time domain can be solved by using ordinary algebraic equations in the frequency domain and complicated convolution integrals can be transformed into simple multiplications. Moreover, signal processing techniques can be quickly evaluated on computers by using discrete versions of the Fourier transform. These valuable features have turned the Fourier analysis into a fundamental signal processing tool in such fields as acoustics, imaging, communications, control, etc. Since there exist very rigorous books dealing with the different branches of the Fourier analysis in the technical literature [1, 2], this section avoids entering into dense studies and just introduces essential concepts regarding the application of Fourier analysis to grid synchronizing power converters.

4.2.1.1 The Fourier Series

Joseph Fourier (1768–1830) presented in 1807 a new analysis technique for solving the differential equations describing heat flow in a metal plate. It was the first version of what is known today as the Fourier Series [3]. This signal analysis technique allows the frequency components of a periodic signal to be obtained by multiplying it by a set of basic functions (sine/cosine) at different frequencies. When a cosine function is multiplied by a unitary cosine at the same frequency, its amplitude (divided by 2) appears as a DC component in the resultant squared function. A similar result is obtained when squaring a sine function, i.e.

$$A\cos(n\omega t)\cos(n\omega t) = A\cos^2(n\omega t) = \frac{A}{2} + \frac{A\cos(2n\omega t)}{2}$$
$$B\sin(n\omega t)\sin(n\omega t) = B\sin^2(n\omega t) = \frac{B}{2} - \frac{B\cos(2n\omega t)}{2}$$
(4.1)

Taking into account the fact that the DC value of the product of two sinusoidal functions with different frequencies is always equal to zero, it is possible to use these unitary sine/cosine basic functions as a sort of probe function for detecting the amplitude and phase-angle of the sinusoidal components of the processed signal at given frequencies. On the other hand, the DC component of the processed signal is conventionally obtained by calculating the mean value of the processed signal over a period. According to this reasoning, Fourier stated that a generic periodic signal $v(t)$ can be expressed by a sum of the following terms:

$$v(t) = a_0 + \sum_{n=1}^{\infty} (a_n \cos(n\omega t) + b_n \sin(n\omega t)) \qquad (4.2)$$

where the different coefficients are calculated by

$$a_0 = \frac{1}{T} \int_0^T v(t)\, dt$$

$$a_n = 2\frac{1}{T} \int_0^T v(t) \cos(n\omega t)\, dt = \frac{1}{\pi} \int_{-\pi}^{\pi} v(\theta) \cos(n\theta)\, d\theta \qquad (4.3)$$

$$b_n = 2\frac{1}{T} \int_0^T v(t) \sin(n\omega t)\, dt = \frac{1}{\pi} \int_{-\pi}^{\pi} v(\theta) \sin(n\theta)\, d\theta$$

According to (4.3), the Fourier series principle can easily be used to implement a selective band-pass filter by multiplying the input signal, v_{in}, by the sine/cosine basic functions at the desired frequency. A diagram of this adaptive band-pass filter is shown in Figure 4.1. Assuming the grid frequency as a constant and well-known magnitude, the order of the harmonic to be extracted at the output of this filter, v', is selected by setting the value of the parameter n. The amplitude and the phase-angle of this frequency component are given by

$$V'_n = V_n \angle \theta_n \begin{cases} V_n = \sqrt{a_n^2 + b_n^2} \\ \theta_n = \arctan \dfrac{b_n}{a_n} \end{cases} \qquad (4.4)$$

In the diagram of Figure 4.1, the mean value of the signals resulting from multiplying v_{in} by the sine/cosine basic functions is obtained by using a low-pass filter (LPF). The cut-off frequency of this LPF is a function of the lowest frequency component of the input signal. Under regular operating conditions, the lowest frequency component of the grid voltage

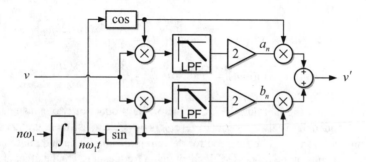

Figure 4.1 Adaptive filter based on Fourier series decomposition

matches the fundamental frequency component, ω_1, usually 50/60 Hz. This fundamental frequency component is actually the variable of interest in most of the applications of the synchronizing power converters to the grid. Therefore, the frequency of the sine/cosine basic functions will be set at ω_1 as well, i.e. $n = 1$. Consequently, the lowest frequency component at the input of the LPF will be at $2\omega_1$, usually 100/120 Hz. It implies that the cut-off frequency of the LPF should be at least one decade lower than $2\omega_1$, around 10/12 Hz – which implies a very slow dynamic response of the system. This cut-off frequency should be even lower if either subharmonics or DC components are present in the input signal.

Using Euler's equation, developed in the early 1800s, the sine/cosine basic functions can be written as

$$\cos(n\omega t) = \frac{e^{jn\omega t} + e^{-jn\omega t}}{2}$$
$$\sin(n\omega t) = \frac{e^{jn\omega t} - e^{-jn\omega t}}{2j} \quad (4.5)$$

Therefore, the Fourier series' coefficients of (4.3) can be also calculated by

$$a_n = \frac{1}{T}\int_0^T v(t)\left(e^{jn\omega t} + e^{-jn\omega t}\right) dt$$
$$b_n = \frac{-j}{T}\int_0^T v(t)\left(e^{jn\omega t} - e^{-jn\omega t}\right) dt \quad (4.6)$$

Defining the complex coefficient c_n from (4.6) as

$$c_n = \frac{1}{2}(a_n - jb_n) = \frac{1}{T}\int_0^T v(t)e^{-jn\omega t} dt \quad (4.7)$$

and developing (4.2), the Fourier series of $v(t)$ can be rewritten as

$$v(t) = a_0 + \sum_{n=1}^{\infty} c_n e^{jn\omega t} + \sum_{n=1}^{\infty} c_n^* e^{-jn\omega t} \quad (4.8)$$

where c_n^* is the conjugate complex of c_n. Taking into account that $a_0 = c_0$ and $c_n^*|_{n<0} = c_n|_{n>0}$, the previous expression can become very simple by changing the range of the summations as follows:

$$v(t) = a_0 + \sum_{n=1}^{\infty} c_n e^{jn\omega t} + \sum_{n=1}^{\infty} c_n e^{-jn\omega t} = \sum_{n=0}^{\infty} c_n e^{jn\omega t} + \sum_{n=-1}^{-\infty} c_n e^{jn\omega t} = \sum_{n=-\infty}^{\infty} c_n e^{jn\omega t} \quad (4.9)$$

This is the compact complex form of the Fourier series, in which positive and negative frequencies are considered.

4.2.1.2 The Discrete Fourier Transform

The expression of (4.9) is very useful for introducing the concept of the Fourier transform in an intuitive way. To do that, we will consider that the pulse function described by the expressions of (4.10) constitutes the repetition pattern of a train of pulses to be processed using (4.9).

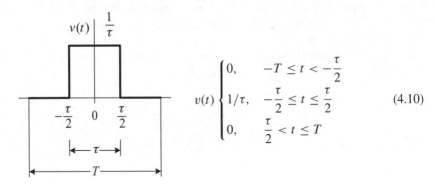

$$v(t) \begin{cases} 0, & -T \leq t < -\dfrac{\tau}{2} \\ 1/\tau, & -\dfrac{\tau}{2} \leq t \leq \dfrac{\tau}{2} \\ 0, & \dfrac{\tau}{2} < t \leq T \end{cases} \qquad (4.10)$$

Complex coefficients of this Fourier series are calculated by solving the integral of (4.7). If we consider that the period T is made gradually longer while the pulse duration τ is kept constant, the complex coefficients calculated by (4.7) become smaller each time, since T appears to be dividing in (4.7) and the value of the integral remains constant. Actually, as T approaches infinity, the signal $v(t)$ becomes aperiodic and all the coefficients calculated by (4.7) are equal to zero.

To allow the time/frequency duality to be applied to the analysis of aperiodic signals, with $T = \infty$, the period T is removed from the denominator of (4.7) and the result is rewritten as follows:

$$V(\omega) = \mathcal{F}[v(t)] = \int_{-\infty}^{\infty} v(t) e^{-j\omega t} \, dt \qquad (4.11)$$

The last expression is known as the Fourier transform and, as long as the integral exists, it allows continuous functions to be obtained in the frequency domain representing continuous periodic and aperiodic functions in the time domain. As an example, Figure 4.2 represents the Fourier transform of the rectangular function described by (4.10).

The Fourier transform of (4.11) has been proved extremely useful when analysing electrical signals and circuits. When the Fourier transform is programmed in a digital signal processor, the integral of (4.11) is implemented by summation of a finite number of samples equally spaced in time. In such cases, the discrete input signal is defined as

$$v[k] = v(t)\delta(t - kT_S) \quad \text{with} \quad k = 0, 1, \ldots, N - 1 \qquad (4.12)$$

where $\delta(x)$ is Dirac's delta function used for sampling, T_S is the sampling period and N is the number of samples to be processed. In this discrete signal, the product NT_S sets the duration of the repetition pattern of the input signal and usually matches with T, the period of the

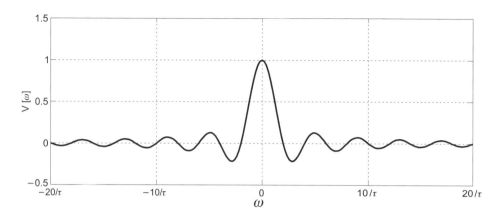

Figure 4.2 Fourier transform of the a rectangular pulse

fundamental frequency component. Therefore, the discrete Fourier transform (DFT) is defined by the following finite summation:

$$V[n] = \sum_{k=0}^{N-1} v[k] \cdot e^{-j2\pi \frac{k}{N} n} \quad \text{with} \quad n = 0, 1, \ldots, N-1 \quad (4.13)$$

It is worth remembering that (4.13) is not a decomposition but a transformation, in which N equally spaced samples in the time domain are transformed into N complex values in the frequency domain representing the finite input signal. The inverse discrete Fourier transformation (IDFT) is defined as

$$v[k] = \frac{1}{N} \sum_{n=0}^{N-1} V[n] e^{j2\pi \frac{n}{N} k} \quad (4.14)$$

Computation of N samples using the DFT algorithm requires N^2 complex multiplications and $N^2 - N$ complex additions. For this reason the DFT did not come into widespread use until the development of microprocessors.

The DFT has been conventionally used to identify the harmonic content of voltages and currents in the power system, mainly with the purpose of calculating power quality indexes or controlling conditioning systems [4]. However, the DFT technique can also be applied to extract the fundamental frequency component of the grid voltage with the aim of synchronizing power converters to such a grid [5]. Applications of this method to converter synchronization [6] and utility power measurement [7] have been reported in the literature. In this technique, a phase error occurs when the sampling of the DFT works asynchronously to the fundamental grid frequency. Strategies based on polynomial regression analysis [7] and the use of time-varying Fourier coefficients [6] have been proposed as suitable strategies to compensate for this phase error.

Based on the inherent symmetry of the complex calculations of the DFT, a new algorithm called the fast Fourier transform (FFT) was presented in 1965 [8]. In the FFT, a mathematical simplification called decimation allows a huge reduction in the computational burden of this algorithm. Computation of N samples with the FFT algorithm requires $(N/2)\log_2 N$ complex multiplications and $N\log_2 N$ complex additions. Although extremely important in itself, the understanding of the mechanism of the FFT algorithm is not an objective of this book. Moreover, there are many references describing the famous FFT 'butterflies' in the literature [1, 2]. Therefore, the implementation details of the FFT algorithm will be skipped over. The FFT algorithm is not suitable for extracting a single-frequency component from the input signal. For this reason it is mainly applied in grid monitoring tasks but not in grid synchronization of power converters.

4.2.1.3 The Recursive Discrete Fourier Transform

The computational burden of the DFT algorithm presented in (4.13) is relatively heavy, so it cannot be calculated by the digital controller of a power converter every sampling period – even when only a single frequency component of the grid voltage is computed. For this reason, a recursive formulation of the DFT algorithm is conventionally used [9, 10]. In this recursive algorithm, the nth frequency component of the input signal at the $[k_S]$ instant is calculated from the value of input signal at the $[k_S]$ instant and the value of the nth frequency component at the $[k_S - 1]$ instant. To explain how it works, the DFT algorithm is computed at the $[k_S - 1]$ and $[k_S]$ instants in order to extract the nth harmonic of the input signal:

$$V[n]\Big|_{k_S-1} = \sum_{k=k_S-N}^{k_S-1} v[k]e^{-j2\pi \frac{k}{N}n} \tag{4.15}$$

$$V[n]\Big|_{k_S} = \sum_{k=k_S-N+1}^{k_S} v[k]e^{-j2\pi \frac{k}{N}n} \tag{4.16}$$

Subtracting (4.15) from (4.16) gives

$$V[n]\Big|_{k_S} = V[n]\Big|_{k_S-1} + v[k_S]e^{-j2\pi \frac{k_S}{N}n} - v[k_S - N]e^{-j2\pi \frac{k_S-N}{N}n} \tag{4.17}$$

Since $e^{-j2\pi \frac{k_S-N}{N}n} = e^{-j2\pi \frac{k_S}{N}n}$, the *recursive discrete* Fourier transform (RDFT) algorithm can be formulated as

$$V[n]\Big|_{k_S} = V[n]\Big|_{k_S-1} + (v[k_S] - v[k_S - N])e^{-j2\pi \frac{k_S}{N}n} \tag{4.18}$$

It is necessary to remember that all the discrete Fourier transforms presented in this section stem from the complex form of the Fourier series formulated by (4.9), which includes positive and negative frequencies. Taking into account that the complex values resulting from the Fourier transform are symmetric with respect to the zero frequency with half the amplitude of the corresponding nth harmonic, such nth harmonic can be reconstructed in the time domain by

$$v[k] = \frac{2}{N}V[n]e^{j2\pi \frac{n}{N}k} \tag{4.19}$$

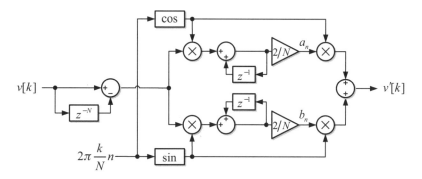

Figure 4.3 Discrete adaptive filter based on the RDFT

Therefore, the RDFT can be applied to implement a discrete adaptive band-pass filter to extract the nth frequency component of the input signal, as shown in Figure 4.3. The amplitude and phase-angle of such a frequency component can be calculated by (4.4). As in all the discrete Fourier analyses, the RDFT gives rise to errors in amplitude and phase-angle estimation when the product NT_S does not match the fundamental frequency of the input signal. An analysis of the phase error occurring in asynchronous conditions allows this drawback to be solved by implementing an extra control loop, which either adjusts the sampling window to match the grid frequency or adds a phase offset to cancel the phase error produced by the RDFT [11].

4.2.2 Grid Synchronization Using a Phase-Locked Loop

A phase-locked loop (PLL) is a closed-loop system in which an internal oscillator is controlled to keep the time of some external periodical signal by using the feedback loop. We are very used to employing this kind of system in our daily lives. Every time we listen to the radio, we tune an internal oscillator to the frequency of a carrier signal on which our favourite radio station is emitting its programmes. In fact, we are used to bringing a local oscillator with us – our own wristwatch. We usually schedule our daily activities according to the signals provided by some clock. When we travel to another country we experiment with a sort of 'phase jump'. One the first things we habitually do on arriving at the new place is to adjust our wristwatch to the new local time so that we can properly follow the daily rhythm of that society.

The PLL techniques are broadly used in fields like communications, computers and modern electronics. They can generate stable frequencies synchronized with external periodical events, recover relevant signals from distorted sources or distribute clock-timing pulses in complex control systems.

A grid-connected power converter perfectly matches the PLL's philosophy since it should work in harmony with the grid. It should phase-lock its internal oscillator to some particular grid power signal in order to generate an amplitude and phase-coherent internal signal that is used by different blocks of the control system. The first grid-connected power converters were based on silicon-controlled rectifiers. These power converters offered a low degree of control and were synchronized to the grid by detecting the zero-crossing of the grid voltage.

The zero-crossing detection method uses comparators for detecting changes in the polarity of the grid voltage. This detection technique presents some drawbacks, such as inaccuracy and detection of multiple zero-crossings in the case of distorted grid voltage. Such drawbacks are even more important in the case of weak grids – grids with a high grid impedance – since their grid voltage is prone to be notably distorted by the harmonics, switching notches and noise. For this reason, modified methods based on comparator circuits with dynamic hysteresis [12], curve-fitting [6] or predictive digital filtering algorithms [11] have been proposed in the literature for cancelling delays in zero-crossing detection and attenuate the adverse effects resulting from the noise and switching notches of the grid voltage. Some of these techniques are relatively complex and their performance is not completely satisfactory when the grid voltage is affected by low-frequency harmonics or remarkable frequency variations.

Currently, grid-connected power converters are based on modern power semiconductor devices operating in switched mode – even at the level of mega-converters, which allow a high degree of control. Advanced synchronous control systems, relying on fast and precise PLLs, are applied to these converters. If a synchronous controller rotating at the fundamental grid frequency is observed from a stationary reference, a relative difference is not expected to exist between the frequency of both the internal variables of the controller and the fundamental grid variables – the effect of harmonics is neglected here. Therefore, the AC grid variables look like DC variables for a properly tuned synchronous controller. As a result, well-known DC controllers can be used to regulate AC magnitudes oscillating at the fundamental grid frequency, which makes the tuning process easier. Moreover, delays introduced by elements acting as the modulator of the power converter and sensors can be compensated by just advancing the phase-angle detected by the PLL. In addition, the PLL provides continuous information about the phase-angle and amplitude of the magnitude of interest, generally the fundamental grid voltage, which allows space vector based controllers and modulators to be implemented, even when working with single-phase signals.

4.2.2.1 Basic Structure of a Phase-Locked Loop

The basic structure of a phase-locked loop (PLL) is shown in Figure 4.4. It consists of three fundamental blocks:

- The *phase detector* (PD). This block generates an output signal proportional to the phase difference between the input signal, v, and the signal generated by the internal oscillator of

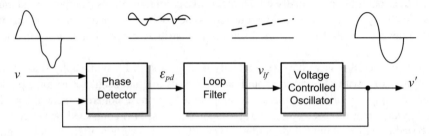

Figure 4.4 Basic structure of a PLL

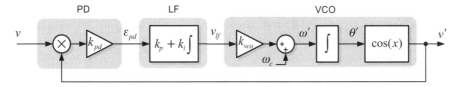

Figure 4.5 Block diagram of an elementary PLL

the PLL, v'. Depending on the type of PD, high-frequency AC components appear together with the DC phase-angle difference signal.
- The *loop filter* (LF). This block presents a low-pass filtering characteristic to attenuate the high-frequency AC components from the PD output. Typically, this block is constituted by a first-order low-pass filter or a PI controller.
- The *voltage-controlled oscillator* (VCO). This block generates at its output an AC signal whose frequency is shifted with respect to a given central frequency, ω_c, as a function of the input voltage provided by the LF.

Different techniques can be used to implement each of the blocks constituting a PLL. A detailed review of these blocks is outside the scope of this book, as it is necessary to consult specialized literature to deepen knowledge of this issue [13]. Basic equations describing the behaviour of an elementary PLL will be developed in the following.

4.2.2.2 Basic Equations of the PLL

The block diagram of an elementary PLL is shown in Figure 4.5. In this case, the PD is implemented by means of a simple multiplier, the LF is based on a PI controller and the VCO consists of a sinusoidal function supplied by a linear integrator.

If the input signal applied to this system is given by

$$v = V \sin(\theta) = V \sin(\omega t + \phi) \tag{4.20}$$

and the signal generated by the VCO is given by

$$v' = \cos(\theta') = \cos(\omega' t + \phi') \tag{4.21}$$

the phase error signal from the multiplier PD output can be written as

$$\begin{aligned}\varepsilon_{pd} &= V k_{pd} \sin(\omega t + \phi) \cos(\omega' t + \phi') \\ &= \frac{V k_{pd}}{2} \left[\underbrace{\sin((\omega - \omega')t + (\phi - \phi'))}_{\text{low-frequency term}} + \underbrace{\sin((\omega + \omega')t + (\phi + \phi'))}_{\text{high-frequency term}} \right]\end{aligned} \tag{4.22}$$

Since the high-frequency components of the PD error signal will be cancelled out by the LF, only the low-frequency term will be considered from now on. Therefore, the PD error signal to be considered in this analysis is

$$\bar{\varepsilon}_{pd} = \frac{Vk_{pd}}{2}\sin((\omega - \omega')t + (\phi - \phi')) \qquad (4.23)$$

If it is assumed that the VCO is well tuned to the input frequency, i.e. with $\omega \approx \omega'$, the DC term of the phase error signal is given by

$$\bar{\varepsilon}_{pd} = \frac{Vk_{pd}}{2}\sin(\phi - \phi') \qquad (4.24)$$

It can be observed in (4.24) that the multiplier PD produces nonlinear phase detection because of the sinusoidal function. However, when the phase error is very small, i.e. when $\phi \approx \phi'$, the output of the multiplier PD can be linearized in the vicinity of such an operating point since $\sin(\phi - \phi') \approx \sin(\theta - \theta') \approx (\theta - \theta')$. Therefore, once the PLL is locked, the relevant term of the phase error signal is given by

$$\bar{\varepsilon}_{pd} = \frac{Vk_{pd}}{2}(\theta - \theta') \qquad (4.25)$$

This equation can be used to implement a small signal linearized model of the multiplier PD. In the locked state, this model represents a zero-order block whose gain depends on the input signal amplitude.

On its part, the averaged frequency of the VCO is determined by

$$\bar{\omega}' = (\omega_c + \Delta\bar{\omega}') = (\omega_c + k_{vco}\bar{v}_{lf}) \qquad (4.26)$$

where ω_c is the centre frequency of the VCO and is supplied to the PLL as a feed-forward parameter dependent on the range of frequency to be detected. Therefore, small signal variations in the VCO frequency are given by

$$\tilde{\omega}' = k_{vco}\tilde{v}_{lf} \qquad (4.27)$$

and variations in the phase-angle detected by the PLL can be written as

$$\tilde{\theta}'(t) = \int \tilde{\omega}' dt = \int k_{vco}\tilde{v}_{lf} dt \qquad (4.28)$$

4.2.2.3 Linearized Small Signal Model of a PLL

Previous equations in the time domain can be effortlessly translated to the complex frequency domain by using the Laplace transform. If it is considered that $k_{pd} = k_{vco} = 1$, the following

Grid Synchronization in Single-Phase Power Converters

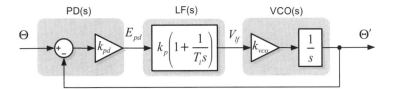

Figure 4.6 Small signal mode of an elementary PLL

expressions are obtained for the signals of interest in the PLL:

- Phase detector:
$$E_{pd}(s) = \frac{V}{2}\left(\Theta(s) - \Theta'(s)\right) \qquad (4.29)$$

- Loop filter:
$$V_{lf}(s) = k_p\left(1 + \frac{1}{T_i s}\right)\varepsilon_{pd}(s) \qquad (4.30)$$

- Controlled oscillator:
$$\Theta'(s) = \frac{1}{s}V_{lf}(s) \qquad (4.31)$$

Therefore, the block diagram of Figure 4.6 can be drawn depicting the small signal model of the PLL. A straightforward analysis of this closed-loop system (with $k_{pd} = k_{vco} = 1$ and $V = 1$) gives the following characteristic transfer functions:

Open-loop phase transfer function:

$$F_{OL}(s) = PD(s) \cdot LF(s) \cdot VCO(s) = k_{in}\frac{k_p\left(1 + \frac{1}{T_i s}\right)}{s} = \frac{k_p s + \frac{k_p}{T_i}}{s^2} \qquad (4.32)$$

Closed-loop phase transfer function:

$$H_\theta(s) = \frac{\Theta'(s)}{\Theta(s)} = \frac{LF(s)}{s + LF(s)} = \frac{K_p s + \frac{K_p}{T_i}}{s^2 + K_p s + \frac{K_p}{T_i}} \qquad (4.33)$$

Closed-loop error transfer function:

$$E_\theta(s) = \frac{E_{pd}(s)}{\Theta(s)} = 1 - H_\theta(s) = \frac{s}{s + LP(s)} = \frac{s^2}{s^2 + K_p s + \frac{K_p}{T_i}} \qquad (4.34)$$

Previous transfer functions allow some preliminary conclusions to be made about the performance of the PLL of Figure 4.5. The open-loop transfer function of (4.32) shows that this PLL is a type 2 system, with two poles at the origin, which means that it is able to track even a constant slope ramp in the input phase-angle without any steady-state error. On its part, the transfer function of (4.33) reveals that the PLL presents a low-pass filtering characteristic in the detection of the input phase-angle, which is a very interesting feature for attenuating the detection error caused by possible noise and high-order harmonics in the input signal. These second-order transfer functions can be written in a normalized way as follows:

$$H_\theta(s) = \frac{2\zeta\omega_n s + \omega_n^2}{s^2 + 2\zeta\omega_n s + \omega_n^2} \quad (4.35)$$

$$E_\theta(s) = \frac{s^2}{s^2 + 2\zeta\omega_n s + \omega_n^2} \quad (4.36)$$

where

$$\omega_n = \sqrt{\frac{K_p}{T_i}} \quad \text{and} \quad \xi = \frac{\sqrt{K_p T_i}}{2}$$

The dynamic response of a second-order system is studied in many books on control systems. The following approximated expression is proposed in reference [14] for estimating the settling time, t_S, measured from the start time to the time in which the system stays within 1% of the steady-state response of a particular second-order system responding to a step input:

$$t_S = 4.6\,\tau \quad \text{with} \quad \tau = \frac{1}{\xi\omega_n} \quad (4.37)$$

This formula can also be used to obtain a rough estimate of the settling time of a system defined by (4.35), and hence the tuning parameters of the PI controller of the PLL of Figure 4.5 can be set as a function of the settling time as follows:

$$K_p = 2\xi\omega_n = \frac{9.2}{t_s}, \quad T_i = \frac{2\xi}{\omega_n} = \frac{t_s \xi^2}{2.3} \quad (4.38)$$

It is worth remarking that the expressions in (4.38) are obtained under the assumption of the unitary input signal, i.e. $V = 1$. Otherwise, these expressions for setting the tuning parameters of the PI controller should be divided by the amplitude of the input signal, V.

Moreover, as commented in reference [14], the expressions resulting from (4.37) should be taken as guides and not precise formulas. They provide a rough estimate of the time response of the system, but it should finally be checked, usually by simulation, in order to verify that the time specifications have been properly met.

4.2.2.4 PLL Response

Figure 4.7 shows some representative plots describing the performance of the PLL of Figure 4.5 when synchronizing with a $100V_{peak}$ single-phase grid affected by both a phase-angle jump ($+45°$) and a frequency jump (from 50 to 45 Hz) at $t = 100$ ms. In this case, the settle time was set to $t_S = 100$ ms with $\xi = 1/\sqrt{2}$.

Figure 4.7(a) shows the grid voltage, Figure 4.7(b) the actual grid frequency in bold trace and the frequency estimated by the PLL in thin trace, Figure 4.7(c) the estimated phase-angle and Figure 4.7(d) the error made in the phase-angle estimation, $\theta - \theta'$. It can be observed in these plots that an oscillatory steady-state error is made in the estimation of the frequency and phase-angle of the input voltage. This error is a consequence of the high-frequency term

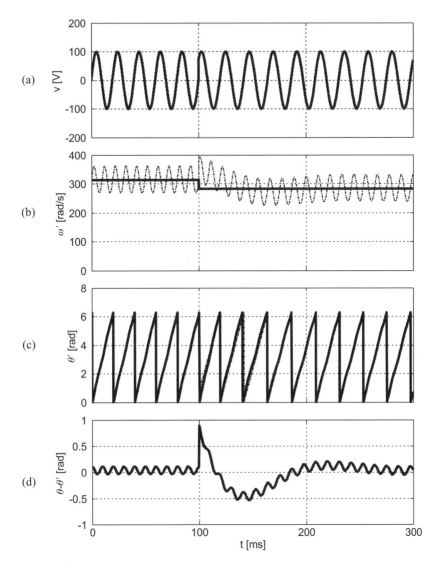

Figure 4.7 Step response of an elementary PLL

existing at the output of the multiplier PD – see (4.22). The amplitude of this oscillatory error can be attenuated by setting a longer settling time for the PLL, which is equivalent to decreasing the bandwidth of the system. In a general case, this bandwidth is given by

$$\omega_{-3\,dB} = \omega_n \left[1 + 2\xi^2 + \sqrt{(1+2\xi)^2 + 1}\right]^{1/2} \qquad (4.39)$$

Plots of Figure 4.7 confirm that the PLL reaches a steady-state response, within a given error margin, after a settling time close to 100 ms. To reduce the magnitude of the settling error it is only necessary to increase the number of constant times in (4.37).

4.2.2.5 Key Parameters of a PLL

Specialized texts propose several key parameters for describing the performance of a PLL. According to reference [13], the main key parameters can be summarized as:

- The hold range $\Delta\omega_H$ is the frequency range at which a PLL is able to keep statically phase-locked. This is calculated as

$$\Delta\omega_H = K_{pd}K_{vco}LF(0) \qquad (4.40)$$

where $LF(0)$ is the DC gain of the loop filter. For a PI controller, $LF(0) = \infty$ and the hold range is only limited by the frequency range of the VCO.

- The pull-in range $\Delta\omega_P$ is the frequency range at which a PLL will always became locked, but the process can become rather slow. This range tends to infinite for the PI loop filter. The time that the PLL needs to be locked when a pull-in process occurs after a variation of the input frequency, $\Delta\omega_{in}$, can be calculated as

$$T_P \approx \frac{\pi^2}{16} \frac{\Delta\omega_{in}^2}{\xi\omega_n^3} \qquad (4.41)$$

- The lock range $\Delta\omega_L$ is the frequency range within which a PLL locks within one single-beat note between the reference frequency and the output frequency. The lock range for the PI loop filter can be approximated to

$$\Delta\omega_L \approx 2\xi\omega_n \approx 2\xi\sqrt{\frac{k_p}{T_i}} \qquad (4.42)$$

and the lock-in time can be calculated by

$$T_L \approx \frac{2\pi}{\omega_n} \qquad (4.43)$$

- The pull-out range $\Delta\omega_{PO}$ is the dynamic limit for stable operation of a PLL. If tracking is lost within this range, a PLL will become phase-locked again after a time longer than the lock-time but shorter than the pull-in time. This range can be calculated as

$$\Delta\omega_{PO} \approx 1.8\omega_n (\xi + 1) \qquad (4.44)$$

4.3 Phase Detection Based on In-Quadrature Signals

As shown in the previous section, the bandwidth of the single-phase PLL of Figure 4.5 should be very low when applied to the detection of power signal parameters in a grid-connected application in order to smooth enough oscillations in the detected frequency and phase-angle. Moreover, the PLL of Figure 4.5 presents an additional drawback relating to the calculation of its key parameters when used for grid-connected applications. Figure 4.8 shows the estimated frequency and phase-angle during the pull-in process that occurs when the 50 Hz grid voltage is suddenly applied to the input of the PLL of Figure 4.5 with no centre frequency supplied to the VCO, i.e. with $\omega_o = 0$. The tuning parameters of the PLL were adjusted to maintain a settle time of $t_s = 100$ ms with $\xi = 1/\sqrt{2}$.

According to (4.41), the pull-in time of this PLL should be 312.7 ms. However, this time is around 2s in the plots shown in Figure 4.8. This difference between the calculated pull-in time and the one observed in simulation also affects the rest of the key parameters of the PLL. Faced with these results, one might question the reliability of formulas presented in the previous section for calculating the key parameters of the PLL. It is worth saying that those formulas are correct. However, they were obtained in a simplified way after making some assumptions. One of these assumptions was that the frequency of the signal to be phase-locked is much higher than the bandwidth of the PLL. Under this assumption, the high-frequency term of the phase error signal provided by the multiplier PD can be neglected when the dominant dynamic response of the PLL is studied. In a grid-connected application, however, the grid frequency is very close to the cut-off frequency of the PLL. When the PLL is locked, the high-frequency oscillations in the phase-angle error signal are only twice the input frequency. For example, in the simulation of Figure 4.8, the PLL bandwidth resulting from the -3 dB cut-off frequency of (4.33) was $\omega_{-3dB} = 21.3$ Hz, while the oscillations to be cancelled out by the LF were at 100 Hz – twice the grid frequency, which was 50 Hz. With these very close frequencies, the assumption about a complete cancellation of the high-frequency term of (4.22) by the LF can no longer be accepted as a valid hypothesis. Therefore, a new PD, different to the simple multiplier PD of Figure 4.5, should be used when designing a PLL for grid-connected applications in order to cancel out oscillations at twice the grid frequency in the phase-angle error signal.

Figure 4.9 shows a PD based on a set of in-quadrature signals. The quadrature signal generator (QSG) of this figure is supposed to be ideal, being able to extract a clean set of in-quadrature signals without introducing any delay at any frequency from a given distorted input signal.

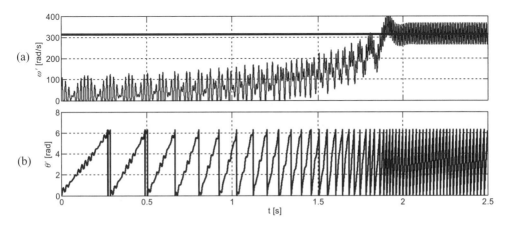

Figure 4.8 Pulling process of a PLL

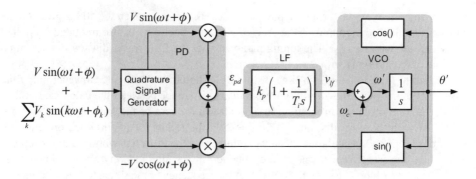

Figure 4.9 Diagram of a PLL with an ideal in-quadrature PD

The phase-angle error signal resulting from this ideal in-quadrature PD is given by

$$\varepsilon_{pd} = V \sin(\omega t + \phi) \cos(\omega' t + \phi') - V \cos(\omega t + \phi) \sin(\omega' t + \phi')$$
$$= V \sin((\omega - \omega')t + (\phi - \phi')) = V \sin(\theta - \theta') \quad (4.45)$$

According to this equation, when the PLL is well synchronized, i.e. with $\omega = \omega'$, the in-quadrature PD does not generate any steady-state oscillatory term, which allows the PLL bandwidth to increase and overcomes aforementioned discrepancies regarding calculation of the PLL key parameters.

As an example, the PLL with the in-quadrature PD of Figure 4.9 is used to synchronize to a $100 V_{peak}/50$ Hz single-phase grid. This PLL was tuned to have a settle time of $t_S = 100$ ms and $\xi = 1/\sqrt{2}$. The frequency estimated by such a PLL during the pull-in process with $\omega_c = 0$ is shown in Figure 4.10(a), where it can be observed that the pull-in time matches so well the 312.7 ms obtained from (4.41). In this example, the lock range and the lock time can be calculated from (4.42) and (4.43) respectively, being $\Delta\omega_L \approx \pm 92$ rad/s and $T_L \approx 96$ ms. Figure 4.10(b) shows the transient response of the frequency detected by the PLL of Figure 4.9 when the input frequency experiences a negative jump of 62.8 rad/s, which is within the frequency lock range. As can be observed in this figure, the settle time of the detected frequency signal matches with the lock time previously calculated. Therefore, it can be concluded that the in-quadrature PD makes possible to design a grid synchronization PLL according to the general design rules applied to PLLs used in other fields.

A review of the trigonometric expression of (4.45) reveals that this is a part of the Park transformation, which is defined by the equation A.19 in Appendix A. Therefore, the diagram of Figure 4.9 can be redrawn as shown in Figure 4.11, where the $\alpha\beta$ to dq transformation block responds to the following transformation matrix:

$$\begin{bmatrix} v_d \\ v_q \end{bmatrix} = \begin{bmatrix} \cos(\theta') & \sin(\theta') \\ -\sin(\theta') & \cos(\theta') \end{bmatrix} \begin{bmatrix} v_\alpha \\ v_\beta \end{bmatrix} \quad (4.46)$$

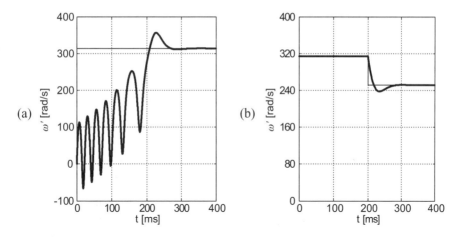

Figure 4.10 Transient response of a PLL with an in-quadrature PD

In Figure 4.11 the voltage-controlled oscillator (VCO) has been removed and a new block called the frequency/phase-angle generator (FPG) has been added to provide the phase-angle for the sinusoidal functions of the Park transformation, which can be considered a sort of synchronous phase detector (PD).

If the input voltage of the PLL is given by

$$v = V \sin(\theta) = V \sin(\omega t + \phi) \qquad (4.47)$$

the output signals from the quadrature signal generator (QSG) can be expressed by the following voltage vector:

$$\boldsymbol{v}_{(\alpha\beta)} = \begin{bmatrix} v_\alpha \\ v_\beta \end{bmatrix} = V \begin{bmatrix} \sin(\theta) \\ -\cos(\theta) \end{bmatrix} \qquad (4.48)$$

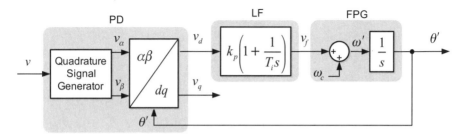

Figure 4.11 PD based on the quadrature signal generator and the Park transformation

Therefore, by substituting (4.48) into (4.46), the output of the PD of Figure 4.11 is given by the voltage vector of the following equation, which will be free of oscillations if the PLL is well tuned to the input frequency, i.e. when $\omega \approx \omega'$:

$$\mathbf{v}_{(dq)} = \begin{bmatrix} v_d \\ v_q \end{bmatrix} = V \begin{bmatrix} \sin(\theta - \theta') \\ -\cos(\theta - \theta') \end{bmatrix} \qquad (4.49)$$

The use of a QSG in the PLL of Figure 4.11 allows a vector approach to be adopted when dealing with a single-phase system. In Figure 4.12, the QSG output signals of (4.48) are represented on an orthogonal and stationary reference frame defined by the $\alpha\beta$ axes, which gives rise to the virtual input vector $\mathbf{v}$. Likewise, the output signals of the Park transformation are represented by the projections of the voltage vector $\mathbf{v}$ on an orthogonal and rotating reference frame defined by the dq axes. If the input voltage is defined by $v_\alpha = V \sin(\theta)$, it can be understood as the projection of the input voltage on the stationary α axis. On the other hand, the angular position of the dq rotating reference frame, θ', is given by the PLL. When the PLL is well tuned to the input frequency ($\omega \approx \omega'$), the virtual input vector and the dq reference frame have the same angular speed.

When the PLL is perfectly locked, one of the axes of the dq reference frame will overlap the virtual input vector $\mathbf{v}$. According to Figure 4.11, the PI regulator of the LF will set the angular position of the dq reference frame to make $v_d = 0$ in the steady state, which means that the input vector $\mathbf{v}$ will rotate orthogonally to the d axis of the rotating reference frame. In the case where the PI regulator is connected to the v_q output of the PD, as shown in Figure 4.13, the virtual input vector $\mathbf{v}$ will rotate, overlapping the d axis of the dq reference frame in the steady state. In such a case, the v_d signal will provide the amplitude of the input voltage vector and the phase-angle detected by the PLL will be in-phase with the virtual input vector $\mathbf{v}$, which means that the detected phase-angle will be 90° lagged with respect to the one of the sinusoidal input voltage, i.e. $\theta' = \theta - \pi/2$.

It is worth highlighting at this point that, thanks to the QSG, the vector interpretation of a single-phase system gives way to thinking about virtual current vectors, which, interacting with a virtual grid voltage vector, will allow regulation of the active and reactive power delivered into a single-phase grid by a power converter. In further explanations in this book,

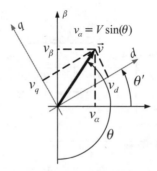

Figure 4.12 Vector representation of the QSG output signals

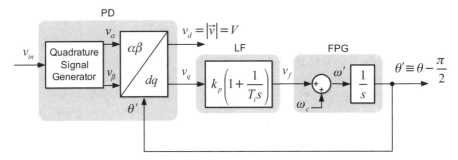

Figure 4.13 PLL with the LF on the q axis of the QSG

it will be assumed that the LF is connected to the v_q output of the PD and, hence, any current vector lying on the d axis of a grid-synchronized dq rotating reference frame will deliver active power into the grid, whereas any current vector on the q axis will deliver reactive power.

4.4 Some PLLs Based on In-Quadrature Signal Generation

Taking into consideration the importance of the QSG in the design of PLLs applied to synchronize with single-phase grids, some relevant techniques to achieve a set of in-quadrature signals, images of the measured single-phase grid voltage, will be presented in the following.

4.4.1 PLL Based on a T/4 Transport Delay

The $T/4$ transport delay technique, with T the fundamental grid frequency period, is probably the easiest way to implement a QSG – see Figure 4.14. The transport delay block can be effortlessly programmed through the use of a first-in-first-out (FIFO) buffer, whose size is set to one-fourth of the number of samples contained in one cycle of the fundamental frequency.

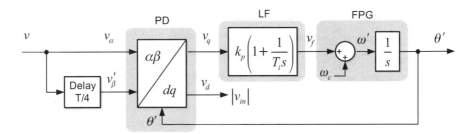

Figure 4.14 PLL based on a T/4 transport delay

The QSG based on the $T/4$ transport delay buffer works well if the input voltage is a purely sinusoidal waveform at the rated grid frequency. If the grid voltage frequency changes in respect to its rated value, the output signals of the QSG will not be perfectly orthogonal, which will give rise to errors in the PLL synchronization.

This QSG technique does not provides any filtering capability, so if the single-phase input voltage is polluted by harmonic components, they will act as a perturbation for the PLL. Moreover, the orthogonal signals generated by the QSG based on a $T/4$ transport delay block will not be really in-quadrature, since each of the frequency components of the input signal had to be delayed by one-fourth of its fundamental period.

4.4.2 PLL Based on the Hilbert Transform

The Hilbert transform, also called a 'quadrature filter', is a fascinating mathematical tool that presents two main features:

1. It shifts $\pm 90°$ the phase-angle of the spectral components of the input signal depending on the sign of their frequency. It is necessary to remember that positive and negative frequencies are considered in the Fourier analysis.
2. It only affects the phase of the signal and has no effect on its amplitude at all.

Therefore, as presented in reference [15], a PLL based on the Hilbert transform can be straightforwardly implemented, as shown in Figure 4.15.

The time domain expression of the Hilbert transform of a given input signal v is defined as

$$H(v) = \frac{1}{\pi} \int_{-\infty}^{\infty} \frac{v(\tau)}{t-\tau} d\tau = \frac{1}{\pi t} * v \qquad (4.50)$$

which describes the convolution product of the function $h(t) = 1/\pi t$ with the signal $v(t)$.

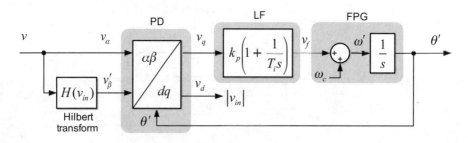

Figure 4.15 PLL based on the Hilbert transform

In the frequency domain, the Hilbert transform can be defined as

$$\mathcal{F}(H(v)) = \mathcal{F}\left(\frac{1}{\pi t}\right)\mathcal{F}(v) = [-j\ sign(\omega)]\mathcal{F}(v) \qquad (4.51)$$

where $\mathcal{F}$ denotes the Fourier transform and $sign(\omega)$ gives the sign of the v frequency. Therefore, in the frequency domain, the Hilbert transform can be understood as a multiplier operator $\sigma_H(\omega) = -j\ sign(\omega)$, which can take the following values:

$$\sigma_H(\omega) = \begin{cases} -j & \text{for} \quad \omega > 0 \\ 0 & \text{for} \quad \omega = 0 \\ +j & \text{for} \quad \omega < 0 \end{cases} \qquad (4.52)$$

Thus, the Hilbert transform has the effect of shifting the phase-angle of positive frequency components by $-90°$. This can easily be proved by calculating the Hilbert transform of the Euler's formula, i.e.

$$H\left(e^{jkt}\right) = \frac{1}{\pi}\int_{-\infty}^{\infty}\frac{e^{j\omega\tau}}{t-\tau}d\tau = -je^{j\omega t}\Big|_{\omega>0} \qquad (4.53)$$

Hence, the Hilbert transform of a sinus input will be given by

$$H(\sin(\omega t)) = H\left(\frac{e^{j\omega t} - e^{-j\omega t}}{2j}\right) = -\frac{e^{j\omega t} + e^{-j\omega t}}{2} = -\cos(\omega t) \qquad (4.54)$$

and the subsequent application of the Hilbert transform will generate a cyclical series of in-quadrature sinusoidal functions as follows:

$$\sin(\omega t) \xrightarrow{H()} -\cos(\omega t) \xrightarrow{H()} -\sin(\omega t) \xrightarrow{H()} \cos(\omega t) \xrightarrow{H()} \sin(\omega t) \qquad (4.55)$$

The Hilbert transform of (4.50) involves convolution of v with $h(t) = 1/\pi t$, which is a noncausal filter and therefore cannot be practically realizable in its current form if v is a time-dependent signal. In many cases, a practical implementation of the Hilbert transform implies that a finite impulse response (FIR) filter, which in addition is made causal by setting a suitable delay, is used to approximate the computation [16]. However, some works have been published for the design of infinite impulse response (IIR) Hilbert transformers as well [17]. Moreover, these approximations can also allow selection of a specific frequency range of interest, removing low and high frequencies, to be subject to the characteristic phase shifting related to the Hilbert transform.

4.4.3 PLL Based on the Inverse Park Transform

The Park transform is used typically as a tool to project an input voltage vector, defined by in-quadrature signals in the $\alpha\beta$ stationary reference frame, on the orthogonal axes of the dq

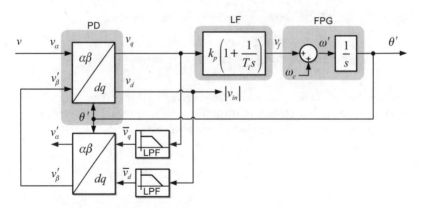

Figure 4.16 PLL based on the inverse Park transform

synchronous reference frame. For a given angular position, the Park transform matrix defined by (4.46) is a linear transformation that can easily be inverted as follows:

$$\mathbf{v}_{(\alpha\beta)} = \begin{bmatrix} v_\alpha \\ v_\beta \end{bmatrix} = \begin{bmatrix} \cos(\theta') & -\sin(\theta') \\ \sin(\theta') & \cos(\theta') \end{bmatrix} \begin{bmatrix} v_d \\ v_q \end{bmatrix} \qquad (4.56)$$

Therefore, an in-quadrature image of a single-phase input signal can be achieved by introducing a filter in a loop consisting of the direct and inverse Park transformations, as presented in reference [16] and shown in Figure 4.16. Although a detailed analysis of this loop is presented in the following, an intuitive explanation of its operation principle can be given if it is assumed that the PLL is well tuned to the input signal frequency. Under such operating conditions, if v_α and v'_β are not in-quadrature, the virtual input vector resulting from these signals will have neither constant amplitude nor rotation speed. Therefore, the v_d and v_q waveforms resulting from the direct Park transform will have oscillations. These oscillations will be attenuated by a low-pass filter (LPF), giving rise to the $\bar{v}_d$ and $\bar{v}_q$ signals. Therefore, the v'_α and v'_β signals resulting from applying the inverse Park transform to $\bar{v}_d$ and $\bar{v}_q$ will be in-quadrature, though v_α and v'_α will not be in-phase if the PLL is not perfectly synchronized. As the PLL locks the phase-angle of the input signal, v_α will become both in-phase with v'_α and in-quadrature with v'_β.

To analyse the QSG of Figure 4.16, we will start by writing the Park transform of (4.46) in terms of the Euler formula, with $\theta' = \omega't$, as follows:

$$\mathbf{v}_{(dq)} = \begin{bmatrix} v_d \\ v_q \end{bmatrix} = \frac{1}{2} \begin{bmatrix} \left(e^{j\omega't} + e^{-j\omega't}\right) & -j\left(e^{j\omega't} - e^{-j\omega't}\right) \\ j\left(e^{j\omega't} - e^{-j\omega't}\right) & \left(e^{j\omega't} + e^{-j\omega't}\right) \end{bmatrix} \begin{bmatrix} v_\alpha \\ v'_\beta \end{bmatrix} \qquad (4.57)$$

By using the Laplace transform, (4.57) can be written in the complex-frequency domain as

$$\begin{bmatrix} V_d(s) \\ V_q(s) \end{bmatrix} = \frac{1}{2} \begin{bmatrix} (V_\alpha(s+j\omega') + V_\alpha(s-j\omega')) & -j(V'_\beta(s+j\omega') - V'_\beta(s-j\omega')) \\ j(V_\alpha(s+j\omega') - V_\alpha(s-j\omega')) & (V'_\beta(s+j\omega') + V'_\beta(s-j\omega')) \end{bmatrix} \qquad (4.58)$$

Therefore, the signals to be applied to the input of the inverse Park transform are given by

$$\begin{bmatrix} \bar{V}_d(s) \\ \bar{V}_q(s) \end{bmatrix} = \frac{\omega_f}{s + \omega_f} \begin{bmatrix} V_d(s) \\ V_q(s) \end{bmatrix} \qquad (4.59)$$

where it is assumed that LPF is a first-order low-pass filter with cut-off frequency ω_f. The inverse Park transform can be written in terms of the Euler formula as

$$v'_{(\alpha\beta)} = \begin{bmatrix} v'_\alpha \\ v'_\beta \end{bmatrix} = \frac{1}{2} \begin{bmatrix} (e^{j\omega' t} + e^{-j\omega' t}) & j(e^{j\omega' t} - e^{-j\omega' t}) \\ -j(e^{j\omega' t} - e^{-j\omega' t}) & (e^{j\omega' t} + e^{-j\omega' t}) \end{bmatrix} \begin{bmatrix} \bar{v}_d \\ \bar{v}_q \end{bmatrix} \qquad (4.60)$$

which gives rise to the following expression in the complex-frequency domain:

$$\begin{bmatrix} V'_\alpha(s) \\ V'_\beta(s) \end{bmatrix} = \frac{1}{2} \begin{bmatrix} (\bar{V}_d(s + j\omega') + \bar{V}_d(s - j\omega')) & j(\bar{V}_q(s + j\omega') - \bar{V}_q(s - j\omega')) \\ -j(\bar{V}_d(s + j\omega') - \bar{V}_d(s - j\omega')) & (\bar{V}_q(s + j\omega') + \bar{V}_q(s - j\omega')) \end{bmatrix} \qquad (4.61)$$

A lengthy but entertaining academic exercise is to substitute (4.59) into (4.61) to arrive at the following transfer function:

$$\frac{V'_\beta}{V_\alpha}(s) = \frac{k\omega'^2}{s^2 + sk\omega' + \omega'^2}; \quad k = \frac{\omega_f}{\omega'} \qquad (4.62)$$

In (4.60), it is possible to realize that

$$v'_\alpha = \frac{1}{\omega'} \frac{d}{dt} v'_\beta \quad \text{and thus} \quad V'_\alpha(s) = \frac{s}{\omega'} V'_\beta \qquad (4.63)$$

Therefore, the following transfer function can be written as well:

$$\frac{V'_\alpha}{V_\alpha}(s) = \frac{sk\omega'}{s^2 + sk\omega' + \omega'^2}; \quad k = \frac{\omega_f}{\omega'} \qquad (4.64)$$

The transfer functions of (4.62) and (4.64) describe the performance of the QSG based on the inverse Park transform. It simultaneously acts as a second-order band-pass filter (v_α to v'_α) and a low-pass filter (v_α to v'_β). The centre frequency of these filters, ω_o, is given by the rotation speed of the dq synchronous reference frame whereas the damping factor, ξ, is given by the k factor, being $k = 2\xi$. Figure 4.17 shows the frequency response of these two transfer functions in the case where $\omega_o = 2\pi \cdot 50$ rad/s and $\omega_f = 2\pi \cdot 70.7$ rad/s, which results in an optimal response with $\xi = 0.707$. It can be observed in this figure how v'_α and v'_β are always in-quadrature (90° shifted) and have the same amplitude in the steady state only if the rotation frequency of the dq reference frame, ω', matches the input frequency ω.

Figure 4.18 shows some representative waveforms resulting from the simulation of the inverse Park transform PLL of Figure 4.16 when, supposing the PLL to be perfectly synchronized, the input signal experiences both a phase-angle jump (+45°) and a frequency jump (from 50 to 45 Hz) at $t = 50$ ms. In this simulation, the LF parameters were set according to (4.38) to achieve a settling time of $t_S = 100$ ms with $\xi = 1/\sqrt{2}$. The in-quadrature signals generated by the QSG based on the inverse Park transform are shown in Figure 4.18(a). The signals resulting from the projection of the virtual input vector on the dq reference frame are shown in Figure 4.18(c). In this plot, the v_d signal matches the amplitude of the input voltage

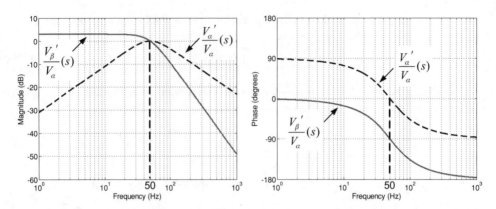

Figure 4.17 Frequency response of the QSG based on the inverse Park transform with $\omega' = 2\pi \cdot 50$ rad/s and $\omega_f = 2\pi \cdot 70.7$ rad/s

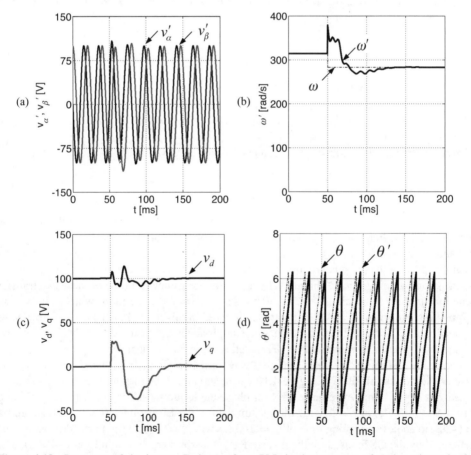

Figure 4.18 Response of the inverse Park transform PLL in the presence of a phase jump in the input signal: (a) in-quadrature signals generated by the QSG, (b) detected frequency, (c) signals in the synchronous reference frame and (d) detected phase-angle

in the steady state, whereas the v_q signal is made equal to zero by the action of the PI controller of the LF. Figure 4.18(d) shows the phase-angle detected by the PLL, which is 90° lagged with respect to the phase-angle of the input voltage in the steady state. This 90° lagging can be appreciated in the vector diagram of Figure 4.12 for the case when the d axis rotates in unison with the virtual voltage vector v.

4.5 Some PLLs Based on Adaptive Filtering

A conventional filter is designed to attenuate a given range of frequencies. The digital implementation of the transfer function of such a filter gives rise to a mathematic algorithm with a series of static coefficients, which are usually set in the design time. In contrast, an adaptive filter is a filter that has the ability to adjust its own parameters automatically according to an optimization algorithm, and their design requires little or no a priori knowledge of the signal to be filtered [18]. In general terms, the optimization algorithm involves the use of a cost function, which sets the performance of the filter (e.g. minimizing a particular noise component of the input) to determine how to modify the filter coefficients to minimize the cost on the next iteration. Adaptive filters have been applied in many control and communication fields, such as systems identification, adaptive and predictive controllers, channel equalization and noise cancellation.

A basic diagram describing the adaptive noise cancelling (ANC) concept is shown in Figure 4.19. In this diagram, the signal to be filtered is applied to the input v. This input signal consists of a primary signal s plus a noise n_0 uncorrelated to the signal s. An auxiliary reference signal n_1, correlated to the noise signal n_0, is applied to the input x. The reference signal n_1 is adaptively filtered to produce an output signal v' that is as close a replica as possible of n_0. This output signal v' is subtracted from the primary input v to produce the output signal e. As a result, the primary noise n_0 is eliminated by cancellation. When the ANC technique is used to cancel out specific frequency components of the input signal, this filtering concept is also called adaptive notch filtering (ANF) [19].

In the digital implementation of an ANC filter, the reference signal x is sampled at a proper sampling period T_S and stored in an N-length buffer to generate the reference vector $\mathbf{x}$. Therefore, at the kth sample, i.e. at the time $t = kT_S$, the reference vector is given by $\mathbf{x}_k = [x_k, x_{k-1}, \ldots, x_{k-N}]$. The elements of the $\mathbf{x}_k$ vector are weighted and summed to give the adaptive filter output v'_k. The most extended adaptation algorithm used to set the weights

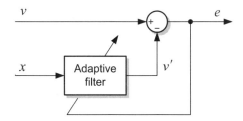

Figure 4.19 Adaptive noise cancelling (ANC) system

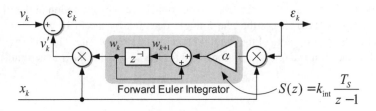

Figure 4.20 Diagram of the LMS algorithm with one adaptive weight in an ANC system

of the adaptive filter, $\boldsymbol{w}_k = [w_k, w_{k-1}, \ldots, w_{k-N}]$, is the least-mean-squares (LMS) algorithm [18]. A simple version of the LMS algorithm at the time k can be given by the following equations:

$$v'_k = \boldsymbol{w}_k^T \cdot \boldsymbol{x}_k \tag{4.65}$$

$$e_k = v_k - v'_k \tag{4.66}$$

$$\boldsymbol{w}_{k+1} = \boldsymbol{w}_k + \alpha e_k \boldsymbol{x}_k \tag{4.67}$$

The LMS algorithm is an iterative gradient-descent algorithm that uses an estimate of the gradient on the mean-square error surface to seek the optimum weight vector at the minimum mean-square error point. The term $e_k \cdot x_k$ represents the estimate of the negative gradient and the adaptation gain α determines the step size taken at each iteration along that estimated negative-gradient direction. A schematic representation of a very simple LMS algorithm with only one weight ($N = 1$) is shown in Figure 4.20. It is worth realizing in this figure that the highlighted area matches a forward Euler discrete integrator $S(z)$ affected by a gain k_{int}, being $\alpha = k_{\text{int}} T_S$.

4.5.1 The Enhanced PLL

The ANC system of Figure 4.20 can be applied to enhance the performance of the multiplier phase detector (PD) of the conventional single-phase PLL. In such an application, the ANC system works as an adaptive notch filter (ANF) in which the grid voltage signal is applied to the input v and a unitary sinusoidal signal, provided by the voltage controller oscillator (VCO) of the PLL, is applied to the input x as a reference signal. The synchronization system resulting from combining an ANF and a conventional single-phase PLL is shown in Figure 4.21 and is known as the enhanced PLL (EPLL) [20].

In the EPLL, the output of the ANF becomes equal to zero as the frequency and phase-angle of the reference signal generated by the VCO, $x = \cos(\theta')$, match those of the input signal v. As a result, signal oscillations at the output of the multiplier PD are completely cancelled out and the input signal phase-angle is properly detected by the conventional PLL. It is worth remarking, however, that there will exist a 90° phase shifting between θ and θ' in the steady state, i.e. $\theta' = \theta - \pi/2$, due to the effect of the multiplier PD. This performance is illustrated in Figure 4.22, where some representative waveforms from the EPLL of Figure 4.21

Grid Synchronization in Single-Phase Power Converters 71

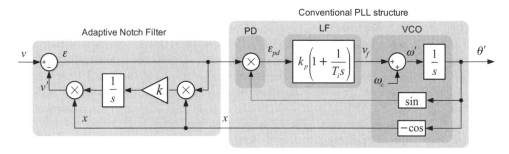

Figure 4.21 Diagram of the enhanced PLL (EPLL)

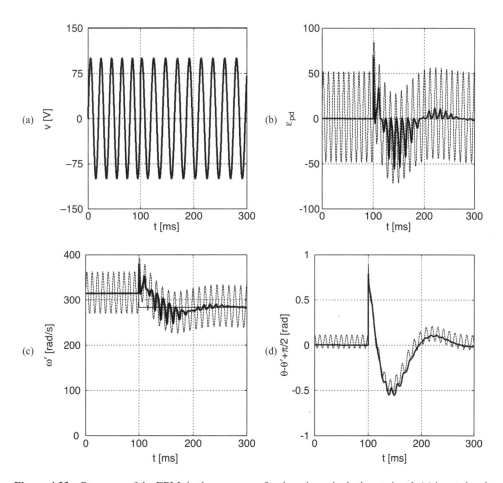

Figure 4.22 Response of the EPLL in the presence of a phase jump in the input signal: (a) input signals v, (b) phase detector output ε_{pd} and (c) detected frequency ω' and (d) error in phase-angle detection $\theta - \theta' + \pi/2$

(in thick continuous line) are overlapped on the waveforms resulting from the conventional PLL of Figure 4.5 (in thin dashed line) when the input signal v experienced both a phase-angle jump ($+45°$) and a frequency jump (from 50 to 45 Hz) at $t = 100$ ms. As can be appreciated in Figure 4.22(b), the output signal of the multiplier PD of the conventional PLL presents steady-state oscillations at twice the grid frequency. Consequently, as shown in Figure 4.22(b) and (c), these oscillations are present on the detected frequency and phase-angle as well. In the EPLL, however, the ANF progressively makes zero the input signal of the multiplier PD as the PLL becomes synchronized. Therefore, as shown in Figure 4.22(b) in thick line, the output of the multiplier PD is equal to zero in the steady state and, as shown in Figure 4.22(b) and (c), the detected frequency and phase-angle are free of oscillations after a transient period.

4.5.2 Second-Order Adaptive Filter

As presented in the previous section, when a single-frequency sinusoidal signal is applied to the input v of the ANC of Figure 4.20, the output error signal ε is equal to zero – after a transient period – only if the frequency and phase-angle of the sinusoidal signal v match those of the sinusoidal reference signal x. There are some applications, however, in which it is interesting that the output error signal ε is equal to zero just when the frequency of v and x are equal –independently of their phase-angle.

Figure 4.23(a) shows a single frequency ANC, namely an ANF, using an LMS algorithm with two adaptive weights [19]. In this filter, two $90°$ shifted sinusoidal signals at the frequency of interest ω are used as reference signals for the adaptive algorithm. Since accumulators of the LMS algorithm can be understood as forward discrete integrators, the discrete system of Figure 4.23(a) can be transformed into the continuous equivalent systems of Figure 4.23(b), which has been arranged according to the standard structure of an adaptive noise cancelling system. In this system, the sine and cosine blocks are integrated into the adaptive filter (AF) structure and the frequency to be filtered, ω', is considered as the reference signal. Moreover, the error signal is affected by a gain $k = \alpha/T_S$, with T_S being the sampling frequency of the original discrete ANF. In the following, the AF of Figure 4.23(b) is analysed in order to obtain its transfer function.

Defining $g = k\varepsilon_v$, the v_d and v_q signals of Figure 4.23(b) can be written as

$$v_d = g\cos(\omega't) = \frac{1}{2}g\left[e^{j\omega't} + e^{-j\omega't}\right] \tag{4.68}$$

$$v_q = g\sin(\omega't) = \frac{1}{j2}g\left[e^{j\omega't} - e^{-j\omega't}\right] \tag{4.69}$$

The A_d and A_q variables, which correspond to the output of the integrators for v_d and v_q, can be expressed in the *Laplace* domain as

$$A_d(s) = \frac{1}{s}v_d(s) = \frac{1}{2s}\left[g(s+j\omega't) + g(s-j\omega't)\right] \tag{4.70}$$

$$A_q(s) = \frac{1}{s}v_q(s) = \frac{1}{j2s}\left[g(s+j\omega't) - g(s-j\omega't)\right] \tag{4.71}$$

Grid Synchronization in Single-Phase Power Converters

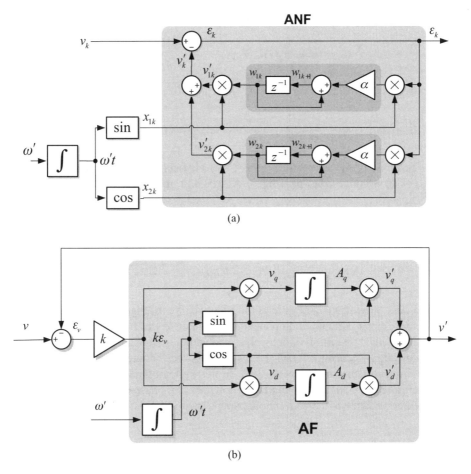

Figure 4.23 (a) ANF based on an LMS algorithm with two adaptive weights and (b) diagram of a second-order AF on the continuous time domain

Likewise the Laplace transforms for the v'_d and v'_q, variables are given by

$$v'_d(s) = \frac{1}{2}\left[A_d(s+j\omega't) + A_d(s-j\omega't)\right]$$

$$= \frac{1}{4(s+j\omega')}\left[g(s) + g(s+2j\omega')\right] + \frac{1}{4(s-j\omega')}\left[g(s) + g(s-2j\omega')\right] \quad (4.72)$$

$$v'_q(s) = \frac{1}{2j}\left[A_q(s+j\omega't) - A_q(s-j\omega't)\right]$$

$$= \frac{1}{4(s+j\omega')}\left[g(s) - g(s+2j\omega')\right] + \frac{1}{4(s-j\omega')}\left[g(s) - g(s-2j\omega')\right] \quad (4.73)$$

Finally, the addition of v'_d and v'_q gives rise to the AF output, v', as follows:

$$v'(s) = v'_d(s) + v'_q(s) = \frac{s}{s^2 + \omega'^2} g(s) \tag{4.74}$$

Consequently, the transfer function of the AF structure of Figure 4.23(b) is given by

$$AF(s) = \frac{v'}{k\varepsilon_v}(s) = \frac{s}{s^2 + \omega'^2} \tag{4.75}$$

Hence, the response of the system of Figure 4.23(b) is defined by two second-order transfer functions, i.e. an adaptive band-pass filter (ABPF) and an adaptive notch filter (ANF), as follows:

$$ABPF(s) = \frac{v'}{v}(s) = \frac{AF(s)}{1 + AF(s)} = \frac{ks}{s^2 + ks + \omega'^2} \tag{4.76}$$

$$ANF(s) = \frac{\varepsilon_v}{v}(s) = 1 - ABPF(s) = \frac{s^2 + \omega'^2}{s^2 + ks + \omega'^2} \tag{4.77}$$

The band-pass filtering characteristic of the adaptive filter of Figure 4.23(b) suggests that it is possible to extract a particular component at the frequency of interest ω' even if the input signal v' is affected by distortion. Moreover, as shown in Figure 4.24, this system can be used as a quadrature signal generator (QSG) by just adding a scaled integrator at the output of the adaptive filtering structure of Figure 4.23(b). In this system, the signals v' and qv' are 90° shifted. Therefore, they can be applied to the input of any phase detector based on QSG to improve the performance of the conventional single-phase PLL.

4.5.3 Second-Order Generalized Integrator

The AF block of Figure 4.24, whose transfer function is given by (4.75), is of special interest when working with sinusoidal signals. Figure 4.25(a) shows the response of an AF block with $\omega' = 2\pi \cdot 50$ rad/s when a unitary step is applied to its input. As expected, the response of the system defined by the transfer function given by (4.75), with two imaginary complex poles placed at $\pm j\omega'$, is like that of a resonator oscillating at the frequency ω'. As further shown, this feature can be of great use for implementing the voltage controlled oscillator (VCO) block in a PLL structure. Figure 4.25(b) shows the response of the AF block defined by (4.75) in the case of applying both a sine and a cosine signal with frequency ω' to its input. As can be

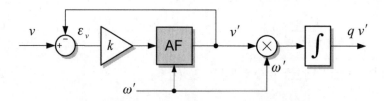

Figure 4.24 QSG based on a second-order AF

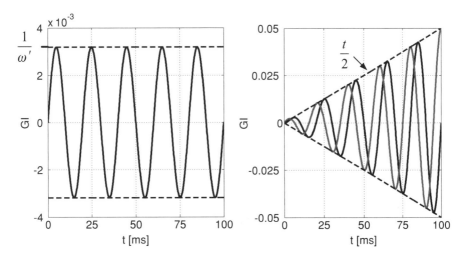

Figure 4.25 Response of the GI ($\omega' = 2\pi \cdot 50$ rad/s) with: (a) unitary step input and (b) unitary sine/cosine input ($\omega = 2\pi \cdot 50$ rad/s)

observed in this figure, the system acts as an amplitude integrator for both sinusoidal input signals. For this reason, the system with the transfer function given by (4.75) is also referred as a 'sinusoidal integrator'.

In the following, a more detailed analysis of this response of Figure 4.25(b) is conducted. As known, the Laplace transforms of the sine and cosine functions are given by

$$L\left[\sin(\omega't)\right] = \frac{\omega'}{s^2 + \omega'^2} \quad (4.78)$$

$$L\left[\cos(\omega't)\right] = \frac{s}{s^2 + \omega'^2} \quad (4.79)$$

Therefore, the time response of the system characterized by (4.75) in the presence of sinusoidal inputs is given by

$$L^{-1}\left(\frac{\omega'}{s^2 + \omega'^2} \frac{s}{s^2 + \omega'^2}\right) = \frac{1}{2} t \sin(\omega't) \quad (4.80)$$

$$L^{-1}\left(\frac{s}{s^2 + \omega'^2} \frac{s}{s^2 + \omega'^2}\right) = \frac{1}{2}\left[\frac{\sin(\omega't)}{\omega'} + t\cos(\omega't)\right] \quad (4.81)$$

As evidenced in (4.81), the system with the transfer function (4.75) does not act as an ideal amplitude integrator for any kind of sinusoidal input signal with frequency ω' but, depending of the phase-angle of the input signal, its output contains a steady-state error – related to the step response previously noted. However, the amplitude of this error is low enough to be considered negligible in most of the applications, which is why the system defined by the transfer function given by (4.75), multiplied by 2, is commonly known as the *generalized integrator* (GI). A detailed study about this issue can be found in [21].

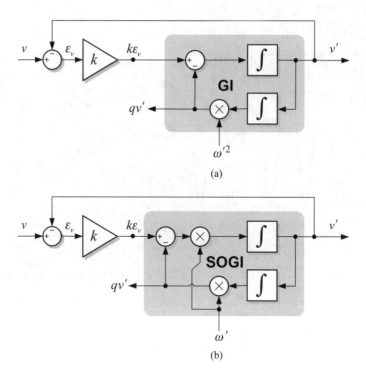

Figure 4.26 (a) Second-order AF based on a GI and (b) second-order AF based on an SOGI (SOGI-QSG)

As the transfer function of the GI provides an infinite gain at the resonance frequency, it allows any steady-state error to be cancelled when controlling sinusoidal signals at such resonance frequency. This interesting feature makes the GI the core of so-called proportional-resonant controllers [22, 23].

As previously shown, the AF structure of Figure 4.23 gives rise to the transfer function of (4.75). However, this is not the only way to implement a GI, for there exist other solutions as reported in the literature, different direct discrete realizations of this resonant system can be used to reduce its computational burden [21, 24]. Figure 4.26(a) shows a diagram of a second-order AF based on very efficient implementation of the GI. The resonant structure forming the GI of Figure 4.26(a) has been used in different systems for grid monitoring and synchronization purposes [25–28]. The characteristic transfer functions of the adaptive filter of Figure 4.26(a) are given by

$$GI(s) = \frac{v'}{k\varepsilon_v}(s) = \frac{s}{s^2 + \omega'^2} \qquad (4.82)$$

$$D(s) = \frac{v'}{v}(s) = \frac{ks}{s^2 + ks + \omega'^2} \qquad (4.83)$$

$$Q(s) = \frac{qv'}{v}(s) = \frac{k\omega'^2}{s^2 + ks + \omega'^2} \qquad (4.84)$$

As expected, the transfer function of (4.82) for the GI is identical to that of (4.75) for the AF. However, the GI implementation of Figure 4.26(a) is much simpler than the AF structure of Figure 4.23. The GI does not uses sine/cosine functions, which typically require large lookup tables that increase computation time and introduce additional quantization noise into the discrete system.

Moreover, as evidenced by the transfer functions of (4.83) and (4.84), the adaptive filtering structure of Figure 4.26(a) generates two 90° shifted output signals, v' and qv', which makes it suitable for implementation of a PLL based on in-quadrature signals generation. These transfer functions also indicate that the bandwidth of the band-pass filter given by (4.83) and the static gain of the low-pass filter of (4.84) are not only a function of the gain k but also depend on the centre frequency of the filter, ω'. This issue can become an inconvenience when designing variable-frequency systems, as is the case of a PLL. This problem can be overcome by modifying the flow diagram of the adaptive filtering system [27, 28]. Another very straightforward solution can be obtained by modifying the structure of the GI itself. The alternative sinusoidal integrator is known as the *second-order generalized integrator* (SOGI) [29] to differentiate it from the conventional GI. The adaptive filtering structure based on the SOGI is shown in Figure 4.26(b) and its characteristic transfer functions are given by:

$$SOGI(s) = \frac{v'}{k\varepsilon_v}(s) = \frac{\omega's}{s^2 + \omega'^2} \quad (4.85)$$

$$D(s) = \frac{v'}{v}(s) = \frac{k\omega's}{s^2 + k\omega's + \omega'^2} \quad (4.86)$$

$$Q(s) = \frac{qv'}{v}(s) = \frac{k\omega'^2}{s^2 + k\omega's + \omega'^2} \quad (4.87)$$

These transfer functions show that the bandwidth of the SOGI-based adaptive filter is not a function of the centre frequency ω' but only depends on the gain k, which makes it suitable for variable-frequency applications. Moreover, the amplitude of the in-quadrature signals, v' and qv', matches the amplitude of the input signal v when the centre frequency of the filter, ω', matches the input frequency, ω. Therefore, if by some mechanism it is provided that $\omega' = \omega$, the SOGI-based filtering structure of Figure 4.26(b) may be considered a very suitable method for quadrature signal generation (QSG). This is the reason why this system was named the SOGI-QSG in reference [29].

To evaluate the time response of the SOGI-QSG of Figure 4.26(b) a sinusoidal input signal $v = V\sin(\omega t)$ is applied to its input. Hence, if it is assumed that $\omega = \omega'$, the output signals of the adaptive filter defined by the transfer functions (4.86) and (4.87) are given by

$$v' = -\frac{V}{\sqrt{1-(k/2)^2}}\sin\left(\omega\sqrt{1-(k/2)^2}\,t\right)e^{-\frac{k\omega}{2}t} + V\sin(\omega t) \quad (4.88)$$

$$qv' = -\frac{V}{\sqrt{1-(k/2)^2}}\cos\left(\omega\sqrt{1-(k/2)^2}\,t - \varphi\right)e^{-\frac{k\omega}{2}t} - V\cos(\omega t) \quad (4.89)$$

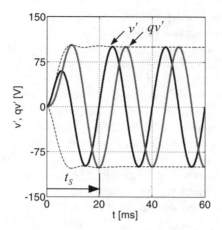

Figure 4.27 Response of the SOGI-QSG ($t_S = 20.7$ ms, $k = \sqrt{2}$, $\omega' = 2\pi \cdot 50$ rad/s)

where

$$\varphi = \arctan \frac{k/2}{\sqrt{1 - (k/2)^2}}$$

As proposed in reference [14], the settling time for a second-order system can be roughly estimated by $t_S = 4.6\tau$. Therefore, since $\tau = 2/k\omega'$ in (4.88) and (4.89), the gain of the SOGI-QSG can be calculated for a given settling time as

$$k = \frac{9.2}{t_S \omega'} \tag{4.90}$$

Figure 4.27 shows the waveforms of (4.88) and (4.89) when the SOGI-QSG parameters are $k = \sqrt{2}$ and $\omega' = \omega = 2\pi \cdot 50$ rad/s. In this case, the settling time is around 20 ms, which matches the value obtained from (4.90). It is worth noting that a gain $k = \sqrt{2}$ implies a damping factor $\xi = 1/\sqrt{2}$, which roughly results in an optimal relationship between the settling time and overshooting in the dynamic response.

4.5.4 The SOGI-PLL

The SOGI-QSG of Figure 4.26(b) can be straightforwardly applied to implement a PLL based on in-quadrature signal generation like the one shown in Figure 4.28, which is known as the SOGI-PLL [29, 30]. This system has a double feedback loop; i.e. the frequency/phase generator provides both the phase-angle to the Park transform and the central frequency to the SOGI-QSG.

Figure 4.29 shows the response of the SOGI-PLL of Figure 4.28 when the 100 V_{peak}/50 Hz single-phase input signal v experienced both a phase-angle jump ($+45°$) and a frequency jump (from 50 to 45 Hz) at $t = 100$ ms. In this simulation, the SOGI-QSG gain was set to $k = \sqrt{2}$, which in theory implies a settling time of 20 ms for this adaptive filter, and the LF parameters

Grid Synchronization in Single-Phase Power Converters

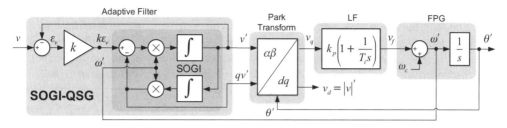

Figure 4.28 Diagram of the SOGI-based PLL (SOGI-PLL)

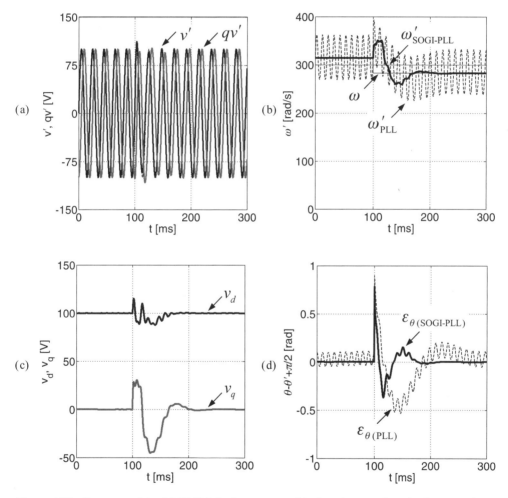

Figure 4.29 Response of the SOGI-PLL in the presence of both a phase-angle and a frequency jump in the input signal: (a) in-quadrature signals generated by the SOGI-QSG, (b) detected frequency, (c) signals in the synchronous reference frame and (d) error in the phase-angle detection $\theta - \theta' + \pi/2$

were calculated according to (4.38) to achieve a settling time of 100 ms in the PLL. As shown in Figure 4.29, the SOGI-QSG and the PLL interact with each other and the resulting response is a combination of the action of both systems. The response of the SOGI-PLL differs from that of the EPLL since in that case the feedback loops for both the adaptive filter and the PLL depended on the same variable, i.e. the detected phase-angle. For this reason, the ANF and the PLL blocks of the EPLL reach steady-state conditions at the same time. In the case of the SOGI-PLL there are two variables involved in the synchronization process, i.e. the SOGI-QSG tuned by using the detected frequency, ω', whereas the PLL is locked to the input phase-angle. As evidenced in Figure 4.29, the response of the SOGI-PLL is very close to the one shown in Figure 4.18 for the inverse Park transform PLL. Actually, it seems logical since, as demonstrated by (4.64), such an inverse Park transform structure behaves as an adaptive filter as well.

Figure 4.29(a) shows the two in-quadrature signals generated by the SOGI-QSG. As appreciated in this plot, the transient response is extended until the grid frequency is newly tuned. Figure 4.29(b) shows the frequency that would be detected by a conventional PLL (ω'_{PLL} in dashed thin line) and the frequency detected by the SOGI-PLL ($\omega'_{SOGI-PLL}$ in thick line), free of steady-state oscillations. Figure 4.29(c) shows the output variables of the Park transform. The v_d signal is given the amplitude of the input voltage and the v_q signal is made equal to zero in the steady state by the action of the phase-locking loop. Figure 4.29(d) shows the error made by a conventional PLL in the detection of the input phase-angle ($\varepsilon_{\theta(PLL)}$ in dashed thin line) and the one made by the SOGI-PLL ($\varepsilon_{\theta(SOGI-PLL)}$ in thick line). As appreciated in this figure, the SOGI-PLL detects the input phase-angle faster than the conventional PLL and with no steady-state oscillations.

4.6 The SOGI Frequency-Locked Loop

In Section 4.5, the SOGI was used to implement an in-quadrature signal generator (QSG), which improved phase-angle detection in a conventional PLL. This PLL locked the phase-angle of its internal oscillator to that of the input signal at the same time as the input frequency was detected, which allowed the SOGI-QSG to remain properly tuned. However, as commented in Section 4.5, the inherent resonant character of the SOGI makes itself work as a voltage-controlled oscillator, which stimulates one to think about designing a simple control loop to auto-adapt the centre frequency of the SOGI resonator to the input frequency and discard the PLL block from the SOGI-PLL structure. This is the main idea supporting the study of the frequency-locked loop (FLL) presented in this section.

The first thing to do in order to make the SOGI-QSG structure auto-tuneable is to analyse the error signal ε_v and study how the centre frequency of the SOGI-QSG might be regulated by using this error signal. The transfer function from the input signal v to the error signal ε_v is given by

$$E(s) = \frac{\varepsilon_v}{v}(s) = \frac{s^2 + \omega'^2}{s^2 + k\omega's + \omega'^2} \qquad (4.91)$$

The transfer function in (4.91) responds to a second-order notch filter, with zero gain at the centre frequency. An interesting feature of this transfer function is that the phase-angle of the

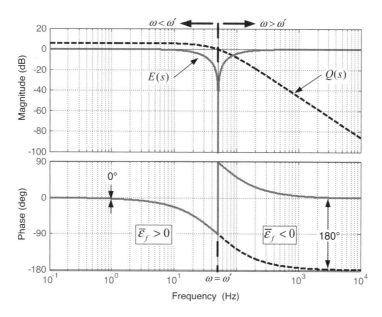

Figure 4.30 Bode diagram of the $E(s)$ and $Q(s)$ in an SOGI-QSG

output signal experiences a jump of 180° when the frequency of the input signal, ω, changes from lower to higher than the SOGI-QSG centre frequency, ω'. This characteristic is used in the following to compare the value of both frequencies.

Figure 4.30 shows the *Bode* diagram of $E(s)$ and $Q(s)$ transfer functions, the latter given by (4.87), to study the relationship between ε_v and qv'. As can be appreciated from this figure, the signals ε_v and qv' are in phase when the input frequency is lower than the SOGI resonance frequency ($\omega < \omega'$) and are in counter-phase in the opposite case, i.e. when $\omega > \omega'$.

Therefore, a frequency error variable ε_f can be defined as the product of qv' and ε_v. As indicated in Figure 4.30, the average value of ε_f will be positive when $\omega < \omega'$, zero when $\omega = \omega'$ and negative when $\omega > \omega'$. This frequency error variable allows a straightforward frequency locking loop (FLL) to be designed, such as the one shown in Figure 4.31. In this loop, an integral controller with a negative gain $-\gamma$ is used to make the DC component of ε_f zero by shifting the centre frequency of the SOGI-QSG, ω', until matching the input frequency, ω. In addition, as shown in Figure 4.31, the nominal value of the grid frequency is added to the FLL output as a feed-forward variable, ω_c, to accelerate the initial synchronization process.

The combination of both the SOGI-QSG and the FLL building blocks according to Figure 4.31 gives rise to a single-phase grid synchronization system named the SOGI-FLL [31]. In the SOGI-FLL, the input frequency is directly detected by the FLL, whereas the estimation of the phase-angle and the amplitude of the input 'virtual vector' can be indirectly calculated by

$$|\boldsymbol{v}'| = \sqrt{(v')^2 + (qv')^2}; \quad \underline{|\boldsymbol{v}'|} = \arctan \frac{qv'}{v'} \qquad (4.92)$$

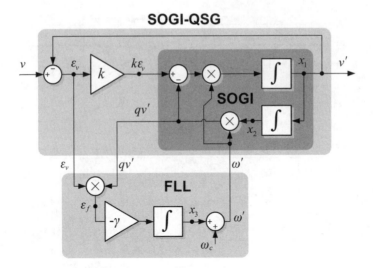

Figure 4.31 Diagram of the SOGI-FLL

Figure 4.32 shows the response of the SOGI-FLL of Figure 4.31 when a 100 V_{peak}/50 Hz single-phase input signal v experienced both a phase-angle jump (+45°) and a frequency jump (from 50 to 45 Hz) at $t = 100$ ms. In this simulation, the SOGI-QSG gain was set to $k = \sqrt{2}$ and the FLL gain was set to $\gamma = 2.22$ – a justification of this value is given in next section. Figure 4.32(b) shows how the SOGI-FLL is able to track the input frequency variation, which allows the SOGI-QSG to remain properly tuned. This can be appreciated from Figure 4.32(a), where the amplitude of the in-quadrature signals matches the input voltage amplitude, even after the input frequency variation. This makes it possible to obtain a correct estimation of the input amplitude and phase-angle by applying (4.92), as shown in Figure 4.32(c) and (d) respectively.

4.6.1 Analysis of the SOGI-FLL

The performance and dynamical response of the SOGI-FLL depends mainly on the appropriate selection of the control parameters k and γ. In this section, the SOGI-FLL equations are briefly analysed in order to set proper values for k and γ to achieve a desired performance in the detection of the amplitude and frequency of the input signal.

From the SOGI-FLL diagram of Figure 4.31, the following space-state equations can be written:

$$\dot{x} = \begin{bmatrix} \dot{x}_1 \\ \dot{x}_2 \end{bmatrix} = \mathbf{A}x + \mathbf{B}v = \begin{bmatrix} -k\omega' & -\omega'^2 \\ 1 & 0 \end{bmatrix} \begin{bmatrix} x_1 \\ x_2 \end{bmatrix} + \begin{bmatrix} k\omega' \\ 0 \end{bmatrix} v \qquad (4.93)$$

$$y = \begin{bmatrix} v' \\ qv' \end{bmatrix} = \mathbf{C}x = \begin{bmatrix} 1 & 0 \\ 0 & \omega' \end{bmatrix} \begin{bmatrix} x_1 \\ x_2 \end{bmatrix} \qquad (4.94)$$

$$\dot{\omega}' = -\gamma x_2 \omega' (v - x_1) \qquad (4.95)$$

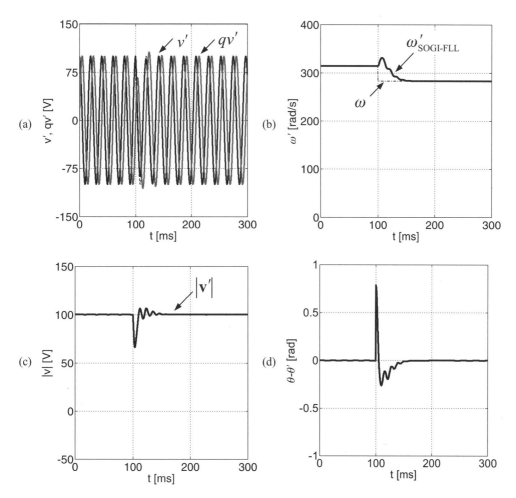

Figure 4.32 Response of the SOGI-FLL in the presence of both a phase-angle and a frequency jump in the input signal: (a) in-quadrature signals generated by the SOGI-QSG, (b) detected frequency, (c) detected input voltage amplitude and (d) error in phase-angle detection $\theta - \theta'$

where $x = [x_1, x_2]^T$ and $y = [v', qv']^T$ are the SOGI-QSG state and output vectors. The state equation describing the behaviour of the FLL is given by (4.95).

Considering the stable operating conditions with the FLL properly tuned, which implies $\dot{\omega}' = 0$, $\omega = \omega'$ and $x_1 = v$, the SOGI-QSG state vector in the steady state is given by

$$\dot{x}\bigg|_{\dot{\omega}'=0} = \begin{bmatrix} \dot{\bar{x}}_1 \\ \dot{\bar{x}}_2 \end{bmatrix} = \begin{bmatrix} 0 & -\omega'^2 \\ 1 & 0 \end{bmatrix} \begin{bmatrix} \bar{x}_1 \\ \bar{x}_2 \end{bmatrix} \qquad (4.96)$$

where the steady-state variables are indentified by a bar over. The eigenvalues of the Jacobian obtained from (4.96) are conjugated complex with a null real part, something that confirms the resonant behaviour of the system, as the steady-state response remains in a periodic orbit at the ω' frequency. Therefore, for a given sinusoidal input signal $v = V\sin(\omega t + \phi)$ the steady-state output vector will be given by

$$\bar{y} = \begin{bmatrix} v' \\ qv' \end{bmatrix} = V \begin{bmatrix} \sin(\omega t + \phi) \\ -\cos(\omega t + \phi) \end{bmatrix} \quad (4.97)$$

If the FLL was intentionally frozen at a frequency ω' different from the input frequency ω, e.g. by making $\gamma = 0$, the SOGI-QSG output vector would still keep in a stable orbit defined by

$$\bar{y}' = V|D(j\omega)| \begin{bmatrix} \sin(\omega t + \phi + \angle D(j\omega)) \\ -\dfrac{\omega'}{\omega}\cos(\omega t + \phi + \angle D(j\omega)) \end{bmatrix} \quad (4.98)$$

where $|D(j\omega)|$ and $\angle D(j\omega)$ can be obtained from (4.86) and are given by

$$|D(j\omega)| = \dfrac{k\omega\omega'}{\sqrt{(k\omega\omega')^2 + (\omega^2 - \omega'^2)^2}} \quad (4.99)$$

$$\angle D(j\omega) = \arctan\dfrac{\omega'^2 - \omega^2}{k\omega\omega'} \quad (4.100)$$

As indicated in (4.98), if the input signal is assumed sinusoidal at the frequency ω and even though $\omega \neq \omega'$, the state variables of the SOGI-QSG keep the following relationship:

$$\ddot{\bar{x}}_1 = -\omega^2 \bar{x}_2 \quad (4.101)$$

Therefore, from (4.93), the steady-state synchronization error signal can be written as

$$\bar{\varepsilon}_v = (v - \bar{x}_1) = \dfrac{1}{k\omega'}\left(\ddot{\bar{x}}_1 + \omega'^2 \bar{x}_2\right) \quad (4.102)$$

and substituting (4.101) into (4.102) the steady-state frequency error signal is given by

$$\bar{\varepsilon}_f = \omega' \bar{x}_2 \bar{\varepsilon}_v = \dfrac{\bar{x}_2^2}{k}\left(\omega'^2 - \omega^2\right) \quad (4.103)$$

The equation in (4.103) proves that the ε_f signal certainly collects information about the error made in the frequency estimation, which makes it suitable to act as the control signal of the FLL. However, this expression is highly nonlinear, which means that linear control analysis techniques cannot be directly applied to set the value of the FLL gain, γ. Hence, some assumptions should be made to determine the performance of the FLL.

In this way, the local dynamics of the FLL can be studied by considering steady-state conditions, namely $\omega' \approx \omega$. In such a case, $\omega'^2 - \omega^2$ can be approximated as $2(\omega' - \omega)\omega'$, and the small signal performance of the FLL can be described as follows:

$$\dot{\omega}' = -\gamma \bar{\varepsilon}_f = \frac{\gamma}{k}\bar{x}_2^2\left(\omega'^2 - \omega^2\right) \approx -2\frac{\gamma}{k}\bar{x}_2^2(\omega' - \omega)\omega' \quad (4.104)$$

Moreover, taking $v = V \sin(\omega t + \phi)$ as the input signal for the SOGI-FLL, the square of the state $\bar{x}_2$ can be written from (4.98) as

$$\bar{x}_2^2 = \frac{V^2}{2\omega^2}|D(j\omega)|^2 \left[1 + \cos(2(\omega t + \phi + \angle D(j\omega)))\right] \quad (4.105)$$

According to (4.99) and (4.100), the terms $|D(j\omega)|$ and $\angle D(j\omega)$ in (4.105) tend to 1 and 0 respectively, as the frequency detected by the FLL locks the input frequency ($\omega' \to \omega$). Hence, in the vicinity of the steady-state operation of the FLL, $\bar{x}_2^2$ will present a DC component equal to $V^2/(2\omega^2)$ plus an AC term oscillating at twice the input frequency. Therefore, the averaged dynamics of the FLL with $\omega' \approx \omega$ can be described by the following equation, where the AC component of $\bar{x}_2^2$ has been neglected:

$$\dot{\bar{\omega}}' = -\frac{\gamma V^2}{k\omega'}\left(\bar{\omega}' - \omega\right) \quad (4.106)$$

Equation (4.106) is very interesting because it clears up the existing relationship between the dynamic response of the FLL, the SOGI-QSG gain and the parameters of the input signal. This equation encourages the value of γ to be normalized according to (4.107), by using feedback variables, in order to achieve a first-order linearized frequency adaptive system like the one shown in Figure 4.33. This linearized system is not dependent on either the grid variables or the SOGI-QSG gain, and its time response is perfectly defined by the value of the gain Γ:

$$\gamma = \frac{k\omega'}{V^2}\Gamma \quad (4.107)$$

The transfer function of the first-order frequency locking loop of Figure 4.33 is given by

$$\frac{\bar{\omega}'}{\omega} = \frac{\Gamma}{s + \Gamma} \quad (4.108)$$

and its settling time can be roughly set as follows:

$$t_{s(FLL)} \approx \frac{4.6}{\Gamma} \quad (4.109)$$

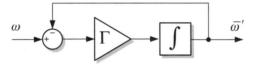

Figure 4.33 Simplified frequency adaptation system of the FLL

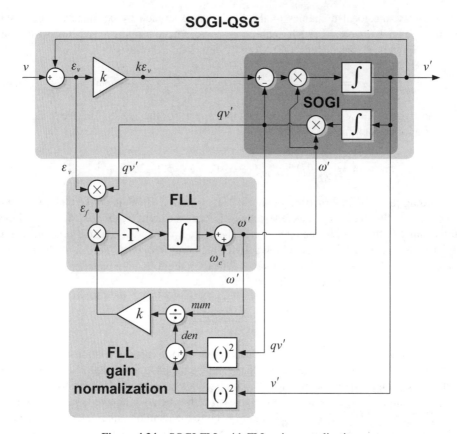

Figure 4.34 SOGI-FLL with FLL gain normalization

A practical implementation of the feedback-based linearized SOG-FLL is shown in Figure 4.34. In this system, the FLL gain is online adjusted by feeding back the estimated grid operating conditions, which guarantees a constant settling time in the frequency estimation independently of the input signal parameters. In this implementation, the square input voltage amplitude is estimated by

$$V^2 = v'^2 + qv'^2 \qquad (4.110)$$

Figure 4.35(a) and (b) shows some representative plots from the simulation of the feedback-based linearized SOGI-FLL of Figure 4.34 when a 100 V_{peak}/50 Hz single-phase input signal v experiences a frequency jump (from 50 to 45 Hz) at $t = 100$ ms (without modifying its amplitude). In this simulation, the SOGI-QSG gain was set to $k = \sqrt{2}$ and the normalized FLL gain was set to $\Gamma = 46$. According to (4.109), this value for the normalized FLL gain entails a settling time in frequency adaptation, $t_{S(FLL)}$, of around 100 ms. As can be appreciated from Figure 4.35(b), the detected frequency fits a first-order exponential response very well with a settle time of 100 ms, which validates the analysis conducted in this section.

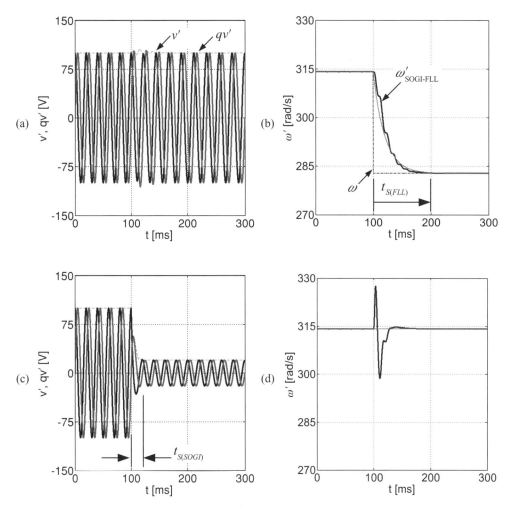

Figure 4.35 Response of the SOGI-FLL with FLL gain normalization: (a),(b) in-quadrature signals and detected frequency when the input signal experiences a frequency jump and (c),(d) in-quadrature signals and detected frequency when the input signal experiences a voltage dip

Figure 4.35(c) and (d) shows the response of the feedback-based linearized SOGI-FLL when the amplitude of the input signal v drops down to 20% of its rated value at $t = 100$ ms (without modifying its frequency). In this case, the values for the control parameters are the same as in the previous case, i.e. $k = \sqrt{2}$ and $\Gamma = 46$. Since the input frequency is kept constant in this simulation, the settling time in the detection of the input voltage amplitude mainly depends on the SOGI-QGS dynamics. As indicated in reference [14], this settling time can be roughly estimated by

$$t_{S(SOGI)} = 4.6\tau, \quad \text{being} \quad \tau = 2/k\omega' \quad (4.111)$$

In this simulation, therefore, the settling time in amplitude detection should be $t_{S(SOGI)} = 20.7$ ms. A zoomed view of Figure 4.35(c) corroborates that, even though the frequency detected by the FLL shows some transient oscillations when the input voltage level varies, the settle time in the response of the SOGI-QSG matches a value calculated by (4.111).

Until now, the SOGI-QSG and the FLL have been studied by considering separated variations in both the amplitude and the frequency of the input signal. However, both systems are interdependent, which means that the global time response of the SOGI-FLL will differ from the one obtained in earlier sections whenever the input signal experiences simultaneous variations in frequency and amplitude. Figure 4.36 shows some representative plots of an SOGI-FLL in which the amplitude of the input signal drops down to 20% of its rated value

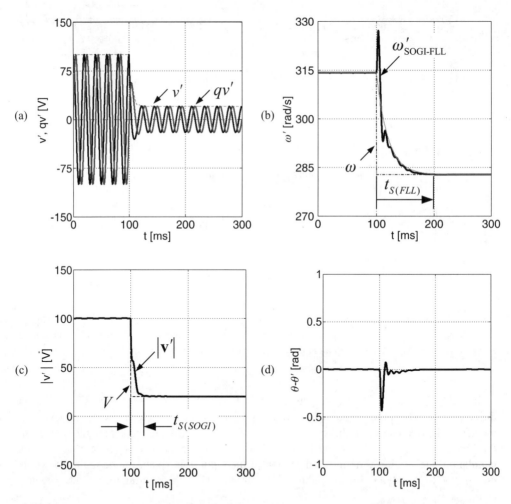

Figure 4.36 Response of the SOGI-FLL with FLL gain normalization when the input signal simultaneously experiences a voltage dip and a frequency jump: (a) in-quadrature signals, (b) detected frequency, (c) detected input voltage amplitude and (d) error in phase-angle detection $\theta - \theta'$

at the same time that its frequency varies from 50 to 45 Hz. As can be appreciated from the plots of Figure 4.36, both the detected input frequency signal and the detected input amplitude signal are coupled to each other, which implies that they differ from the simplified waveforms previously studied. However, for certain conditions, it can still be assumed as valid that the settle times of the SOGI and the FLL match those calculated by (4.109) and (4.111). In this regard, an analysis based on the simulation demonstrated that, for $k = \sqrt{2}$, the settle times for the SOGI and the FLL had to be in the range of $t_{S(FLL)} \geq 2t_{S(SOGI)}$ to ensure enough different constant times in both subsystems and to be able to apply (4.109) and (4.111) satisfactory to calculate settling times.

4.7 Summary

In this chapter, the paradigm of synchronizing in a single-phase system was discussed. Different techniques for quadrature signal generation (QSG) are presented and compared in terms of dynamic performances. The novel QSG based on a second-order generalized integrator (SOGI) proves to be a very good solution, easy to design and implement, and able to perform close to a delay-free filter with the required bandwidth.

This is a very important feature model in the application of synchronization of grid-connected converters as the voltage of the grid is typically affected by disturbances and harmonics.

The second challenge is to make the synchronization process adaptive to frequency changes that may occur in the grid system, e.g. during the power balancing events. Moreover, some grid codes require operation within a frequency range of 47–52 Hz, posing a big challenge for the synchronization loop.

Some classical solutions based on adaptive filtering techniques are presented along with the novel frequency-locked loop (FLL) technique, which is able to achieve frequency adaptability using a single algorithm without any use of trigonometric functions. The FLL feature also measured stability over transients from a physical point of view, the frequency change being much slower than the grid voltage phase-angle. The FLL is a nonlinear system and a practical design method was proposed in this chapter.

References

[1] Stade, E., *Fourier Analysis*, John Wiley & Sons, Ltd, 2005. ISBN 978-0-471-66984-5.
[2] Brigham, E., *Fast Fourier Transform and Its Applications*, Prentice-Hall, 1988. ISBN 0133075052.
[3] Sakakibara, Y. and Gleason, A., *Who Is Fourier? A Mathematical Adventure*, Language Research Foundation, Transnational College of Lex Tokyo, April 1995. ISBN 10: 0964350408.
[4] Asiminoaei, L., Teodorescu, R., Blaabjerg, F. and Borup, U., 'A Digital Controlled PV-Inverter with Grid Impedance Estimation for ENS Detection'. *IEEE Transactions on Power Electronics*, **20**(6), November 2005, 1480–1490.
[5] Dolen, M. and Lorenz, R. D., 'Industrially Useful Means for Decomposition and Differentiation of Harmonic Components of Periodic Waveforms'. In *Proceedings of the IEEE Industry Applications Society Annual Meeting*, 2000, pp. 1016–1023.
[6] Begovic, M. M., Djuric, P. M., Dunlop, S. and Phadke, A. G., 'Frequency Tracking in Power Networks in the Presence of Harmonics', *IEEE Transactions on Power Delivery*, **8**(2), April 1993, 480–486.
[7] Nedeljkovic, D., Nastran, J., Vocina, D. and Ambrozic, V. 'Synchronization of Active Power Filter Current Reference to the Network'. *IEEE Transactions on Industrial Electronics*, **46**(2), April 1999, 333–339.

[8] Cooley, J.W. and Tukey, J.W., 'An Algorithm for the Machine Calculation of Complex Fourier Series'. *Mathematical Computations*, **19**, 1965, 297–301.
[9] Vainio, O., Ovaska, S.J. and Polla, M., 'Adaptive Filtering Using Multiplicative General Parameters for Zero-Crossing Detection'. *IEEE Transactions on Industrial Electronics*, **50**(6), December 2003, 1340–1342.
[10] Dolen, M. and Lorenz, R. D., 'Industrially Useful Means for Decomposition and Differentiation of Harmonic Components of Periodic Waveforms'. In *Proceedings of the IEEE Industry Applications Society Annual Meeting*, 2000, pp. 1016–1023.
[11] McGrath, B. P., Holmes, D. G. and Galloway, J., 'Improved Power Converter Line Synchronisation Using an Adaptive Discrete Fourier Transform (DFT)'. In *Power Electronics Specialists Conference*, 2002. pesc. 02. 2002 IEEE 33rd Annual, Vol. 2, 2002, pp. 821–826.
[12] Wall, R.W., 'Simple Methods for Detecting Zero Crossing'. In *Industrial Electronics Society, IECON '03. The 29th Annual Conference of the IEEE*, Vol. 3, 2–6 November 2003, pp. 2477–2481.
[13] Best, R. E., *Phase-Locked Loops: Design, Simulation, and Applications*, 5th edition, New York: McGraw-Hill Professional, 2003. ISBN 0071412018.
[14] Franklin, G.F., Powell, J. D. and Emami-Naeini, A., *Feedback Control of Dynamic Systems*, 4th edition, Prentice Hall, 2002. ISBN 0130323934.
[15] Saitou, M., Matsui, N. and Shimizu, T., 'A Control Strategy of Single-Phase Active Filter Using a Novel *d-q* Transformation'. In *Proceedings of the Industry Applications Conference 2003 (IAS'03)*, Vol. 2, October 2003, pp. 1222–1227.
[16] Silva, S. M., Lopes, B. M., Filho, J.C., Campana, R. P. and Bosventura, W.C., 'Performance Evaluation of PLL Algorithms for Single-Phase Grid-Connected Systems'. In *Proceedings of the Industry Applications Conference 2004 (IAS'04)*, Vol. 4, October 2004, pp. 2259–2263.
[17] Rader, C, M. and Jackson, L. B., 'Approximating Noncausal IIR Digital Filters Having Arbitrary Poles, Including New Hilbert Transformer Designs, Via Forward/Backward Block Recursion'. *IEEE Transactions on Circuits and Systems I: Regular Papers*, **53**(12), December 2006, 2779–2787.
[18] Haykin, S.S., *Adaptive Filter Theory*. Upper Saddle River, NJ: Prentice Hall, 2002.
[19] Widrow, B., Glover Jr, J. R., McCool, J. M., Kaunitz, J., Williams, C.S., Hearn, R.H., Zeidler Jr, J. R., Dong, E. and Goodlin, R.C., 'Adaptive Noise Cancelling: Principles and Applications'. *Proceedings of the IEEE*, **63**(12), December 1975, 1692–1716.
[20] Karimi-Ghartemani, M. and Iravani, M. R., 'A Nonlinear Adaptive Filter for Online Signal Analysis in Power Systems: Applications'. *IEEE Transactions on Power Delivery*, **17**, April 2002, 617–622.
[21] Yuan, X., Merk, W., Stemmler, H. and Allmeling, J., 'Stationary Frame Generalized Integrators for Current Control of Active Power Filters with Zero Steady-State Error for Current Harmonics of Concern under Unbalanced and Distorted Operating Conditions'. *IEEE Transactions on Industrial Applications*, **38**(2), March/April 2002, 523–532.
[22] Teodorescu, R., Blaabjerg, F., Liserre, M. and Loh, P.C., 'Proportional-Resonant Controllers and Filters for Grid-Connected Voltage-Source Converters'. *Electric Power Applications, IEE Proceedings*, **153**(5), September 2006, 750–762.
[23] Zmood, D.N. and Holmes, D. G., 'Stationary Frame Current Regulation of PWM Inverters with Zero Steady-State Error'. *IEEE Transactions on Power Electronics*, **18**(3), May 2003, 814–822.
[24] Padmanabhan, M., Martin, K. and Peceli, G., *Feedback-Based Orthogonal Digital Filters: Theory, Applications, and Implementation*, Norwell, MA: Kluwer Academic Publishers, 1996.
[25] Burger, B. and Engler, E., 'Fast Signal Conditioning in Single Phase Systems'. In *European Conference on Power Electronics and Applications (EPE'01)*, PP00287, Graz, August 2001.
[26] de Brabandere, K., Loix, T., Engelen, K., Bolsens, B., Van den Keybus, J., Driesen, J. and Belmans, R., 'Design and Operation of a Phase-Locked Loop with Kalman Estimator-Based Filter for Single-Phase Applications'. In *Proceedings of the IEEE Industrial Electronics Conferece (IECON'06)*, November 2006, pp. 525–530.
[27] Mojiri, M. and Bakhshai, A., 'An Adaptive Notch Filter for Frequency Estimation of a Periodic Signal'. *IEEE Transactions on Automatic Control*, **49**(2), February 2004, 314–318.
[28] Mojiri, M., Karimi-Ghartemani, M. and Bakhshai, A., 'Time Domain Signal Analysis Using Adaptive Notch Filter'. *IEEE Transactions on Signal Processing*, **55**(1), January 2007, 85–93.
[29] Rodriguez, P., Teodorescu, R., Candela, I., Timbus, A.V., Liserre, M. and Blaabjerg, F., 'New Positive-Sequence Voltage Detector for Grid Synchronization of Power Converters under Faulty Grid Conditions'. In *Proceedings of the IEEE Power Electronics Special Conference (PESC'06)*, June 2006, pp. 1–7.

[30] Ciobotaru, M., Teodorescu, R. and Blaabjerg, F., 'A New Single-Phase PLL Structure Based on Second Order Generalized Integrator'. In *Proceedings of the IEEE Power Electronics Special Conference (PESC'06)*, June 2006, pp. 1–7.

[31] Rodriguez, P., Luna, A., Candela, I., Teodorescu, R. and Blaabjerg, F., 'Grid Synchronization of Power Converters Using Multiple Second Order Generalized Integrators'. In *Proceedings of the IEEE Industrial Electronics Conference (IECON'08)*, November 2008, pp. 755–760.

5

Islanding Detection

5.1 Introduction

A higher penetration of distributed power generation systems (DPGSs), involving both conventional and renewable technologies, is changing the power system face. There is a clear evolution towards active grids that could include a significant amount of storage systems, could work in island mode and could be connected through flexible transmission systems. This complex scenario will put different requirements on the DPGS units depending on their size and on their level of integration with the power system. Thus the monitoring of the grid condition will always be a crucial feature of the DPGS units at every level. The detection of a possible island condition will always be important in a power system with a significant amount of DPGS.

Typically in a low-power DPGS like, for example, PV systems, this feature is defined as an 'anti-islanding requirement' in order to highlight the request of the utility operator, as pointed out in Chapter 2, that the DPGS should disconnect in case the main electric grid should cease to energize the distribution line. A higher power DPGS, typically wind plants, have completely different requirements and generally benefits of communication systems and a supervisory control that interact with the utility operator in view of making the DPGS contribute to the stability of the grid. Here the latest grid codes require low-voltage ride-through capability, meaning that they should stay connected during grid faults, which is quite opposite for the PV systems. Hence islanding detection can be considered a requirement only for low-power DPGSs. However, as already pointed out, the power system is evolving and the future scenario may consider the presence of a smart micro-grid (SMG) usually operated in connection to distribution grids but with the capability of automatically switching to a stand-alone operation if faults occur in the main distribution grid, and then reconnecting to the grid. Since it is not possible to predict the level of connection and the reliability of the information exchange among the different players of this future scenario, the detection of islanding can be considered as an important feature, requested in some cases and optional in others.

In this chapter islanding detection will be treated with attention paid to the consequences of an uncontrolled islanding (amplitude and frequency variation of the grid voltage, which are usually the first signs of the island condition) and to the performances of the islanding detection methods: reliability, selectivity and minimum perturbation. Ideally, the methods should be able to detect the island condition in every grid condition, strong or weak, with a limited or high penetration of DPGS – this property can be defined as reliability; the method should also be able to discriminate between a condition of islanding and a simple perturbation of the grid – this property can be defined as selectivity; finally, the method should degrade the grid power quality as little as possible in order to make suitable the parallel operation of several DPGS – this property can be defined as minimum perturbation.

In the following the nondetection none (NDZ) will be defined with reference to the effects on the grid voltage amplitude and frequency of uncontrolled island operation, which can be used as the basic means to detect the island operation. Then the methods will be reviewed, classifying them in passive, and hence only based on the measurements of the electric grid quantities, and active, and hence based also on the deliberate and periodic perturbation of the electric grid to test its presence. Next a final comparison is made with reference to the actual standards and codes and to the industrial use. Some novel relevant active methods are described in more detail with the deliberate purpose of promoting their implementation in practice.

5.2 Nondetection Zone

The reliability of islanding detection methods can be represented by the nondetection zone (NDZ), defined in the power mismatch space (ΔP versus ΔQ) at the point of common coupling (PCC) where the islanding is not detectable and there is potential for parasitic trips [1].

Figure 5.1(a) shows the typical interconnection of a PV inverter to the utility and local load and Figure 5.1(b) the balance of the power in the system. ΔP is the real power output of the grid, ΔQ is the reactive power output of the grid, P_{DG} is the real power output of the PV, Q_{DG} is the reactive power output of the PV, P_{load} is the real power of the load and Q_{load} is the reactive power of the load. Hence the power balance is

$$P_{load} = P_{DG} + \Delta P \qquad (5.1)$$

$$Q_{load} = Q_{DG} + \Delta Q \qquad (5.2)$$

If $P_{load} = P_{DG}$ there is not a mismatch between the power produced by the PV system and the power produced by the utility and in the same way if $Q_{load} = Q_{DG}$ there is no reactive power mismatch between the PV system and the utility.

The behaviour of the system at the time of utility disconnection will depend on ΔP and ΔQ at the instant before the switch opens to form the island [2]. If the resonant frequency of the RLC load is the same as the grid line frequency the linear load does not absorb or consume reactive power. Active power is directly proportional to the voltage. After the disconnection of the grid, the active power of the load is forced to be the same with the power generated by the PV system; hence the grid voltage changes into

$$V' = KV \qquad (5.3)$$

where

$$K = \sqrt{\frac{P_{DG}}{P_{load}}} \qquad (5.4)$$

Islanding Detection

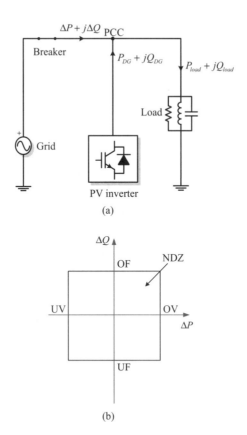

Figure 5.1 (a) Interconnection of the grid inverter to the utility and the load and (b) the nondetection zone (NDZ)

When $P_{DG} > P_{load}$ there is an increase in the amplitude of the voltage and if $P_{DG} < P_{load}$ there is a decrease in the amplitude. Reactive power is tied to the frequency and amplitude of voltage:

$$Q'_{load} = Q_{DG} = \left(\frac{1}{\omega' L} - \omega' C\right) V'^2 \quad (5.5)$$

In this way is possible to calculate the islanding pulsation (ω'):

$$\omega' = \frac{-\frac{Q_{DG}}{CV'^2} + \sqrt{\left(\frac{Q_{DG}}{CV'^2}\right)^2 + \frac{4}{LC}}}{2} \quad (5.6)$$

The grid is subject to numerous disturbances, such as voltage dips, overvoltage, harmonic distortion and frequency variations. It is necessary to set an islanding protection immune to

Table 5.1 Maximum variation of grid voltage and frequency EN 50160

Value	Minimum	Maximum
Frequency	$f_{min} = 49$ Hz	$f_{max} = 51$ Hz
Voltage	$V_{min} = 0.9$ p.u.	$V_{max} = 1.1$ p.u.

these disturbances. The grid voltage and frequency limits according to EN 50160 (requirements for public low-voltage distribution grid) are given in Table 5.1.

The worst case for islanding detection is represented by a condition of balance of the active and reactive power in which there is no change in amplitude and frequency, i.e. $\Delta P = 0$ and $\Delta Q = 0$. It is straightforward that a small ΔP results in an insufficient change in voltage amplitude and a small ΔQ results in an inadequate change in frequency to effectively disconnect the PV and prevent islanding.

It is possible to calculate the NDZ area from the mismatches of active and reactive power and to set the values of the threshold for frequency and amplitude of the voltage (Figure 5.1(b)). The probability that ΔP and ΔQ fall into the NDZ of OUV/OUF (over/under voltage and frequency) can be significant. Because of this concern, the standard over/under voltage and frequency protective devices alone are generally considered to be insufficient anti-islanding protection and thus they must be combined with other islanding detection methods, as explained in the following.

5.3 Overview of Islanding Detection Methods

Three main approaches can be employed for islanding detection:

- Grid-resident detection.
- External switched capacitor detection.
- Inverter-resident detection.

First two methods methods require either a communication system through the power line or an external switched capacitor at the PCC in order to detect the islanding condition accurately, which increases the system complexity and its economical costs.

The grid-resident methods are based on the communication between the grid and PV inverters and are completely different from the other inverter-resident techniques. In fact, a transmitter (T) is installed near the line protection switch and a receiver (R) is positioned in the PCC in the proximity of the inverter, as shown in Figure 5.2(a) [3]. The system uses a PLCC (power line carrier communications) line whose support is the power grid. Under normal conditions of operation a specific frequency signal is sent to the receiver using the energized power lines.

When a PV inverter cannot 'hear' the signal any more islanding is detected. The main problems remain the high cost of broadcasting equipment and the fact that such a method requires the collaboration of grid operators to be implemented. The same goal can be achieved with a dedicated line of communication. This method is very good for islanding detection because it is independent from power flows and it does not present an NDZ. The evolution of

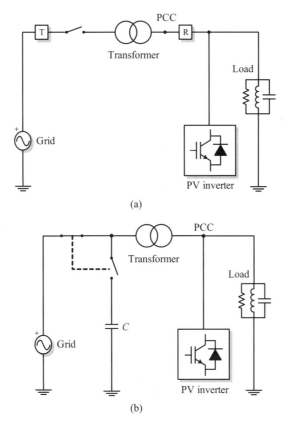

Figure 5.2 (a) Grid-resident islanding detection methods based on communication and (b) switched external capacitor detection

grid resident methods could be done using a SCADA (supervisory control and data acquisition) system even if the penetration of this communication systems in the low voltage distribution grid is limited to smart metering [4].

These methods have not been commercialized yet because of the high costs of installation, but the situation could change if PV penetration increases. They may be the ultimate islanding detection methods, also working perfectly in the case of multiple inverters operating in parallel, one of the hardest goals to achieve by all other methods.

External switched capacitor (ESC) *detection* is based on the concept that an external capacitor periodically switched on in parallel with the grid would produce a zero-crossing delay proportional to the grid impedance. The method has been successfully implemented some years ago as a separate device connected at the PCC in parallel with PV inverters to comply with an ENS standard (like, for example, the product ENS 32 by UfE GmbH). The VDE 0126-1-2006 has relaxed the impedance detection and an anti-islanding compliance can be achieved by software algorithms (inverter-resident methods). The ESC method can be used in applications with multiple inverters in parallel where the islanding detection function can be implemented separately using only one device.

Inverter-resident detection relies exclusively on software implementation inside the PV inverter control platform and can use:

- Passive methods.
- Active methods.

The *passive methods* are based on detection of a change of a power system parameter (typically amplitude, frequency, phase or harmonics of the voltage) caused by the power mismatch after the disconnection. Passive methods have nonzero NDZ and are typically combined with active methods to improve the reliability.

The *active methods* generate a disturbance in the PCC in order to force a change of a power system parameter that can be detectable by the passive methods. With active methods, the NDZ can be significantly reduced; however, they have the potential to affect power quality and to generate instability in the grid, especially if more inverters are connected in parallel.

5.4 Passive Islanding Detection Methods

The passive methods are based on monitoring grid parameters that typically change during islanding. In the following the most relevant passive islanding detection methods are described. Some of the classical methods are described with a more overview character while some novel methods are described in more detail.

5.4.1 OUF–OUV Detection

All grid-connected PV inverters are required to have an over/under frequency (OUF) and over/under voltage (OUV) protection window that causes the PV inverter to cease supplying power to the utility grid if the values in the PCC are out of this window (see Figure 5.2(b)). As described in Chapter 3, the typical ranges for the voltage–frequency working window are $+10/-15\%$ in voltage and ± 1 Hz in frequency around the nominal values (in some countries there can be slightly different values).

Voltage and frequency monitoring is typically used in order to trip the inverter in case of OUV or OUF and thus islanding detection is achieved. However, when the amount of power mismatch at the PCC due to islanding is very small it may fail to detect islanding as voltage and frequency variations could be too small to hit OUF or OUV. The worst case for islanding detection is represented by a condition of balance of the active and reactive power in which there is no change in amplitude and frequency, i.e. $\Delta P = 0$ and $\Delta Q = 0$. The minimum values for ΔP and ΔQ that would hit the OUF or OUV can be determined analytically as:

For OUF:

$$q\left(1-\left(\frac{f}{f_{\min}}\right)^2\right) \leq \frac{\Delta Q}{P_{DG}} \leq q\left(1-\left(\frac{f}{f_{\max}}\right)^2\right) \qquad (5.7)$$

For OUV:

$$\left(\frac{V}{V_{\max}}\right)^2 - 1 \leq \frac{\Delta P}{P_{DG}} \leq \left(\frac{V}{V_{\min}}\right)^2 - 1 \tag{5.8}$$

Thus the NDZ can be precisely determined, but in most cases this method is considered to be insufficient as anti-islanding protection complying with the PV standards (e.g. IEEE 1574).

Another important factor for the reliability of this method is the accuracy in monitoring the voltage and frequency. The grid is subject to numerous disturbances as voltage dips, overvoltage, harmonic distortion and frequency variations. Fast and high accuracy voltage monitoring techniques showing robustness over grid background harmonics and disturbances are reported in reference [5], where different phase-locked loop (PLL) structures are analysed, or in reference [6], where a frequency-adaptive PLL scheme is proposed. These are described in detail in Chapter 4.

5.4.2 Phase Jump Detection (PJD)

This method observes the phase difference between the inverter terminal voltage and its output current, which typically occurs during islanding due to a reactive power mismatch. In contrast with the OUF method, the phase can change much faster than the frequency, so much faster island detection is theoretically possible. This method can be easily implemented using the zero-cross detection synchronization methods, where the phase of the current is updated once every zero-crossing of the voltage and thus it is possible to detect the eventual phase jump. However, as nowadays fast PLL are typically implemented for more robust synchronization, this phase jump can become negligible since the PLL forces the current to resynchronize with the voltage after islanding, minimizing the phase jump. A possible solution is to implement two PLLs, one fast for synchronization (e.g. with a settling time in the range of 50–100 ms) and one very slow for islanding detection (e.g. with a settling time of 1–2 s). As the slow PLL will 'filter out' the phase jump, it is possible to detect the phase jump by comparing the phase calculated by the two PLL. However, it is still difficult to choose the correct threshold for obtaining a reliable islanding detection as phase jumps in the grid voltage occur quite frequently due to, for example, switching reactive loads (capacitor banks, induction motors, etc.) and can result in nuisance trips. Also for three-phase DG systems, which are allowed to work with nonunity PF, like, for example, wind turbines, this method could lead to nuisance trips.

As with all passive methods, the NDZ cannot be zero as during islanding without any power mismatch the phase of the voltage will not change.

5.4.3 Harmonic Detection (HD)

The DG interface inverter, even if perfectly controlled to work as an ideal current source, produces harmonics due to switching (high order harmonics), dead-time and semiconductor voltage drop (even harmonics) or ripple of the DC link voltage (odd harmonics). These harmonics are maintained low to comply with standard regulations (e.g. IEEE 1574 *THD* < 5 %) by means of hardware solutions (typically filters) or compensation mechanisms embedded in the control algorithm. However, they generate an amount of voltage harmonics that depends on the level of grid impedance. As the grid impedance is usually low, these voltage harmonics are quite low and difficult to detect.

In the islanding mode, the grid impedance is now replaced by the load impedance, which can also be much higher with respect to the grid impedance (at least on order of magnitude for a low-voltage DG), so the harmonics level in the voltage will be increased significantly and can be used as an indicator for islanding detection. If the electric grid is highly polluted or weak (high grid impedance) and the DG grid inverter is optimally controlled the harmonic level can decrease. In reference [2] two thresholds are proposed: one related to the distortion that can be expected in the grid voltage and the second related to the distortion that will be produced during islanding by either of the inverter harmonics amplified by the local load or nonlinear loads. If the system distortion is outside these two thresholds it means that the system is in the islanding condition because the harmonic distortion is either too low or too high. Either the THD of the voltage or the amplitude of the most important harmonics (3rd, 5th, 7th, 9th, 11th) can be used as an indicator.

The main advantage of this method is that it is one of the few passive methods that can theoretically reduce the NDZ to zero because it does not depend on the power mismatch between active and reactive powers produced by the PV-system and absorbed by the loads. However, there are also few shortcomings of this method:

- Connection/disconnection of nonlinear loads may change the harmonic condition and can be interpreted as an event of islanding.
- No-load transformers, which are known for generating high amounts of 3rd harmonic.
- Some DG inverters may increase the voltage background distortion in an effort to inject 'clean' currents.

The main difficulties related to the application of this method are in the choice of the parameters that should be evaluated (harmonics or indexes that combines some of them) for islanding detection and in the choice of the thresholds. In fact, it is not easy to discriminate between the harmonic pollution created by the grid, by the loads and by the DG unit, and hence to guarantee not only the islanding detection but also to avoid a false trip. The main challenge to engineers is to make the method selective with respect to normal change in the harmonic situation that may appear as islanding conditions. Some studies have highlighted the relation between the DC link ripple of the DG unit, the grid impedance and the amount of harmonic generated by the DG unit [2] in order to give some guidelines for tuning the method and its thresholds. However, most of the studies are related to the choice of the indicators, with the aim to increase the robustness of the HD (minimize the NDZ and maximize the selectiveness of the method).

Three approaches are proposed in the following: the first is suitable for three-phase DG because it is based on an unbalance evaluation; the second is based on the Kalman filter estimation of the 3rd, 5th and 7th harmonic variations; and the third is based on the wavelet and is an attempt to evaluate a portion of the frequency spectrum that is subjected to larger variations in the case of the islanding condition.

The HD method modified by taking into account the system unbalance is the result of the consideration that the detection method in general can fail if the load has strong low-pass characteristics, which occurs for loads with a high value of the quality factor Q or when the inverters have high-quality, low-distortion outputs. In order to increase the reliability of this method several other indicators, such as, for example, voltage unbalance, can be used.

Thus in reference [7] the averaged *THD* of the current (5.9) is combined with information about voltage unbalance, obviously applicable for three-phase DPGS. Thus, VUB_{avg}, the variation in the voltage unbalance (5.10), and V_{avg}, the three-voltage amplitude variation (5.11), are defined. All these parameters are averaged over one cycle and are used together with the one-cycle variation (ΔTHD, ΔVUB_{avg} and ΔV_{avg}) in order to increase the sensitivity of this method.

$$THD_{avg}[\%] = \int_0^T \left[\frac{\sqrt{\sum_{h>1} I_h^2}}{I_1} \times 100 \right] dt \qquad (5.9)$$

$$VUB_{avg}[\%] = \int_0^T \left[\frac{\max(V_a, V_b, V_c) - \min(V_a, V_b, V_c)}{\frac{1}{3}(V_a + V_b + V_c)} \times 100 \right] dt \qquad (5.10)$$

where V_a, V_b, V_c denote RMS values of phase voltages and T is the grid period

$$V_{avg}[V] = \frac{\pi}{3\sqrt{2}} \int_0^T [\max(v_a, v_b, v_c) - \min(v_a, v_b, v_c)] dt \qquad (5.11)$$

where v_a, v_b, v_c denote instantaneous values of phase voltages.

At every sampling time, this method calculates the THD_{avg} of phase a current, the averaged unbalance VUB_{avg} and V_{avg}, the averaged amplitude variation. In the first, it checks whether the V_{avg} is lower than the preset value (0.5 p.u.). If it is, then the trip signal is generated immediately. This will be a typical islanding due to a large variation in the loading for the DG. Otherwise, the method checks the other monitoring parameters THD_{avg} and VUB_{avg}. If there is still no islanding condition detection, the algorithm tests the one-cycle variation of these variables according to the rule:

$$\left(\Delta THD_{avg} > 75\,\%\right) \text{ OR } \left(\Delta THD_{avg} < -100\,\%\right) \text{ AND } \left(\Delta VUB_{avg} > 50\,\%\right) \text{ OR } \left(\Delta VUB_{avg} < -100\,\%\right) \qquad (5.12)$$

If the logical condition (5.12) is true for more than one cycle, the trip signal is generated, signalling the islanding condition due to little variation in the loading for DG. Thus a highly sensitive islanding detection can be achieved.

The method, described in reference [8], based on the use of the Kalman filter estimation of the 3rd, 5th and 7th harmonic variations, exploits the natural sensitivity to disturbances of a grid voltage sensorless control to highlight the islanding condition, as shown in Figure 5.3(a). The algorithm evaluates not only the absolute value of the grid voltage harmonics but also the variation of the spectrum power density (energy). Often DPGS controllers adopt harmonic compensators that produce a voltage contribution to the control action used to null the harmonic voltage drop on the grid filter and as a consequence the current harmonics. The contribution of the harmonic compensators can be considered as an estimate of the background distortion and compared with the measured one (see Figure 5.3(b)).

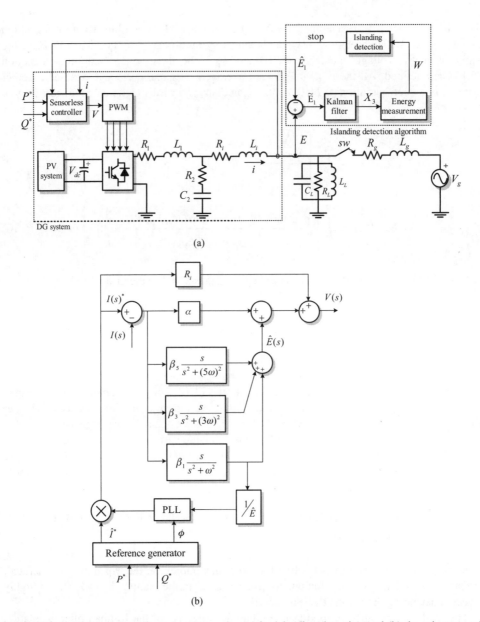

Figure 5.3 (a) Voltage sensorless algorithm adopted for islanding detection and (b) the scheme used for voltage harmonic estimation

The third method reviewed in this chapter in the category of the HD method is a wavelet-based detection method, described in reference [9], which can detect the islanding condition from local measurements of PCC voltage and current signals, as in the case of passive methods, but evaluates the high-frequency components injected by the PV inverter, which depend on the characteristics of the employed pulse width modulator, LCL filter and current controller, to

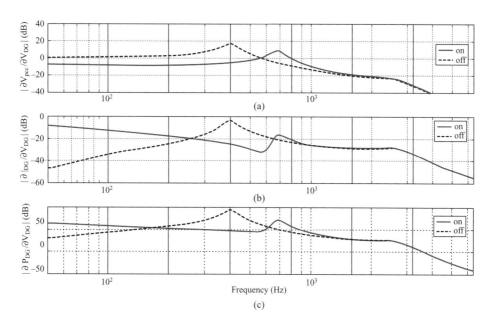

Figure 5.4 Frequency behaviour of the DG unit with the LCL filter when it is connected to the grid (on) and when it is working in the islanding condition (off)

reveal the islanding condition, as done by active methods. As shown in Figure 5.4, the spectrum of the PV inverter output power suffers a small variation, after the islanding operation mode, over a continuous and relatively wide frequency band. Passive harmonic detection methods, based on DFT, do not allow these variations to be detected due to their low resolution, which depends on the number of selected harmonics. As a consequence, wavelet filter banks are proposed for tracking purposes of such spectrum variations in a properly selected frequency band.

5.4.4 Passive Method Evaluation

In order to determine the islanding condition more accurately, the different individual detection methods can be operated simultaneously. The reliability of passive methods is limited as there will always be a nonzero NDZ for small power unbalances. Thus passive methods are often combined with active methods. A performance comparison of the passive methods has been reported in Table 5.2 as outlined in references [2] and [11].

Table 5.2 Comparison among passive anti-islanding methods

Method	NDZ	Trip time (power balance)
OUV	$-17\% \leq \Delta P \leq 24\%$	Not applicable
OUF	$-5\% \leq \Delta Q \leq 5\%$	Not applicable
PJD	$-5\% \leq \Delta Q \leq 5\%$	Not applicable
HM	Absent	It can be less than 200 ms

5.5 Active Islanding Detection Methods

The central concept of active methods relies on the generation of small perturbations at the output of the PV inverter generating small changes in one of the power system parameter (frequency, phase, harmonics, P, Q). The targeted actions are:

- *Frequency* drifts enough to activate OUF protection in due course.
- *Voltage* drifts enough to activate OUV protection in due course.
- *Grid impedance* estimation and thus indirectly to detect the islanding.
- *PLL- based* estimation.
- *Negative sequence detection.*

This concept has generated a very large number of ideas, materialized in publications and patents, as the ways of implementing it can be very different. Generally, the active methods, usually in combination with passive methods, lead to islanding without NDZ, as required by IEEE 1574 or VDE 0126-1-2006. Comprehensive evaluation of the most known techniques is reported in references [1], [12] and [13].

In the following the main concepts of the most relevant active AI methods are described.

5.5.1 Frequency Drift Methods

These methods target the grid frequency drift by disturbing the frequency reference with, for example, a positive feedback. As long as the grid is present, it is obvious that the frequency cannot be drifted, but when the grid is disconnected, the disturbance will be able to drift the frequency until it hits OUF protection. Several implementations exist as in the following.

5.5.1.1 Active Frequency Drift (AFD)

The output current waveform is slightly distorted, presenting a zero current segment for a drift-up operation [14]. This is done by forcing the current frequency to be slightly ($\delta f = 0.5 - 1.5$ Hz) higher than the voltage frequency in the previous cycle and keeping the inverter current equal to zero from the end of its negative semi-cycle to the positive zero-crossing of the voltage (as shown in Figure 5.5).

The so-called 'chopping factor' for AFD is defined as

$$cf = \frac{2T_z}{T} = \frac{\delta f}{f + \delta f} \tag{5.13}$$

where T_z is the zero time of the AFD signal and T is the period of the utility voltage.

The inverter reference and the phase for this method in the steady state are

$$i^* = \sqrt{2}I \sin\left[2\pi \left(f + \delta f\right)\right] t$$

$$\theta_{AFD} = \pi f T_z = \frac{\pi \delta f}{f + \delta f} \tag{5.14}$$

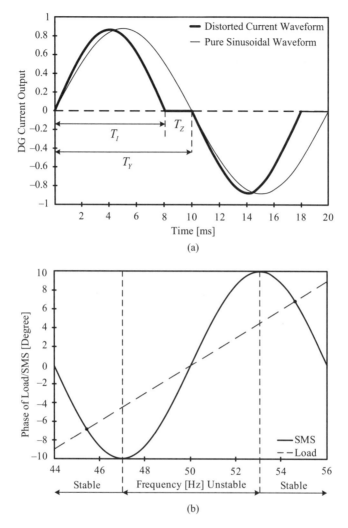

Figure 5.5 (a) Current shape in AFD and (b) phase–frequency dependence in SMS

Thus, there is a continuous trend to change the frequency but the grid presence will prevent this. In an island, the frequency will be drifted away and eventually tripping of the OUF protection. According to reference [15], the NDZ cannot be reduced to zero and it will be dependent on the quality factor of the LC load (Q) and of δf being very close to the OUF–OUV method for high Q loads.

5.5.1.2 Slip-Mode Frequency Shift (SMS)

In this scheme, a *positive feedback is* applied to the phase of PCC voltage to destabilize the inverter by changing the short-term frequency [1]. The phase angle of the inverter is made a function of the frequency, as depicted in Figure 5.5(b). The phase response curve of the inverter is designed such that the phase of the inverter increases faster than the phase of the (RLC) load with a unity-power factor in the region near the utility frequency. If the utility is tripped and the frequency of PCC voltage is distorted, the inverter phase response curve increases the phase error and, hence, causes instability in the frequency. This instability further amplifies the perturbation of the frequency of PCC voltage and the frequency is eventually driven away until it hits OUF protection.

The inverter reference and the phase for this method in the steady state are

$$i^* = \sqrt{2}I \sin(2\pi ft + \theta_{SMS})$$
$$\theta_{SMS} = \theta_m \sin\left(\frac{\pi}{2}\right) \frac{f_i - f}{f_m - f} \quad (5.15)$$

where f_m is the frequency at which the maximum phase shift θ_m occurs. For this method it is possible to obtain zero NDZ for a given Q by choosing f_m and θ_m that satisfy the condition [15]:

$$\frac{\theta_m}{f_m - f} \geq \frac{12Q}{\pi^2} \quad (5.16)$$

For $Q = 2.5$ (as in IEEE 929-2000) for $f_m - f = 3$ Hz, a typical $\theta_m = 10°$ results.

5.5.1.3 Sandia Frequency Shift (SFS)

Also called active frequency drift with positive feedback (AFDPF), this is an extension of the AFD method and is another method that utilizes positive feedback. In this method, it is the frequency of voltage at PCC to which the positive feedback is applied. To implement the positive feedback, the 'chopping fraction' from AFD is made to be a function of the error in the line frequency [1]:

$$cf = cf_0 + k(f - f_n) \quad (5.17)$$

where k is the acceleration gain, cf_0 is the chopping factor when there is no frequency error and $f - f_n$ is the difference between the estimated frequency and nominal value.

When connected to the utility grid, minor frequency changes are detected and the method attempts to increase the change in frequency; however, the stability of the grid prevents any change. When the utility is disconnected and as f increases, the frequency error increases, the chopping fraction increases and the PV inverter also increases its frequency. The inverter thus acts to reinforce the frequency deviation, and this process continues until the frequency reaches the threshold of the OUF. According to reference [15] the NDZ can be reduced to zero for $Q < 4.8$ by choosing $cf_0 = 0.05$ and $k = 0.01$. This method will distort the current, but for these values the THD of the current can be kept under the 5 % limit.

This method is typically combined with the complementary method SVS (Sandia voltage shift) [1] to maximize the islanding performance.

5.5.1.4 Active Frequency Drift with Pulsating Chopping Factor (AFDPCF)

In reference [10] an improved SFS method is reported where the chopping factor, instead of depending on a gain, has an alternating pulse shape leading to faster frequency drift during islanding:

$$cf = \begin{cases} cf_{max} & \text{if } T_{cfmax\_on} \\ cf_{max} & \text{if } T_{cfmin\_on} \\ 0 & \text{otherwise} \end{cases} \quad (5.18)$$

where cf_{max} and cf_{min} are the maximum and minimum values of cf respectively, as depicted in Figure 5.6(a).

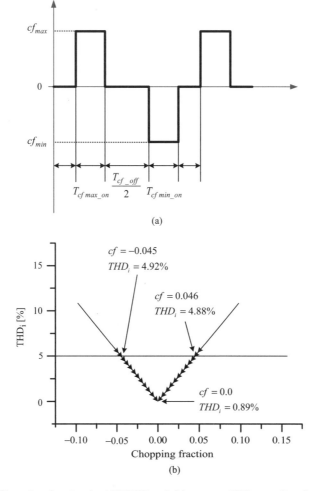

Figure 5.6 (a) Chopping fraction in AFDPCF and (b) current *THD* as a function of the chopping fraction

Actually, the frequency is pushed to increase in one period and then to decrease in the second period. The positive and negative values of the chopping factor can be set using an analytical calculation by imposing a certain grid current *THD* as required by standards. Figure 5.6(b) depicts the dependence of current *THD* on the chopping factor. Thus, complying with the power quality standard can be guaranteed.

Also the potential for parallel operation is higher than the previous AFD base method. The same authors have recently proposed in reference [16] a combination of AFD and SMS where both frequency and current starting phase are changed using software PLL technology. Thus a robust method is obtained with more design flexibility, leading to customized NDZ with minimal *PQ* degradation. The design procedure complexity is still being increased and needs to be clarified.

5.5.1.5 GE Frequency Shift (GEFS)

This is another frequency drift AI method based on positive feedback like SFS. Here the reactive current reference is augmented with a positive feedback derived from the frequency estimation, with proper filtering and gain in order to maintain the stability [17]. Increasing the reactive current reference will lead to higher reactive power, which in the case of islanding on the RLC load will further increase the frequency, causing it to be quickly pushed outside the OUF limits, as shown in Figure 5.7(a).

As typically the frequency positive feedback is implemented in combination with the voltage positive feedback method, both methods are illustrated in Figure 5.7(a).

By injecting continuous feedback signals into the grid, which has a spectral content confined to a very narrow band around the fundamental frequency, the *THD* degradation is negligible. As a result, the positive feedback gain for GEFS is not limited by the power-quality constraint as in the case of SFS, leading to the possibility of obtaining a very small NDZ. The GEFS also has good potential for parallel connection of inverters.

In reference [18] an interesting approach is introduced where the magnitude of the phase shift between the voltage and current (typically null) is modulated with a 1 Hz value. When grid power is available, the phase shift has no more effect than a slow variation of the current around the average value, and does not change the power being delivered to the grid. When grid power is lost and the load is resonating, the phase modulation throws the resonance off-balance and modifies the resonant voltage frequency. This change of voltage and/or frequency is detected by the OUF or OUV protection and immediately initiates a shutdown signal to the inverter system.

5.5.1.6 Reactive Power Variation (RPV)

The concept here is to add a harmonic disturbation signal (typically low frequency) in the reference of the reactive current i_q^*. In the presence of the grid, this disturbance will try to modulate the voltage frequency with the disturbing one but will not be able to due to the stiff character. In the islanding situation, the voltage will depend linearly on the current and the frequency variations will be present and can be detected. In reference [19], 1 Hz – 1 % harmonic current is added to the reactive current reference with very reliable detection that is not sensitive to the grid impedance. A frequency deviation detector is used to count the half-periods between the zero-crossing that deviates from the rated frequency. After a

Islanding Detection

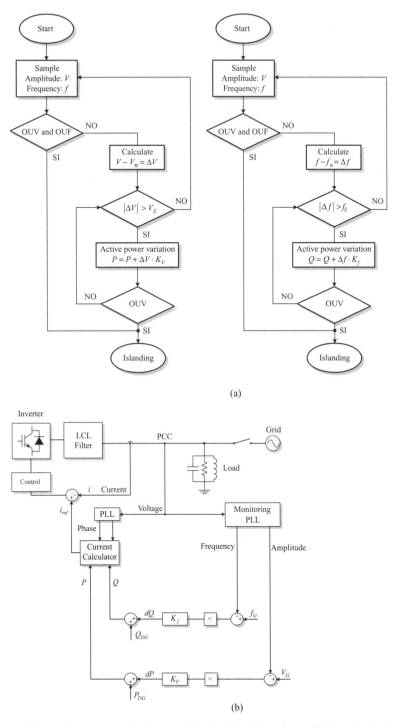

Figure 5.7 (a) Frequency and voltage positive feedback flowchart and (b) control scheme.

predetermined count the trip signal is generated. Shorter detection times can be obtained by increasing the frequency of the harmonic current, if desired, but keeping its amplitude low.

In reference [20], even lower and slower 0.5 Hz – 0.33 % harmonic current is added to the reactive current reference. The detection mechanism is based on the fact that during islanding a small phase change between the current and voltage will be produced, and this is effectively detected by the passive PJD method measuring the phase difference of two PLLs: one fast for the current and one slow for the voltage. This islanding method creates almost undetectable power quality degradation.

In reference [21], step variations of Q in the range of 2.5 % of the rated kVA are used instead of harmonic distortions. The periodical alteration of the reactive power component makes a phase difference between the output voltage and current of the system during islanding. The phase difference induces an increase or decrease of the frequency of load voltage that eventually hits the OUF protection.

An interesting analysis of grid frequency variation using RPV is reported in reference [22], where an analytical approach is used to design RPV optimally for specific harmonic requirements.

5.5.2 Voltage Drift Methods

These methods target the grid voltage drift either by positive feedback of the current or by varying the reactive power Q. As long as the grid is present, the voltage cannot be drifted, but when the grid is disconnected, the disturbance will be able to drift the voltage until it hits OUV protection. Several implementations exist as in the following.

5.5.2.1 Sandia Voltage Shift (SVS)

In this method the amplitude of the voltage acts as a positive feedback to the current reference [1]. Thus, if there is a decrease in the amplitude of voltage at the PCC (usually it is the RMS value that is measured in practice), the PV inverter reduces its current output and thus its power output. If the utility is connected, there is little or no effect on the voltage when the power is reduced. When the utility is absent and there is a reduction in voltage, there will be a further reduction in the amplitude, as dictated by the Ohm's law response of the (RLC) load impedance to the reduced current. This additional reduction in the amplitude of voltage leads to a further reduction in PV inverter output current, leading to an eventual reduction in voltage that can be detected by the UVP. It is possible to either increase or decrease the power output of the inverter, leading to a corresponding OVP or UVP trip. It is, however, preferable to respond with a power reduction and a UVP trip as this is less likely to damage load equipment.

The same concept with slightly different implementation has been developed by GE as a voltage islanding scheme, as shown in Figure 5.7(b). Useful design guidelines for the gain and band-pass filter of the positive feedback signals are reported in reference [17].

5.5.3 Grid Impedance Estimation

Especially for compliance with the old VDE 0126 or the first option in VDE 0126-1-1 (ENS), where a specific increase of 0.5 Ω in the grid impedance should be detected at a perfect balanced local load, more complex methods aiming at accurate impedance estimation have been developed. The concept is that a certain disturbance, such as harmonic injection or *PQ* variation, is used to estimate the grid impedance based on the response of the grid. The main methods are as follows.

5.5.3.1 Harmonic Injection (HI)

HI is based on injection of noncharacteristic harmonic current and extraction of the resultant voltage harmonic, which is dependent on the grid impedance at that frequency. This assumes that these frequencies are normally not present in the grid voltage so the voltage detected at this frequency will be only the voltage drop over the grid impedance.

In reference [23], first two equidistant injection frequencies (40 Hz and 60 Hz for a 50 Hz grid) are used and linear extrapolation is used to estimate the grid impedance at these frequencies and then interpolate for 50 Hz. Another method using only one frequency (75 Hz) is reported in reference [24]. A further development of this method is reported in reference [25], where higher frequencies (400–500 Hz) and zero-crossing synchronization of harmonic injection lead to less grid disturbance, as shown in Figure 5.8.

The grid-connected inverter is used directly to inject the harmonic current by adding a harmonic voltage to the inverter voltage reference, as can be seen in Figure 5.9, where θ_{PLL} represents the grid phase angle as provided by the PLL, I_g the actual grid current, I_g^* the grid current reference, V_{h12} the injected voltage harmonics and V_{PWM}^* the inverter voltage reference.

The voltage and current harmonics are estimated using an algorithm based on removing the fundamental frequency and filtering using tuned resonant filters. The noncharacteristic frequencies are chosen not to interact with the current controller in the case when the proportional resonant current controller with harmonics compensation is used and should not be near the output filter resonance frequency.

Extrapolating the estimated impedance at 400 Hz and 600 Hz to the low grid frequency leads to some limited accuracy in the estimation of the absolute value due to nonlinearity of the impedance. However, for the detection of a 0.5 Ω jump, as required by VDE 0126-1-2006, it can work satisfactorily.

The processing algorithm proposed is shown in Figure 5.9(b). The grid parameters are calculated by solving the following set of equations:

$$\begin{cases} Z_1^2 = R_g^2 + \omega_1^2 L_g^2 \\ Z_2^2 = R_g^2 + \omega_2^2 L_g^2 \end{cases} \tag{5.19}$$

$$L_g = \sqrt{\frac{Z_1^2 - Z_2^2}{\omega_1^2 - \omega_2^2}} \tag{5.20}$$

$$R_g = \sqrt{\frac{\omega_1^2 Z_2^2 - \omega_2^2 Z_1^2}{\omega_1^2 - \omega_2^2}} \tag{5.21}$$

$$Z_g = \sqrt{R_g^2 + \omega_g^2 L_g^2} \tag{5.22}$$

where ω_1, ω_2 denote injected harmonic frequencies; Z_1, Z_2 are the impedances calculated for ω_1, ω_2; and R_g, L_g are the resistive and inductive parts of the grid.

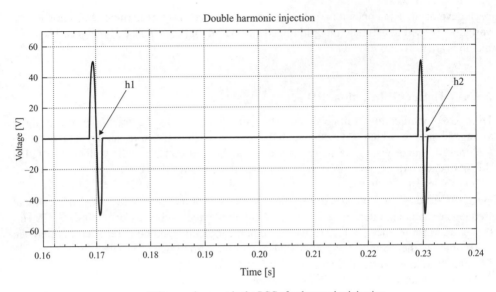

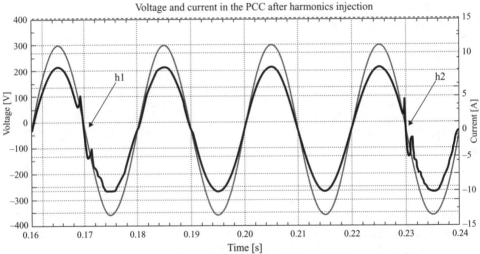

Figure 5.8 Principle of double harmonic signals injection

Another interesting work is reported in reference [26] where the existing voltage harmonic compensator is used to estimate the grid impedance indirectly. Basically, finding the amount of voltage to compensate a specific harmonic is equivalent to measuring the grid impedance at that frequency.

HI has been successfully used as an islanding detection method complying with ENS requirements, but it exhibits an inaccurate estimation and nuisance trips when multiple inverters are operating in parallel. In reference [27] a method is proposed to avoid concurrent injection with multiple inverters.

Islanding Detection

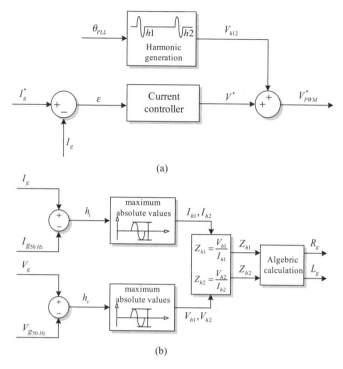

Figure 5.9 Double harmonic injection: (a) injection and (b) processing

5.5.3.2 Grid Impedance Estimation by Active Reactive Power Variation (GIE-ARPV)

This method is based on the fact that the grid impedance can be calculated by using two stationary working points, as shown in Figure 5.10, and solving the voltage Kirchhoff law. This is because normally there are two unknowns in the circuit, the grid impedance and the voltage at the power supply terminals. The latter can be eliminated if two sets of measurements are available, as expressed below.

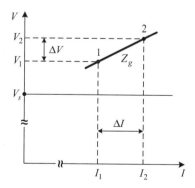

Figure 5.10 Grid impedance estimation by using two stationary working points

The voltages, V_1 and V_2, represent the voltage at the PCC (V_{PCC}) for the working points 1 and 2. By subtracting V_2 from V_1, the unknown variable V_g is avoided:

$$\begin{cases} V_1 = I_1 \cdot Z_g + V_g \\ V_2 = I_2 \cdot Z_g + V_g \end{cases} \quad (5.23)$$

$$V_1 - V_2 = Z_g (I_1 - I_2) \Leftrightarrow \Delta V = Z_g \cdot \Delta I \quad (5.24)$$

The relation of the grid impedance Z_g can be written as

$$Z_g = R_g + j\omega L_g = \frac{V_1 - V_2}{I_1 - I_2} \quad (5.25)$$

Furthermore, the expressions of the resistance R_g and inductance L_g are given in as

$$\begin{cases} R_g = \mathrm{Re}\left(\dfrac{V_1 - V_2}{I_1 - I_2}\right) \\ L_g = \dfrac{1}{\omega}\mathrm{Im}\left(\dfrac{V_1 - V_2}{I_1 - I_2}\right) \end{cases} \quad (5.26)$$

$$\begin{cases} R_g = \dfrac{\Delta V_d \Delta I_d + \Delta V_q \Delta I_q}{\Delta I_d^2 + \Delta I_q^2} \\ L_g = \dfrac{\Delta V_q \Delta I_d - \Delta V_d \Delta I_q}{[\Delta I_d^2 + \Delta I_q^2]\omega} \end{cases} \quad (5.27)$$

It can be seen that the calculation algorithm for the grid impedance is less complicated in contrast with harmonic injection. The only problem is how to make the inverter move to different stationary working points.

In reference [28] small pulse variations of P are used to determine the resistive part and small Q variations to determine the inductive part of the grid impedance. A voltage control loop is employed to avoid flickering. An improvement of this method is reported in reference [29] where it is demonstrated that if the grid converter is providing voltage control at PCC by droop control it creates natural P and Q variations necessary for the grid impedance estimation.

5.5.4 PLL-Based Islanding Detention

This method, reported in reference [30], takes advantage of the existing PLL structure responsible for the synchronization of the inverter output current with the grid voltage and is based on the deliberate alteration of the derived angle of the inverter current angle (θ_{PLL}). In particular, a sinusoidal signal (σ_{inj}) synchronized with the cycle is added to θ_{PLL} in order slightly to modify the inverter current phase. A feedback signal is then extracted from the voltage at the PCC

Islanding Detection

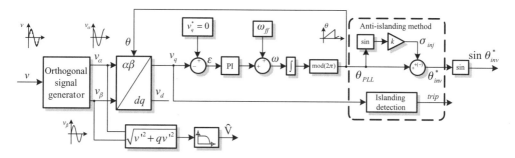

Figure 5.11 PLL-based islanding method

(namely from v_q) as a consequence of the injected signal σ_{inj}. The signal injection can be done with either a positive or negative sign, as shown in Figure 5.11.

The injected signal is defined by

$$\sigma_{inj} = k \sin \theta_{PLL} \qquad (5.28)$$

where the gain k is used to choose the amount of disturbance needed for the islanding detection. The resulting phase angle reference for the current becomes

$$\sin \theta^*_{inv} = \sin(\theta_{PLL} + \sigma_{inj}) = \sin(\theta_{PLL} + k \sin \theta_{PLL}) \qquad (5.29)$$

which, after some transformation, can be approximated by

$$\sin \theta^*_{inv} = \sin \theta_{PLL} + \frac{k}{2} \sin 2\theta_{PLL} \qquad (5.30)$$

Thus, for small values of k (e.g. for $k < 0.05$), the addition of $k \sin \theta_{PLL}$ to θ_{PLL}, as provided by the PLL, is equivalent to the addition of a second-harmonic signal ($A \sin 2\theta_{PLL}$) without affecting the amplitude, as shown in Figures 5.12 and 5.13.

The response of the grid will be a voltage second harmonic with amplitude related to the grid impedance value, as shown in Figure 5.14.

This feedback signal can be extracted from the voltage at PCC after the Park transform (from v_q), which, due to the transformation, will become the first harmonic and can be extracted using a tuned resonant filter. The signal processing is described in Figure 5.15, where λ_{Amp} is the amplitude of the feedback signal extracted from v_q, $\lambda_{AmpAvg50}$ is the 50 Hz average of the signal λ_{Amp}, $\lambda_{AmpAvg5}$ is the 5 Hz average of the signal $\lambda_{AmpAvg50}$ and δ represents the difference between the actual value of $\lambda_{AmpAvg50}$ and the delayed value of $\lambda_{AmpAvg5}$.

When the islanding occurs, this is also reflected in the grid impedance value. The change in the grid impedance value is then detected in the feedback signal in the form of δ. Then, following the algorithm shown in Figure 5.15, a trip signal is generated.

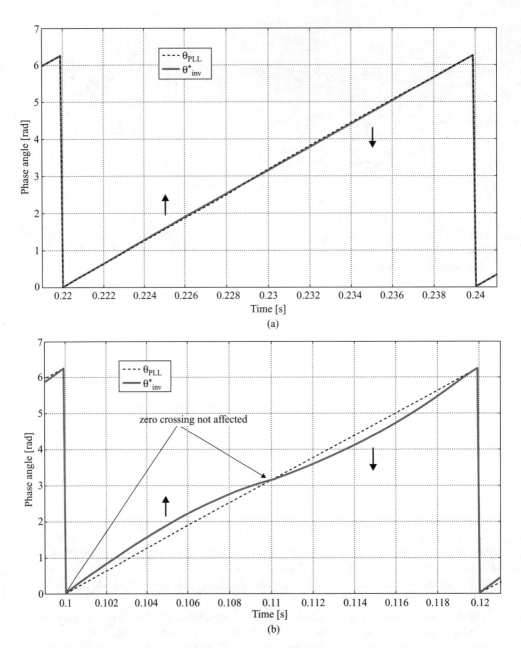

Figure 5.12 (a) The effect of the injected signal (σ_{inj}) for the inverter phase angle reference (θ^*_{inv}) at normal operation and (b) when the amplitude of the injected signal is 10 times larger than necessary

Islanding Detection

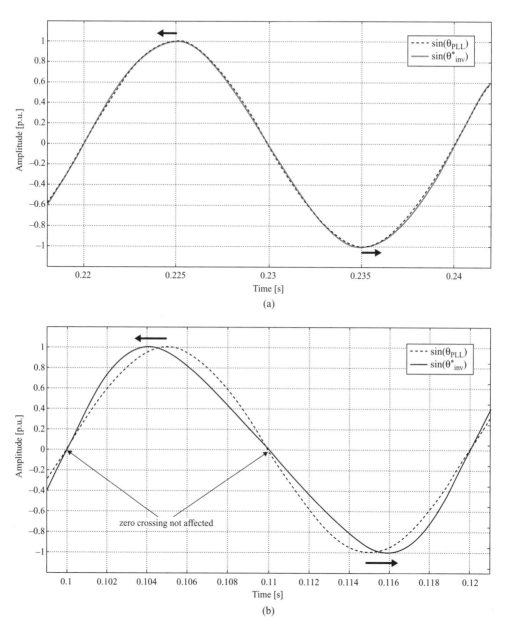

Figure 5.13 (a) The effect of the injected signal (σ_{inj}) for the grid current reference in p.u. ($\sin \theta_{inv}^*$) at normal operation and (b) when the amplitude of the injected signal is 10 times larger than necessary in order to illustrate the method

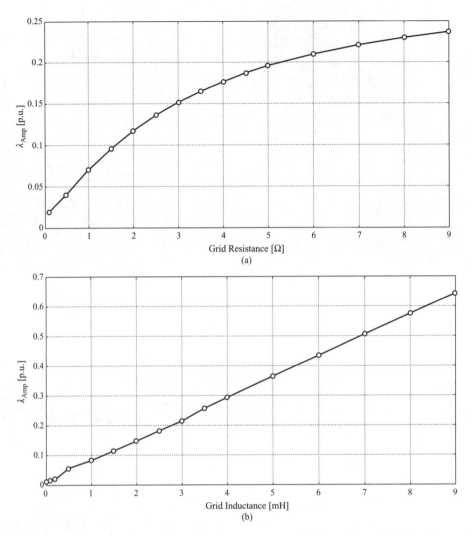

Figure 5.14 The relation between the feedback signal and the grid impedance: (a) grid resistance R_g and (b) grid inductance L_g

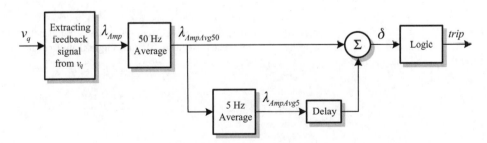

Figure 5.15 Signal processing for islanding detection

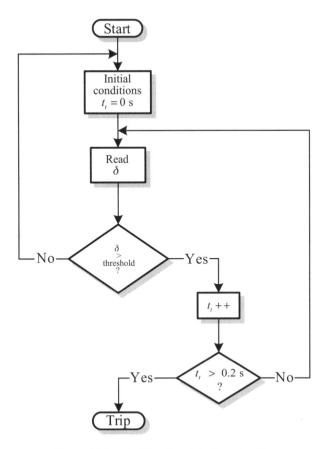

Figure 5.16 The islanding detection algorithm

The flowchart of the islanding detection algorithm is presented in Figure 5.16. The algorithm is described as follows:

- The component δ is compared to a threshold value.
- When δ is larger than the threshold value a timer is started.

The time value of the timer is named t_t and its initial condition is 0 s. If the δ remains larger than the threshold value for more than 0.2 s the inverter is tripped. Otherwise, if δ becomes smaller than the threshold value within 0.2 s, the timer is reset to 0 s. The threshold value and the tripping time can be chosen in accordance with one of the standards IEEE 929-2000, IEEE 1547.1-2005 or VDE 0126.1.1.

5.5.5 Comparison of Active Islanding Detection Methods

A synthetic comparison of the most relevant active islanding deduction methods, from the above described ones, is shown in Table 5.3, with the focus on reliability, power quality

Table 5.3 Performance comparison of active AI methods

AI active method	Reliability	Maintaining power quality	Suitability for parallel inverter operation	Potential for standardization
Active frequency drift (AFD)	Medium, as it is not able to eliminate the NDZ	Low, as it introduces low harmonics	Low, as it cannot handle concurrent detections	Low, as more likely for the AFD with positive feedback
Slip mode frequency shift (SMS)	Medium, as it is not able to eliminate the NDZ	Medium, as PF is affected but no harmonics are injected	Low, as it cannot handle concurrent detections	Low
Sandia frequency shift (SFS+SVS)	High, as it can eliminate NDZ but is susceptible for parasitic trips	Medium, as with continuous drifting the PQ can be affected	Medium, as it can work with parallel inverters but PQ can be affected	Medium, as more likely for the improved AFDPCF or GEFS
Active frequency drift with pulsating chopping factor (AFDPCF)	Medium high, as usually it leads to longer detection times as AFDPF but stability is controlled	High, as it introduces harmonics but THD can be controlled	High, but limited in case one inverter is increasing frequency while another is decreasing it	High, as both stability and THD degradation can be controlled
General Electric frequency shift (GEFS)	Very high, as NDZ can be eliminated and stability can be controlled	Very high, as practically no influence on THD	High, but not unlimited; according to GE more research is needed	High, as no degradation of THD is made
Reactive power variation (RPV)	Moderately high, as it can eliminate NDZ	High, as no harmonics are injected, only the PF can be slightly reduced	Low, as frequency changes at PCC can be also caused by other inverters	Low, due to low suitability for parallel operation
Grid impedance estimation–harmonic injection (GIE-HI)	High, as NDZ can be eliminated; there is potential for parasitic trip depending on grid impedance	Medium, depending on time between injections	Low, as readings can be influenced during parallel injection	Low, as ENS has now softened the requirements and accepted IEEE 1547 as an alternative
Grid impedance estimation–external capacitor switching (GIE-ESC)	High, as it can eliminate NDZ	Low, as it introduces low harmonics	Low for inverter level implementation; it can be implemented as one unit for a group of inverters	Low, as for inverter level cannot compete with software anti-islanding methods
Grid communication (GC)	Very high, as long as communication is good; it does not depend on PQ mismatch	High (no influence on PQ)	Very high, just as it is dependent on communication reliability	Moderately high on long terms due to cost issues

degradation, suitability for parallel inverter operation and the potential for standardization issues [13].

5.6 Summary

In this chapter the important function of islanding detection required for all grid-connected PV inverters is addressed.

Passive methods based on either voltage–frequency accurate monitoring or advanced filtering or harmonic deduction are evaluated in terms of the nondetection zone (NDZ).

Especially for active methods, many innovative methods have been reported based on the frequency drift, voltage drift, grid impedance estimation or PLL. All these methods are basically an introduction to small disturbances in the grid in order to deduct the grid presence.

The active methods are discussed and compared in terms of the performance in deduction (NDZ, reliability and suitability for parallel inverter operation).

References

[1] Bower, W. and Ropp, M., 'Evaluation of Islanding Detection Methods for Utility-Interactive Inverters in Photovoltaic Systems'. SANDIA Report SAND2002-3591, Albuquerque, NM: Sandia National Labs, November 2002. Online ordering: http://www.doe.gov/bridge.

[2] De Mango, F., Liserre, M., Dell'Aquila, A. and Pigazo, A., 'Overview of Anti-islanding Algorithms for PV Systems. Part I: Passive Methods'. In *Proceedings of the 12th International Power Electronics and Motion Conference*, August 2006, pp. 1878–1883.

[3] Ropp, M., Larson, D., Meendering, S., MacMahon, D., Ginn, J., Stevens, J.,W. Bower, W., Gonzalez, S., Fennell, K. and Brusseau, L., 'Discussion of Power Line Carrier Communications-Based Anti-islanding Scheme Using a Commercial Automatic Meter Reading System'. In *Proceedings of the 4th IEEE World Conference on Photovoltaic Energy Conversion*, May 2006, pp. 2351–2354.

[4] Liserre, M., Sauter, T. and Hung, J.Y. "Integrating Renewable Energy Sources into the Smart Power Grid Through Industrial Electronics" *IEEE Industrial Electronics Magazine*, **4**(1), March 2010, 18–37.

[5] Blaabjerg, F., Teodorescu, R., Liserre, M. and Timbus, A.V., 'Overview of Control and Grid Synchronization for Distributed Power Generation Systems'. *IEEE Transactions on Industrial Electronics*, **53**(5), October 2006, 1398–1409.

[6] Rodriguez, P., Luna, A., Ciobotaru, M., Teodorescu, R. and Blaabjerg, F., 'Advanced Grid Synchronization System for Power Converters under Unbalanced and Distorted Operating Conditions'. In *32nd IEEE Annual Conference on Industrial Electronics*, November 2006, 5173–5178.

[7] Jang, S.-I. and Kim, K.-H., 'An Islanding Detection Method for Distributed Generations using Voltage Unbalance and Total Harmonic Distortion of Current'. *IEEE Transactions on Power Delivery*, **19**(2), April 2004, 745–752.

[8] Liserre, M., Pigazo, A., Dell'Aquila, A. and Moreno, V. M., 'An Anti-Islanding Method for Single-Phase Inverters Based on a Grid Voltage Sensorless Control'. *IEEE Transactions on Industrial Electronics*, **53**(5), October 2006, 1418–1426.

[9] Pigazo, A., Moreno, V. M., Liserre, M. and Dell'Aquila, A., 'Wavelet-Based Islanding Detection Algorithm for Single-Phase Photovoltaic (PV) Distributed Generation Systems'. In *Proceedings of the 2007 IEEE International Symposium on Industrial Electronics*, June 2007, pp. 2409–2413.

[10] Jung, Y., Choi, J., Yu, B., So, J. and Yu, G., 'A Novel Active Frequency Drift Method of Islanding Prevention for the Grid-Connected Photovoltaic Inverter'. In *IEEE 36th Power Electronics Specialists Conference, PESC '05*, 2005, pp. 1915–1921.

[11] Ye, Z., Kolwalkar, A., Zhang, Y., Du, P. and Walling, R., 'Evaluation of Anti-islanding Schemes Based on Nondetection Zone Concept'. *IEEE Transactions on Power Electronics*, **19**(5), September 2004, 1171–1176.

[12] De Mango, F., Liserre, M. and Dell'Aquila, A., 'Overview of Anti-islanding Algorithms for PV Systems. Part II: Active Methods'. In *Proceedings of the 12th International Power Electronics and Motion Conference*, August 2006, pp. 1884–1889.

[13] Petrone, G., Spagnuolo, G., Teodorescu, R., Veerachary, M. and Vitelli, M., 'Reliability Issues in Photovoltaic Power Processing Systems'. *IEEE Transactions on Industrial Electronics*, **55**(7), July 2008, 2569–2580.

[14] Ropp, M. E., Begovic, M. and Rohatgi, A., 'Analysis and Performance Assessment of the Active Frequency Drift Method of Islanding Prevention'. *IEEE Transaction on Energy Conversion*, **14**(3), September 1999, 810–816.

[15] Lopes, L. A. C. and Sun, H., 'Performance Assessment of Active Frequency Drifting Islanding Detection Methods'. *IEEE Transactions on Energy Conversion*, **21**(1), March 2006, 171–180 .

[16] Yu, B., Jung, Y., So, J., Hwang, H. and Yu, G., 'A Robust Anti-islanding Method for Grid-Connected Photovoltaic Inverter'. In *Proceedings of the 2006 IEEE 4th World Conference on Photovoltaic Energy Conversion*, Vol. 2, May 2006, pp. 2242–2245.

[17] Ye, Z., Walling, R., Garces, L., Zhou, R., Li, L. and Wang, T., 'Study and Development of Anti-Islanding Control for Grid-Connected Inverters'. Report NREL/SR-560-36243, Golden, CO: National Renewable Energy Laboratory, May 2004.

[18] Hudson, R.M., Thorne, T., Mekanik, F., Behnke, M.R., Gonzalez, S. and Ginn, J., 'Implementation and Testing of Anti-islanding Algorithms for IEEE 929-2000 Compliance of Single Phase Photovoltaic Inverters'. In *Photovoltaic Specialists Conference, 2002. Conference Record of the 29th IEEE*, 19–24 May 2002, pp. 1414–1419.

[19] Hernandez-Gonzalez, G. and Iravani, R., 'Current Injection for Active Islanding Detection of Electronically-Interfaced Distributed Resources'. *IEEE Transactions on Power Delivery*, **21**(3), July 2006, 1698–1705.

[20] Istvan, V., Attila, B. and Sandor, H., 'Sensorless Control of a Grid-Connected PV Converter'. In *12th International Power Electronics and Motion Control Conference, EPE-PEMC 2006*, August 2006, pp. 901–906.

[21] Jeong, J.B., Kim, H.J., Ahn, K.S. and Kang, C.H., 'A Novel Method for Anti-Islanding Using Reactive Power'. In *27th International Telecommunications Conference, INTELEC '05*, September 2005, pp. 101–106.

[22] Choe, G.-H., Kim, H.-S., Kim, H.-G., Choi, Y.-H. and Kim, J.-C., 'The Characteristic Analysis of Grid Frequency Variation under Islanding Mode for Utility Interactive PV System with Reactive Power Variation Scheme for Anti-Islanding'. In *37th IEEE Power Electronics Specialists Conference, PESC '06*, 18–22 June 2006, pp. 1–5.

[23] United States Patent, 'Method and Apparatus for Measuring the Impedance of an Electrical Energy Supply System'. US Patent 6,933,714 B2, 23 August 2005.

[24] Asiminoaei, L., Teodorescu, R., Blaabjerg, F. and Borup, U., 'A Digital controlled PV-Inverter with Grid Impedance Estimation for ENS Detection'. *IEEE Transactions on Power Electronics*, **20**(6), November 2005, 1480–1490.

[25] Ciobotaru, M., Teodorescu, R. and Blaabjerg, F., 'On-line Grid Impedance Estimation Based on Harmonic Injection for Grid-Connected PV Inverter'. In *IEEE International Symposium on Industrial Electronics, ISIE 2007*, 4–7 June 2007, pp. 2437–2442.

[26] Bertling, F. and Soter, S., 'A Novel Converter Integrable Impedance Measuring Method for Islanding Detection in Grids with Widespread Use of Decentral Generation'. In *Proceedings of the International Symposium on Power Electronics, Electrical Drives, Automation and Motion*, 2006, pp. 503–507.

[27] Timbus, A.V., Teodorescu, R., Blaabjerg, F. and Borup, U., 'Online Grid Impedance Measurement Suitable for Multiple PV Inverters Running in Parallel'. In *21st Annual IEEE Applied Power Electronics Conference and Exposition, APEC '06*, 19–23 March 2006, p. 5.

[28] Ciobotaru, M., Teodorescu, R., Rodriguez, P., Timbus, A. and Blaabjerg, F., 'Online Grid Impedance Estimation for Single-Phase Grid-Connected Systems Using PQ Variations'. In *IEEE Power Electronics Specialists Conference, PESC 2007*, 17–21 June 2007, pp. 2306–2312.

[29] Timbus, A.V., Teodorescu, R. and Rodriguez, P., 'Grid Impedance Identification Based on Active Power Variations and Grid Voltage Control'. In *IEEE Industry Applications Conference, 42nd IAS Annual Meeting. Conference Record of the 2007 IEEE*, 23–27 September 2007, pp. 949–954.

[30] Ciobotaru, M., Agelidis, V. and Teodorescu, R., 'Accurate and Less-Disturbing Active Anti-islanding Method Based on PLL for Grid-Connected PV Inverters'. In *IEEE Power Electronics Specialists Conference, PESC 2008*, 15–19 June 2008, pp. 4569–4576.

6

Grid Converter Structures for Wind Turbine Systems

6.1 Introduction

The connection of a wind turbine (WT) to the grid is a delicate issue. In fact, the stochastic power production of large-power wind turbines or wind-parks could create problems to the transmission line designed for constant power and to the power system stability. This important issue justifies the concerns related to increasing penetration of wind energy within the power system [1]. However, if the wind power plant (single large-power wind turbine or wind-park) would behave as a classical power source by allowing a decision to be made on how much power to inject and when, the main limitation to its use would cease to exist. The use of wind forecasting can help the management of a power system with a high penetration of wind power but cannot transform the wind system in a traditional power plant. A possible solution to the problem is in the use of suitable energy storage. However, this solution is not yet practical even if it will be part of the future power system scenario [2]. On the other hand, the increased use of power electronics, especially on the grid side, in connection with the control of the pitch angle of the blades can partially relive the problem, allowing the WT power plant to behave similarly to a conventional power plant. In this sense it should be noticed that the introduction of power converters in a variable-speed wind turbine has been mainly associated with the possibility of controlling the generator and as a consequence of controlling the active power in order to maximize the power extraction, leaving its limitation to the mechanical control of the blade angle (passive or active). Then the active power control has been viewed as a means to exercise the wind turbine system in a similar way to a traditional power plant, e.g. by providing reserve capability by means of delta control (i.e. producing less power than what is available in order to have the possibility of providing an indirect reserve). However, it is the use of a grid converter that gives to the modern wind turbine system (WTS) the capability of managing the reactive power exchange and allowing its participation in the voltage regulation.

In this chapter the focus is on the grid converter structures adopted in the WTS. The structures are classified as reduced power (for doubly fed generators) and full power. The

---

Grid Converters for Photovoltaic and Wind Power Systems Remus Teodorescu, Marco Liserre, and Pedro Rodríguez
© 2011 John Wiley & Sons, Ltd

latter are further divided into single cell and multicell. In fact, in order to achieve an efficient and reliable management of higher power, two possibilities are given: to use high-power converters topologies (e.g. neutral point clamped) or to use several medium-power converters connected as series or parallel cells. In the chapter attention is paid also to the grid converter control structures, leaving to the following chapters the task of going into more detail. This introductory chapter on WTSs opens the second section of the book, introducing the topics of the following chapters dealing with grid regulations, grid monitoring and synchronization, grid converter control and control under grid faults.

6.2 WTS Power Configurations

The basic power configuration of a wind turbine system is made of two parts: a mechanical part and an electrical one (see Figure 6.1). The first subsystem extracts the energy from the wind and makes the kinetic energy of the wind available to a rotating shaft; the second subsystem is responsible for the transformation of the electrical energy, making it suitable for the electric grid. The two subsystems are connected via the electric generator, which is an electromechanical system and hence transforms the mechanical energy into electrical energy [3–7].

This description highlights the fact that there are three stages used to optimize the extraction of the energy from the wind and adapt it: one mechanical, one electromechanical and another one electrical. The first stage may regulate the pitch of the blades, the yaw of the turbine shaft and the speed of the motor shaft. The second stage can have a variable structure (pole pairs, rotor resistors, etc.), an external excitation and/or a power converter that adapts the speed or the torque of the motor shaft and the waveforms of the generator voltages/currents. The third stage adapts the waveforms of the grid currents. Power electronics converters may be present in the second and/or third stages [8].

This chapter focuses on the third stage. In Figure 6.2 a classification of the possible power converter solutions is reported [9, 10].

The main step that has led to controllable power electronics in a wind turbine has been made with the doubly fed induction generator (Figure 6.3(a)), where a wound rotor is fed by a back-to-back system with a rated power of 30 % of the system power. However, in this case the speed range is quite limited ($-30\% + 30\%$) and the slip rings are needed in order to connect the converter on the rotor. The gear is still needed and the speed regulation via the rotor is used only to optimize power extraction from the wind.

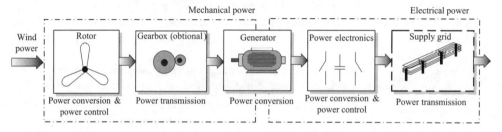

Figure 6.1 Basic power conversion wind turbine system

Grid Converter Structures for Wind Turbine Systems 125

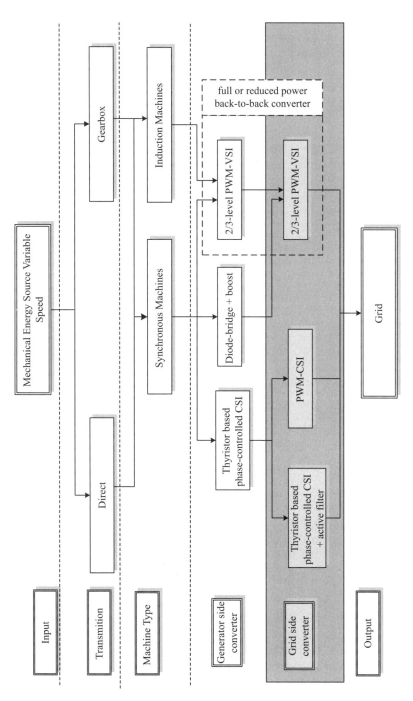

Figure 6.2 Scenario of the power conversion structures for variable-speed wind turbine systems

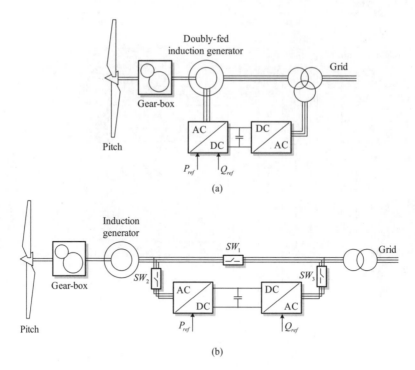

Figure 6.3 Reduced power back-to-back converter options: (a) a doubly fed induction generator with back-to-back connected to the rotor and (b) an induction generator with back-to-back connection only when the system is working at half-power or for reactive power compensation

It is worth noting that this is the first configuration allowing partial control on the grid electrical quantities. In fact, acting on a back-to-back converter it is possible to vary the injected active and reactive power [11]. In particular the rotor-side converter controls the rotor current in order to control the active and reactive power injected into the grid. The current-controlled rotor-side inverter can be seen as a controlled current source in parallel with the DFIG magnetization reactance. If in parallel to these two elements a Thevenin equivalent is substituted, the DFIG model will match the model of a synchronous generator and the active/reactive power control will be straightforward [11]. Moreover, the previously described approach allows the DFIG with a back-to-back converter to be treated with the same theory used to describe the behaviour of the grid converter, which is the subject of this book.

The DFIG system contributes to the definition of the short-circuit power because the stator is directly coupled to the grid. This means that during a fault in the grid, high currents are generated by DFIG and this is an advantage for the coordination of the protections that can detect the fault because of the consequent overcurrent. On the other hand this may limit the capacity of the DFIG system to stay connected to the grid if needed and to reduce the power injection acting as a rolling capacity in the grid to be used to restore the system stability after the fault unless a crowbar is adopted in order to limit to safe level current and voltages in the rotor circuit where the back-to-back power converter is used [12].

Grid Converter Structures for Wind Turbine Systems

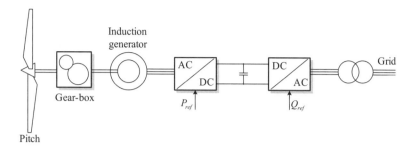

Figure 6.4 Full-power back-to-back converter with an induction generator

A further step in the improvement of the grid-side behaviour of the wind turbine system is made with the use of a squirrel cage induction generator and a reduced scale back-to-back power converter (Figure 6.3(b)). The back-to-back converter is only connected in two cases:

- At medium and low power the converter is used to optimize the power extraction and transfer to the grid (SW_1 open, SW_2 and SW_3 closed).
- At full power only the grid-side converter is connected to perform harmonic and reactive power compensations (SW_2 open, SW_1 and SW_3 closed).

The use of a full-power back-to-back converter (Figure 6.4) leads to an induction generator completely decoupled from the grid, and as a consequence this system has a complete rolling capacity being able to actively contribute to the limitation of the effects of grid faults and to the restoration of the normal grid operation after the fault. However the system does not contribute to the short-circuit power because the grid converter limits the fault current and as a consequence the protection coordination should be redesigned. This system can completely be at stand-by and operate in an island [13]. However, the gear is still needed and the power converter is full-scale.

A similar system can be obtained using an unsynchronized synchronous generator (Figure 6.5). This topology is termed 'synchronous' as the generated frequency is synchronous with the rotor rotation. However, because the generated frequency is not synchronized with the grid frequency, power electronics are necessary. The generator voltage is rectified with a fully controlled converter or with a diode-bridge plus a dc/dc, in case of permanent magnet

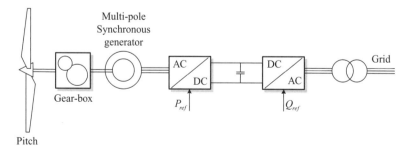

Figure 6.5 Full-power back-to-back converter with a synchronous generator

generator, or with a diode-bridge plus a converter controlling the excitation, in case of generator with independent excitation. Then a fully controlled inverter is adopted to connect the system to the grid. Hence a full-scale back-to-back power converter is needed and a reduced scale converter for the excitation may be used.

In case a multi-pole gnerator is used the gear-box is not necessary. It may therefore be an ideal solution if the WT has to be installed in an extreme environmental condition characterized by a very low temperature that may challenge the maintenance of the gearbox. Some producers of large-power wind turbines prefer to use reduced gearboxes. These reduced gearboxes are more reliable because they involve less rotating components and the inverter is integrated in the nacelle, allowing full control of the active/reactive power produced by the generator.

The use of a synchronous generator with full-power back-to-back converters appears to be the most successful configuration for the near future, gaining the doubly fed generator actual market share.

6.3 Grid Power Converter Topologies

There are many demands on power converter topologies in wind turbine systems. The main ones are: reliability, minimum maintenance, limited physical size/weight and low power losses. The AC/AC conversion can be direct or indirect. In the indirect case there is a DC link that connects two converters performing AC/DC and DC/AC conversions, while in the direct case the DC link is not present. The advantage of the indirect conversion is the decoupling between the grid and generator (compensation for nonsymmetry and other power quality issues) while its major drawback is the need for major energy storage in the DC link (reduced lifetime and increased expenses). However, the DC storage and consequent decoupling between the generator and grid side can give an advantage to indirect conversion over a direct conversion in the case of low-voltage ride-through and for providing some inertia in the power transfer from the generator to the grid.

The main advantage of the direct conversion, such as the matrix converter topology, is that it is a one-stage power conversion (and hence without intermediate energy storage). Moreover, it presents several advantages, such as the thermal load of the power devices is better compared to others, there is less switching losses than two-level back-to-back VSI as well as better harmonic performance on the generator side than two-level back-to-back VSI (and maybe a lower switching frequency). However, these advantages are balanced by many and well-known disadvantages, such as the fact that this is not a proven technology requiring a higher number of components (hence more conduction losses) and a more complex control part. Moreover, the grid filter design is more complex and there is not a unity voltage transfer ratio. Also, the absence of a DC link storage (generally the less reliable part of the converters and the most subject to maintenance) makes this solution attractive, especially for offshore wind turbine systems characterized by difficult maintenance. It has been the subject of a patent in the case of a doubly fed induction generator.

6.3.1 Single-Cell (VSC or CSC)

The grid converter topologies can be classified into voltage-stiff (voltage-fed or voltage-source) and current-stiff (current-fed or current-source) ones respectively, indicated with the acronyms VSC and CSC (Figure 6.6). A third option is represented by the Z-source converter employing, on the DC side, an impedance network with capacitors and inductors [14]. Depending on the main power flow direction they are named rectifiers or inverters, or in case they can work with both power flows they are bidirectional. Then they can be classified as phase-controlled (typically using thyristors and natural commutation synchronized with the grid voltage) or PWM using forced commutated devices. Grid converters for distributed power generation need to work as inverters, but they can benefit from bidirectional power flow in order to pre-charge the DC link.

In the case of the VSC a relatively large capacitor feeds the main converter circuit, a three-phase bridge. Six switches are used in the main circuit, each composed traditionally of a power transistor and a free-wheeling diode to provide bidirectional current flow and a unidirectional voltage blocking capability. The VSC needs both AC and DC passive elements. The passive elements, such as capacitors or inductors, have both storage and filtering functions. The operation of the VSC is connected with the use of a DC capacitive storage instead of a DC inductive storage. The DC capacitor is charged to a certain voltage. This voltage ensures the basic function of the VSC: the VSC can control the AC current through the switching. Then, through the AC current control, the VSC can change the DC value as in active rectifier and filter applications. This can easily be understood from the power balance. Once it is assumed that there are no losses in the operation, the AC active power is transformed to DC power through the VSC. The control of AC active power could be done through control of the AC current amplitude; then the change in AC active power causes the DC power to change, resulting in a charge or discharge of the DC capacitor.

The process of the DC capacitor charge–AC current control–DC voltage control is a virtuous circle that is based on the possible storage of energy due to the DC capacitor.

Then the filtering action, necessary because of the PWM, is done both on the DC side and on the AC side. The passive elements are charged/discharged during the switching period, ensuring smoothing of the AC currents and of the DC voltage. This filtering action is also the basis of the control performed. In fact, the dynamic of the AC current/DC voltage controls depends on the time constants of the two filtering stages. Generally the overall design, which should include filtering and control issues, is a trade-off between high filtering and fast dynamic.

Considering the example of a industrial inverter used in electric drives, if all the energy stored in the AC passive stage is considered, it is less than 5 % of all the energy stored.

The VSC is widely used. It has the following features:

- The AC output voltage cannot exceed the DC voltage. Therefore, the VSC is a buck (step-down) inverter for DC/AC power conversion and is a boost (step-up) rectifier (or boost converter) for AC/DC power conversion. In case the available DC voltage is limited (e.g. in the case of a direct-driven synchronous generator with a diode bridge rectifier) an additional DC/DC boost converter is needed to obtain the proper DC voltage that allows the VSC to operate properly with the grid. The additional power converter stage increases the system cost and lowers efficiency.
- The upper and lower devices of each phase leg cannot be gated on simultaneously either by purpose or by EMI noise. Otherwise, a shoot-through would occur and destroy the devices. This is a serious issue for the reliability of these converters. Dead-time to block both upper and lower devices has to be provided in the VSC, which causes waveform distortion.

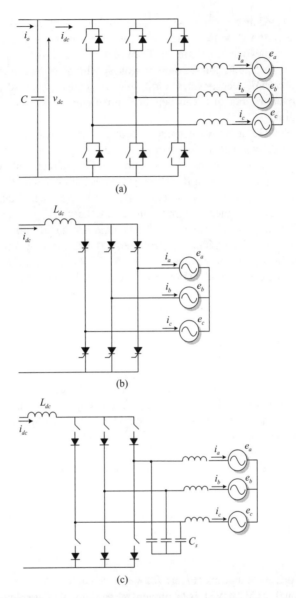

Figure 6.6 Grid converter in the case of indirect type conversion: (a) forced-commutated VSI, (b) phase-controlled line-commutated converter and (c) forced-commutated CSI

- An output high-order filter is needed for reducing the ripple in the current and complying with the harmonic requirements. This causes additional power loss and control complexity.

The traditional CSC had more limited application. A DC current source feeds the main converter circuit. The DC current source can be a relatively large DC inductor fed by a source. Six switches are used in the main circuit, each composed traditionally of a semiconductor switching device with reverse block capability, such as GTO and SCR or a power transistor with a series diode, to provide unidirectional current flow and bidirectional voltage blocking. Operation of the current-source converters requires a constant current source, which could be maintained by either a generator-side or a grid-side converter. Generally, the grid-side converter controls the DC link current based on the assumption of a stiff grid. However, the actual DC link current is determined by the power difference of both sides. The power disturbances of the generator output, mainly due to the disturbances of wind speed, are not simultaneously reflected by the grid-side converter control. This results in a large overshoot or undershoot of the DC link current, which may further affect the stability of the whole system.

The CSC has the following features:

- The AC output voltage has to be greater than the original DC voltage that feeds the DC inductor. Therefore, the CSC is a boost inverter for DC/AC power conversion and the CSC is a buck rectifier (or buck converter) for AC/DC power conversion. For grid converter applications this is a clear advantage.
- At least one of the upper devices and one of the lower devices has to be gated and maintained at any time. Otherwise, an open circuit of the DC inductor would occur and destroy the devices. The open-circuit problem of EMI noise is a major concern for the reliability of these converters. The overlap time for safe current commutation is needed in the CSC, which also causes waveform distortion.
- The main switches of the CSC have to block reverse voltage, which requires a series diode to be used in combination with high-speed and high-performance transistors such as IGBTs. This prevents the direct use of low-cost and high-performance IGBT modules and IPMs. In the following the single cell power converters solutions based on VSC or CSC topologies for medium power and high power wind turbines are reviewed.

6.3.1.1 Medium-Power Converter

Medium-power wind turbine systems of 2 MW are still the best seller on the market and their power level can still allow a good design trade-off to be found using single-cell topologies with just six switches forming a bridge. This solution can be full power or reduced power in the case of a doubly fed induction generator or a converter working only in low wind conditions.

In all the cases forced commutated converters allow better control of the injected power and harmonics. Between the forced commutated converters the preferred solution is the VSI. Particularly in the case where the VSI is adopted, as usual, on the generator side, the resulting configuration is called back-to-back (Figure 6.7).

The two-level back-to-back VSI is a proven technology that employs standard power devices (integrated), but power losses (switching and conduction losses) may limit the use in higher power systems.

The alternative can be the use of CSCs (Figure 6.8), which have three main advantages [15]:

- A portion of the needed DC link inductance is realized by exploiting the cable length and, if necessary, a proper cable layout, which can be possible if the generator with the first convert is located in the nacelle and the grid-side CSI is located at the tower base. Moreover, in the case of a wind-park a DC grid can be adopted and the consequent DC cables can be long enough to provide the needed inductance.

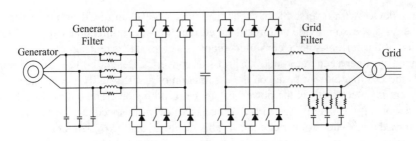

Figure 6.7 Two-level back-to-back PWM-VSI

- The DC link reactor provides natural protection against short-circuit faults and therefore the fault ride-through strategy required by the grid code can be integrated easily into the system.
- A small filter is required on the AC side to cope with the standards in terms of harmonic requirements.

Although the DC link current can be maintained at the highest level to obtain the best dynamic response, the fastest response is not always useful in this application since the output power is regulated to have slow changes rather than fast transients that may cause power system instability. In wind applications, the maximum power generated from a wind turbine is proportional to the cube of the wind speed. In order to extract more energy from the wind, the system requires variable-speed operating capability and the generated power varies in a wide range as the wind speed changes. It is beneficial for a MW system to minimize the DC link current if the power input is reduced. On the other hand, maintaining a high DC link current at lower power input requires a significant amount of shoot-through states in the CSC, causing more conduction loss on the devices and reducing the system efficiency.

For large wind energy applications, the capability of the power factor control or voltage regulation at the grid side is required by the grid codes. When a CSC is connected to the grid, filter capacitors at the grid side result in constant leading reactive power. In a traditional CSC-based drive system, an offline PWM method – selected harmonic elimination (SHE) – is normally used at the grid side due to the capability of eliminating a number of unwanted low-order harmonics. However, the reactive power at the line side is not fully controlled. A unity power factor can be achieved by phase-shifting the modulating signals according to the converter operating point, which is not straightforward for line-side active and reactive power control.

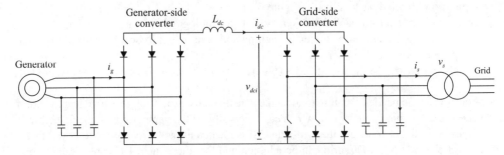

Figure 6.8 Two-level back-to-back PWM-CSI

Grid Converter Structures for Wind Turbine Systems

6.3.1.2 High-Power Converter (NPC)

In case the power level increases over 2 MW a multilevel solution (Figure 6.9) such as the three-level voltage source converter [16] is a known technology that allows lower rating for the semiconductor devices and lower harmonic distortion to the grid (or lower switching losses/smaller grid filter). However, the conduction losses are still high due to the number of devices in series through which the grid current flows and a more complex control is needed to balance the DC link capacitors.

6.3.2 Multicell (Interleaved or Cascaded)

Another option to increase the overall power of the system is to use more power converter cells in parallel or in cascade. In both cases the power-handling capability increases while the reliability if computed in terms of the number of failures decreases and the number of system outages increases. In fact, the modularity implies redundancy that allows the system to continue to operate if one of the cells fails. Moreover, the multicell option allows a reduced number of cells to be used, with consequent reduced losses, in low wind conditions when the produced power is low.

Typically the power cells are connected in parallel on the grid side to allow interleaving operation (as described in Chapter 12). The PWM patterns are shifted in order to cancel PWM side-band harmonics. In this way the size of the grid filter can be considerably reduced.

Figure 6.10 reports a back-to-back converter fed by a six-phase generator and connected in parallel and interleaved on the grid side [17], while Figure 6.11 shows an n-leg diode bridge fed by a synchronous generator producing a high DC voltage shared among several grid/converters connected in parallel and interleaved on the grid side [24].

Similar options can also be achieved with CSC topologies, forming the well-known 12-pulse converter in the case where the CSC is phase-controlled [18]. The DC/AC conversion can be performed by the two series-connected current-source inverters (and) independently supplied by two equal secondaries of a Y–Y transformer. Both inverters require components with a bidirectional voltage blocking capability whereas a unidirectional current-carrying capability is sufficient because the DC link current does not reverse its sign.

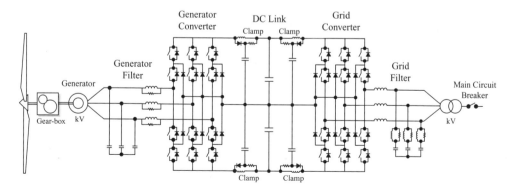

Figure 6.9 Three-level back-to-back PWM VSI

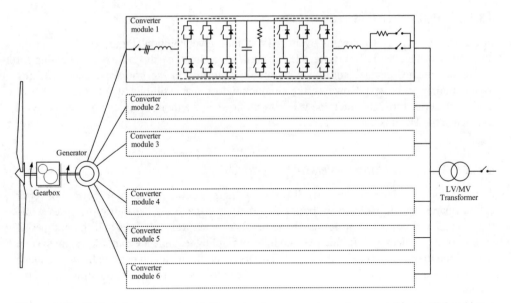

Figure 6.10 Back-to-back converters fed by a six-phase generator and connected in parallel and interleaved on the grid side

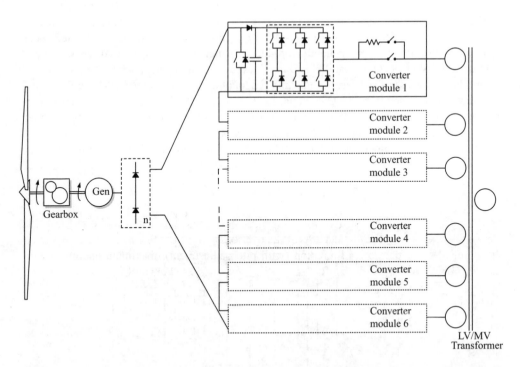

Figure 6.11 An n-leg diode bridge fed by a synchronous generator producing a high DC voltage shared among several grid/converters connected in parallel and interleaved on the grid side

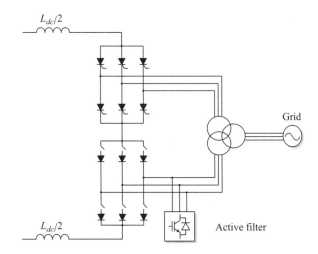

Figure 6.12 Thyristor-based phase-controlled CSI + active filter

The CSI are connected in series on the DC side and in parallel on the AC side to reduce the ripple in the DC current, operate with a higher DC voltage and double the power-carrying capability. The CSCs can be controlled by the well-known phase-control technique with opposite phase-control angles so that the fundamental power factor of the grid current at the transformer primary is guaranteed to be unity at any load condition. This operating mode requires one of the two inverters (bridge in this case) to use fully controllable switches. Then an active filter is adopted to clean the grid current (see Figure 6.12).

6.4 WTS Control

Controlling a wind turbine involves both fast and slow control dynamics. Overall the power has to be controlled by means of the aerodynamic system and has to react based on a set-point given by a dispatched centre or locally with the goal to maximize the power production based on the available wind power. The two subsystems (electrical and mechanical) are characterized by different control goals but interact in view of the main aim: the control of the power injected into the grid. The electrical control is in charge of the interconnection with the grid and active/reactive power control, and also of the overload protection. The mechanical subsystem is responsible for the power limitation (with pitch adjustment), maximum energy capture, speed limitation and reduction of the acoustical noise. The two control loops have different bandwidths and hence can be treated independently.

The power controller should also be able to limit the power both with mechanical and electrical braking systems, since redundancy is specifically requested by the standards. The general scheme of the wind turbine control with different features is reported in Figure 6.13.

Below maximum power production the wind turbine will typically vary the speed proportionally with the wind speed and keep the pitch angle fixed. A pitch angle controller limits the power when the turbine reaches nominal power. The control of the generator-side converter is in charge of extracting the maximum power from the wind. The control of the grid-side converter

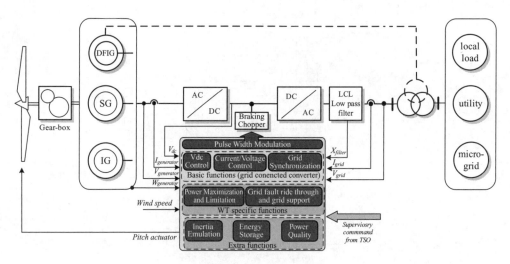

Figure 6.13 Wind turbine control structure

is simply just keeping the DC link voltage fixed. Internal current and voltage loops in both converters are used. The state variables of the LCL filter are controlled for stability purposes. Then there is the grid fault ride-through and the support to the grid voltage restoration after the fault that is a Wind Turbine specific function discussed in Chapter 10. Wind turbine extra functions are: the inertia emulation that is a control function aiming at emulate the relation between active power and frequency normally present in generator with a large inertia, discussed in Chapter 7; the energy storage refers to the possibility to store energy in the inertia of the generator, in the dc-link or to use additional storage to smooth the power output; power quality refers to the possibility to use the grid converter of the WT to provide benefits in terms of grid power quality.

6.4.1 Generator-Side Control

The control of the generator is done in view of the main goal: maximizing the power extraction and limiting the power braking the wind turbine. These two goals result in a torque or speed command for the generator control. In the following the different controllers depending on the generator side are briefly analysed since this is not the main focus of the book.

6.4.1.1 Squirrel Cage Induction Generator Control

The squirrel cage induction generator with a full-power forced commutated back-to-back converter (Figure 6.14) was often chosen by wind turbine manufacturers for low-power stand-alone systems, but recently it has been used for high-power wind turbines as well. A third option already shown in Figure 6.3(b) is to upgrade the fixed-speed WT to a variable-speed WT with a back-to-back converter of reduced power size (50 %), which should only be used during low wind conditions to optimize the power transfer and when it is needed to compensate for the reactive power (only using the grid converter), but is bypassed during high-speed conditions.

Grid Converter Structures for Wind Turbine Systems

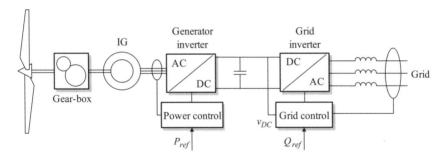

Figure 6.14 Induction generator with a full-scale back-to-back converter

The machine flux and rotor speed or electric torque is controlled via a field orientated control (FOC) or direct torque control (DTC), even if this last option is seldom adopted in WTSs.

6.4.1.2 Synchronous Generator Control

One of the most adopted wind turbine solutions employing a synchronous generator includes a passive rectifier and a boost converter to boost the voltage at low speed. The topology is shown in Figure 6.15. The generator is controlled via the current control of the boost converter, but in

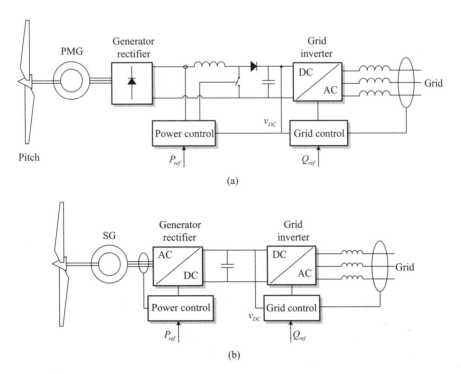

Figure 6.15 Synchronous generator with: (a) diode bridge + VSI and (b) back-to-back converter

this way it is not possible to control selectively the harmonics in the current and the phase of the fundamental current with respect to the generator electromotive force. Filters are usually adopted on the generator side to mitigate especially the 5th and 7th harmonics. This solution is one of the most adopted industrial solutions, especially in the case of direct-driven gearless multiphase wind turbine systems.

The solution displayed in Figure 6.15(b) employs a full-power back-to-back converter. In this case the generator control is usually a standard FOC where the current component that controls the flux can be adapted to minimize the core losses and the reference speed is adapted to optimize the power injection into the grid [19].

6.4.1.3 Doubly Fed Induction Generator Control

The first controller to adapt the rotor speed of an induction generator and try to achieve maximum power extraction and limitation of the mechanical stress on the drivetrain has been the slip control. This was a simple method used to vary the speed of the generator acting on its rotor resistance. The change in the rotor resistance leaves unchanged the synchronous speed while the slope of the machine characteristic varies. The dynamic slip control works below the rated power, the generator acts just like a conventional induction machine and above rated power and the resistors in series with the rotor circuit are adjusted trying to keep the power at the rated value. An interesting alternative is to use a diode bridge plus a transistor in order to vary the apparent resistance of the rotor circuit. In this way the resistors will remain the same and the transistor will be driven in order to change the apparent resistance seen by the rotor circuit. The resulting speed control range is very limited (5–10 % above the synchronous speed) but the method is used in conjunction with a mechanical control acting on the wind turbine blade pitch.

Obviously the use of additional rotor resistors causes additional losses proportional to the speed slip. In a 2 MW wind turbine based on rotor resistance control, a slip of 5 % will result in a rotor power of 95 kW (losses).

It is possible to recover the power losses using the Scherbius drive [20], where the diode bridge is connected to a converter used to inject the slip power into the grid. The system is also called oversynchronous cascade control. An evolution of this system is the doubly fed induction generator equipped with a back-to-back power converter allowing bidirectional power flow.

The doubly fed induction generator equipped with voltage source power converters connected to the grid side and to the rotor side (Figure 6.16) is one of the most adopted solutions in wind turbine systems. This is the case even if it seems that most of the new projects in wind turbine systems are abandoning this solution mainly for the need to comply with LVRT requirements of the standards and grid codes.

The control of the WT (Figure 6.16) is organized such that below maximum power production the wind turbine will typically vary the speed in proportion to the wind speed and keep the pitch angle fixed. At very low winds the speed of the turbine will be fixed at the maximum allowable slip in order not to have overvoltage. A pitch angle controller will limit the power when the turbine reaches nominal power. The generated electrical power is found by controlling the doubly fed generator through the rotor-side converter. The control of the grid-side converter simply keeps the DC link voltage fixed.

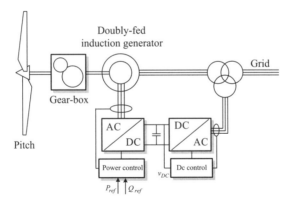

Figure 6.16 Doubly fed induction generator control

A 'crowbar' system can be adopted to ride-through grid faults (see Chapter 10). The three-phase rotor winding is thus short-circuited via the closed crowbar switch, which results in the same behaviour as a standard induction generator.

The power produced by the turbine P_{mecc} follows two paths (stator and rotor):

$$P_{mecc} = P_s - s P_s \qquad (6.1)$$

where P_s is the stator power and s is the slip. It is obvious that the doubly fed generator injects power into the grid both during oversynchronous ($s > 0$) and subsynchronous operation ($s \langle 0$).

The doubly fed induction generator control is different from the control of a standard induction generator [21–23]. In fact, the control is developed on the basis of a power perspective point of view. The machine stator is connected directly and continuously to the grid and exchanges active and reactive power with it. Acting on the rotor current is possible to control the active and reactive power injected by the stator into the grid.

If the machine equations are rewritten in a dq frame oriented with the stator flux, it results in [23]

$$\begin{cases} P_s = -a\left(v_s i_{rq}\right) \\ Q_s = v_s \left(\dfrac{v_s}{b} - i_{rd}\right) \end{cases} \qquad (6.2)$$

where v_s is the stator and grid voltage and a and b are coefficients that depend on the machine parameters. Hence i_{rq} can be used to control the stator active power (and indirectly the grid power since the grid power will be the sum of the stator power and rotor power, i.e. the stator power multiplied for the slip) and i_{rd} can be used to control the overall reactive power.

6.4.2 WTS Grid Control

All the previous reported converter structures employ a grid converter, which in most cases is a VSC. Its principle of operation is similar to the principle of operation of a synchronous generator or of a transmission line (neglecting capacitive coupling). Basically the VSC controls the active and reactive power transfers acting on the amplitude and phase of the produced voltage, as shown in Figure 6.17. In Figure 6.17(a) the case is reported when there is no power produced by the WTS and a small power is absorbed to keep the DC link voltage at its rated value. The VSC is working as a rectifier and the absorbed active power compensates the losses in the overall converter. In Figure 6.17(b) the case is reported when the WTS injects only active power, while in Figure 6.17(c) and (d) the case in which the grid converter is working as a STATCOM is shown, and similarly to Figure 6.17(a) there is no active power injection and hence it is expected that a small amount of power will be drained from the grid to compensate for the losses. This amount is not reported in Figure 6.17(c) and (d) for the sake of simplicity. Figure 6.17(e) and (f) reports the working conditions in which the WTS injects both active and reactive power.

The power transfer between two sections of a short line can be studied using complex phasors, as shown in Figure 6.17 for a mainly inductive grid filter with $X \gg R$, showing that

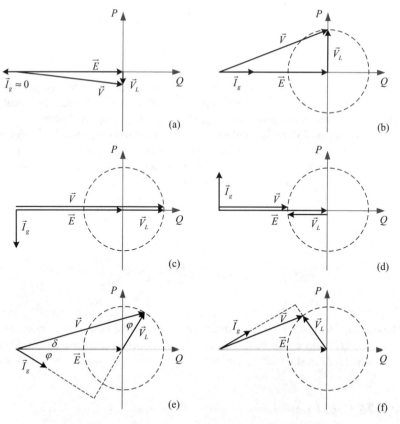

Figure 6.17 Different power transfers achieved by the grid converter in the different operating conditions (V_L is the voltage drop across the grid filter)

the voltage drop V_L is perpendicular to the exchanged current. In this case R may be neglected. If also the power angle δ is small, then $\sin \delta \cong \delta$ and $\cos \delta \cong 1$:

$$P \cong \frac{EV}{X\delta} \qquad (6.3)$$

$$Q \cong \frac{E(E-V)}{X} \qquad (6.4)$$

where E, P, Q denote respectively the voltage, the active power and the reactive power of the grid and V is the voltage of the VSC, (6.3) and (6.4) show that the active power injection depends predominantly on the power angle, whereas the reactive power injection depends on the voltage difference $E - V$.

Equations (6.3) and (6.4) can also be used to explain how the WTS can provide ancillary services influencing the voltage and frequency of the grid with active and reactive power injection. The phasor schemes of Figure 6.17 should be modified while considering the grid voltage E to be not stiff but also dependent on the voltage drop due to the distribution line and the transformer impedances seen at the point of connection, as shown in Figure 6.18. Hence active power can be used to regulate the angle or the frequency of the grid voltage, whereas the reactive power can be used to control the amplitude of the grid voltage. Thus by adjusting the active power and the reactive power, frequency and amplitude of the grid voltage can be influenced.

All the previous reported WTS topologies employ a grid converter, the only difference being between the case where they employ a full power converter or a reduced power converter. Moreover, the grid converter control is slightly different in the case where a doubly fed induction generator is used since in that case the rotor-side converter also determines the reactive power exchange of the overall system. In that case, as already explained, the current controlled rotor converter behaves like a current source that is connected in parallel to the magnetizing reactance of the doubly fed induction generator. If this is parallel it is transformed into a Thevenin equivalent and the rotor plus the back-to-back converter can be seen as a virtual grid converter exchanging active and reactive power with the grid but the capability of the exchanging reactive power is limited. Figure 6.19 compares the capability of active and reactive power handling of a full-power converter and a reduced-power converter (doubly fed induction generator).

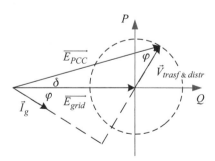

Figure 6.18 Influence of active and reactive power injection by the WTS at the point of common coupling

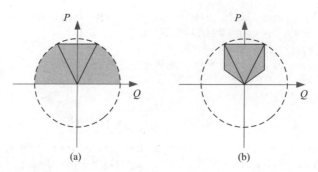

Figure 6.19 Power handling capability in the case of: (a) a full-power converter and (b) a reduced-power converter (the inner triangle shows the normal range of operation with 5 % reactive power injected/absorbed)

6.5 Summary

The aim of the chapter has been to introduce WTSs with a focus on the different topologies, converter structures and control main goals. In the next chapters the grid converter operation will be discussed in more detail and the low-voltage ride-through strategies will be analysed.

References

[1] 'Working with Wind, Integrating Wind into the Power System'. *IEEE Power and Energy Magazine*, **3**(6), November/December 2005.
[2] Pepermans, G., Driesen, J., Haeseldonckx, D., Belmans, R. and D'haeseleer, W., 'Distributed Generation: Definition, Benefits and Issues'. *Energy Policy*, **33**(6), April 2005, 787–798. ISSN 0301-4215, DOI: 10.1016/j.enpol.2003.10.004.
[3] Lubosny, Z., *Wind Turbine Operation in Electric Power Systems Advance Modeling*, Berlin–Heidelberg: Springer, 2003, Hardcover, 259 pages. ISBN 3-540-40340-X.
[4] Ackermann, T., *Wind Power in Power Systems*, John Wiley & Sons, Ltd., 2005. ISBN 0-470-85508-8.
[5] Heier, S., *Grid Integration of Wind Energy Conversion Systems*, John Wiley & Sons, Ltd, 1998.
[6] Hansen, L.H., Madsen, P.H., Blaabjerg, F., Christensen, H.C., Lindhard, U. and Eskildsen, K., 'Generators and Power Electronics Technology for Wind Turbines'. In *Proceedings of IECON '01*, Vol. **3**, 2001, pp. 2000–2005.
[7] Petru, T. and Thiringer, T., 'Modelling of wind turbines for power system studies'. *IEEE Transactions on Power Systems*, **17** (4), November 2002, 1132–1139.
[8] Hansen, L.H., Helle, L., Blaabjerg, F., Ritchie, E., Munk-Nielsen, S., Bindner, H., Sørensen, P. and Bak-Jensen, B., 'Conceptual Survey of Generators and Power Electronics for Wind Turbines', 2001, Risø-R-1205 (EN). ISBN 87-550-2745-8.
[9] Blaabjerg, F., Teodorescu, R., Liserre, M. and Timbus, A.V., 'Overview of Control and Grid Synchronization for Distributed Power Generation Systems'. *IEEE Transactions on Industrial Electronics* **53**(5), October 2006, pp. 1398–1408.
[10] Carrasco, J. M., Franquelo, L. G., Bialasiewicz, J.T. , Galván, E., Guisado, R.C. P., Prats, Á. M., León, J. I. and Moreno-Alfonso, N., 'Power-Electronic Systems for the Grid Integration of Renewable Energy Sources: A Survey'. *IEEE Transactions on Industrial Electronics*, **53**(4), August 2006, pp. 1002–1016.
[11] Machowski, J., Bialek, J. and Bumby, J., *Power System Dynamics: Stability and Control*, John Wiley & Sons, Ltd, 2008, ISBN 10:0470725583.
[12] Hansen, A. G., Michalke, G., Sørensen, P., Lund, T. and Iov, F. 'Co-ordinated voltage control of DFIG wind turbines in uninterrupted operation during grid faults'. *Wind energy*, **10**(1), August 2006, 51–68.

[13] Teodorescu, R. and Blaabjerg, F., 'Flexible Control of Small Wind Turbines with Grid Failure Detection Operating in Stand-Alone and Grid-Connected Mode'. *IEEE Transactions on Power Electronics*, **19** (5), 2004, 1323–1332.

[14] Peng, F. Z., 'Z-source inverter'. *IEEE Transactions on Industry Applications*, **39** (2), March/April 2003, 504–510.

[15] Dai, J., Xu, D. D. and Wu, B., 'A Novel Control Scheme for Current-Source-Converter-Based PMSG Wind Energy Conversion Systems'. *IEEE Transactions on Power Electronics*, **24** (4), April 2009, 963–972.

[16] Faulstich, A., Steakle, J.K. and Wittwer, F., 'New Medium-Voltage Inverter Design with Very High Power Density Medium Voltage Converter for Permanent Magnet Wind Power Generators up to 5 MW'. In *EPE 2005*, Dresden, September 2005.

[17] Andresen, B. and Birk, J., 'A High Power Density Converter System for the Gamesa G10 × 4.5 MW Wind Turbine'. In *EPE 2007*, Aalborg, September 2007.

[18] Tenca, P., Rockhill, A.A., Lipo, T.A. and Tricoli, P., 'Current Source Topology for Wind Turbines with Decreased Mains Current Harmonics, Further Reducible via Functional Minimization'. *IEEE Transactions on Power Electronics*, **23**(3), May 2008, 1143–1155.

[19] Chinchilla, M., Arnaltes, S. and Burgos, J.C., 'Control of Permanent-Magnet Generators Applied to Variable-Speed Wind-Energy Systems Connected to the Grid'. *IEEE Transactions on Energy Conversion*, **21** (1), March 2006, 130–135.

[20] Leonhard, W., *Control of Electrical Drives*, Springer, 1997.

[21] Pena, J.C.C. and Asher, G. M., 'Doubly Fed Induction Generator Using Back-to-Back PWM Converters and Its Application to Variable Speed Wind-Energy Generation'. *IEE Proceedings on Electric Power Application*, 1996, 231–241.

[22] Petersson, A., Harnefors, L. and Thiringer, T., 'Evaluation of Current Control Methods for Wind Turbines Using Doubly-Fed Induction Machines'. *IEEE Transactions on Power Electronics*, **20** (1), January 2005, 227–235.

[23] Tang, Y. and Xu, L., 'A Flexible Active and Reactive Power Control Strategy aor a Variable Speed Constant Frequency Generating System'. *IEEE Transactions on Power Electronics*, **10** (4), July 1995, 472–478.

[24] Schreiber, D., 'Power converter circuit arrangement for generators with dynamically variable power output' - US Patent App. 10/104,474, 2003.

7

Grid Requirements for WT Systems

7.1 Introduction

In order to smooth the effects of wind power penetration over the power system stability and power quality, new grid interconnection requirements, called Grid Codes (GCs), have been developed by Transmission System Operators (TSOs) in different countries. GCs with different requirements for Distribution System Operators (DSOs) have also been developed, but as in most of the countries TSOs are responsible for controlling the power balance and because large wind farms are planned to be installed in the near future, in this chapter only GCs for TSOs are considered. There may also be technical requirements that are not referred to in the grid code, but which apply to the project through the connection agreement or the power purchase agreement or in some other way.

The grid codes are very important as due to them:

- The TSO can maintain the security of power dispatch regardless of the generation technology used.
- The amount of project-specific technical negotiation with the TSO can be reduced.
- Wind turbine manufacturers can design their equipment in the knowledge that the requirements are clearly defined and will not change without warning or consultation.

Due to the political separation of ownership of generation and system operators, the technical requirements governing the relationship between generation and system operators need to be more clearly defined. The introduction of renewable generation has often complicated this process significantly, as these generators have physical characteristics that are different from the directly connected synchronous generators used in large conventional power plants. In some countries, this problem has caused significant delays in the development of appropriate grid code requirements for wind generation.

The GC is developed typically by the system operator, often overseen by the energy regulatory body or government. The requirement modification process should be transparent and

---

Grid Converters for Photovoltaic and Wind Power Systems Remus Teodorescu, Marco Liserre, and Pedro Rodríguez
© 2011 John Wiley & Sons, Ltd

include consultation with generation, system users, equipment suppliers and other affected parties. Typically draft GCs are released before new ones in order to get feedback.

The general message of the GC is that the wind power plants (WPP) should behave as much of possible in the same way as conventional power plants based on large synchronous generators (SG) in both normal operation and during faults. The technology of synchronous generators is very well established, with a long history in the background, and they are able to support the transient stability of the grid by offering inertia, resynchronizing torque, oscillation damping, reactive power generation, short-circuit capability and fault ride-through (FRT). These features allow the SG to comply with the TSO grid connection requirements, which is why today we have a quite stable grid operation worldwide. Typical GC requirements are steady-state and dynamic active and reactive power capability, continuously acting frequency and voltage control and fault ride-through (FRT) capability.

As the grid characteristics are very different worldwide, as, for example, stiffer in Western Europe and weaker in Australia, New Zealand and India, the GCs are also quite different and are updating at a fast pace every few years in order to cope with the explosive wind penetration development.

The wind turbine manufacturers are constantly challenged by the new grid codes as they need to adapt their technology to comply with them and the development times for a wind turbine has to be short in such a market dominated by strong competition. The challenge is how the new modern technology WPP can achieve the new features imposed by the GCs.

7.2 Grid Code Evolution

The first generation of fixed-speed grid-connected wind turbines in the 1980s employed cage induction generators with switched capacitor banks in order to maintain a high power factor. During the 1990s the doubly fed induction generator (DFIG) was successfully introduced where typically a 0.3 p.u. converter controlled the rotor voltage. This technology has the ability to control reactive power within some boundaries and by adding a chopper in the DC link, FRT can be achieved. The GCs reacted by being updated with more requirements, like, for example, a wider range of reactive power control and deeper FRT where the TSO required the WPP to stay connected to the PCC during and after a grid fault, of course within some time window. Another wind turbine technology introduced by Enercon and Siemens in the early 2000s is by using the full-scale converter (FSC) with either induction (Siemens) or synchronous (Enercon) generators. By fully processing the power, a wider range of reactive power control can be obtained as well as grid support during the fault. Today both the DFIG and FSC technologies can offer compliance with the most demanding GCs but there is a strong trend to migrate towards FSC as future GCs can challenge DFIG compliance, especially in the case of asymmetrical grid faults.

Therefore WT technology is an important factor that influences GC evolution. Another factor is the continuously increasing size of the WT and especially of wind farms. Today offshore farms in the range of hundreds of MW are usual and some companies are talking about GW farms (e.g. Borkum and Kriegers Flak). These large power plants call for new requirements in the GC in terms of FRT and active participation in the frequency and voltage control in order not to compromise grid stability.

Thus the TSOs are involved in a continuous process of adapting the GC in order to be able to increase the wind penetration, mainly driven by political goals, without compromising the security of the grid. This is a process involving both TSOs, WT manufacturers and wind farm developers. The typical scenario is that TSOs are reluctant to change the grid operation and control and thus require through GCs the WPP to behave as close as possible to the conventional plants, which may not only be impossible for some of the characteristics or too costly to realize. The WT manufacturers and wind farm developers are interested in more clear and standardized requirements. It will be for the future to see how this process will evolve.

As GCs have evolved mainly in the countries with already or planned high wind power penetration (Denmark, Germany, Spain, UK, Ireland, US and China), this chapter discusses mainly those GCs being considered most relevant from a technical point of view. In Table 7.1,

Table 7.1 GCs in countries with high wind power penetration

Country	TSO	Title	Date	www
Denmark	Energinet.dk	Grid Connection of Wind Turbines to Networks with Voltages above 100 kV, Regulation TF 3.2.5	December 2004	www.energinet.dk
Germany	E.ON, EnBW, Vattenfall, RWE	Transmission Code 2007	August 2007	www.vde.com
		Ordinance on System Services by Wind Energy Plants – SDLWindV	2009	www.erneuerbare-energien.de
Spain	Red Electrica	Resolution-P.O.12.3, Response Requirements against Voltage Dips in Wind Installations	March 2006	www.ree.es
			October 2008	www.res.es
		Draft of Annex of P.O.12.2, restricted to the technical requirements of wind power and photovoltaic facilities		
UK	NGET	The Grid Code, issue 4	June 2009	www.nationalgrid.com
Ireland	EIRGRID	Grid Code, version 3.1	April 2008	www.eirgrid.com
US	FERC	FERC Order 661, Interconnection with Wind Energy	June 2005	www.ferc.gov
	WECC		July 2009	www.wecc.biz
		WECC-060, Low Voltage Ride Through Standard. Revision of 2005 edition (draft)		
China	CEPRI	Revised National Grid Code (draft)	July 2009	www.dwed.org

the current GCs from these countries are listed with information about getting these public documents from the public domain onto the internet.

7.2.1 Denmark

Denmark has the world's highest wind penetration of ca. 20 % (in terms of annual energy consumption covered by WPP) and the current government has planned a challenging 50 % renewable energy penetration by 2025, mostly by adding 3 GW offshore wind power plants to the existing 3 GW by 2008 [1]. As pioneers in wind power penetration, Denmark was the first country to adapt the grid interconnection requirements for WPP in the early 1990s. Currently, the unique system operator is Energinet.dk and the actual GC issued in 2004 is:

- 'Grid Connection of Wind Turbines to Networks with Voltages above 100 kV', Regulation TF 3.2.5, elaborated by Eltra and Elkraft System [2].

In particular, the Danish GC applies only to wind farms and imposes the existence of a wind farm controller (WFC) featuring different types of active power regulation:

- Absolute production constraint.
- Delta production constraint.
- Balance regulation.
- Stop regulation.
- Power gradient constrainer.
- System protection.
- Frequency control.

All these functions ensure that the wind farms can actively participate in the frequency regulation process like any other conventional plant.

7.2.2 Germany

Gemany accumulated around 24 GW of installed wind power by 2008 and with plans to increase this to in excess of 35 GW by 2013 [1], mostly offshore, will remain the biggest wind country in the EU. Germany is divided into four control areas managed by the following TSOs: EnBW, E.ON, Vattenfall and RWE.

The following documents are relevant:

- 'Transmission Code' issued in 2007 by all four system operators [3].
- 'Ordinance on System Services by Wind Energy Plants' (System Service Ordinance – SDLWindV) – Draft [4], an update of Transmission Code 2007 issued in 2009.
- 'Requirements for Offshore Grid Connections in the E.ON Netz Network' by E.ON, 2008 [5].

The recent Ordinance on System Services – SDLWindV updates the national regulation Transmission Code 2007, aiming to boost the security and stability of the networks with high wind energy penetration by primarily offering incentives in the form of a system bonus of:

- 0.5 eurocents/kWh to all new wind farms installed after 30 June 2010.
- 0.7 eurocents/kWh for old wind farms installed before 1 January 2009.

This complies with the new features in terms of voltage control based on reactive power and reactive current support of voltage recovery after faults. The active power control and LVRT requirements defined by Transmission Code 2007 remain unchanged. As a novelty, the term 'wind energy plant' is defined for more specificity of these requirements.

7.2.3 Spain

With over 16 GW of wind power accumulated by 2008 and with plans for over 27 GW by 2013 [1], mostly onshore, Spain will be a very important wind country in Europe, surpassed only by Germany. Although it is a large country, only one TSO operates in Spain and this is Red Electrica.

The most relevant documents are:

- Resolution-P.O.12.3, 'Response Requirements against Voltage Dip in Wind Installations', March 2006 [6].
- Annex of P.O.12.2, restricted to the technical requirements of wind power and photovoltaic facilities (draft), October 2008 [7].

Two years later after the P.O.12.3 had imposed LVRT down to 0 V and limitation of active and reactive power during faults, a specific procedure for verification, validation and certification was published by the Spanish Wind Association AEE [8]. A technical assessment of this document with practical test cases is performed in reference [9].

The recent draft of Annex of P.O.12.2 [7] applies for new wind or PV plants bigger than 10 MW installed after 1 January 2011 and requires for the first time a voltage controller for defining the reactive power support during voltages outside the normal range. Also, as a recommendation for the future, this document defines the possible requirement of inertia emulation and power oscillation damping (POD), two features that require some active power storage, which is very much a reminder of the synchronous generator technology.

7.2.4 UK

With over 3 GW cumulative installation by 2008 and with very ambitious offshore plans to reach over 14 GW by 2013 [1], the UK is another important player in the European wind market.

The relevant document elaborated by the TSO NGNET in 2009 is:

- 'The Grid Code', issue 4 [10].

Here, the wind farms are divided into embedded and nonembedded systems. Nonembedded generating units are directly connected to the TS whereas embedded units only have an indirect connection to the TS by another user (e.g. a wind farm or distribution system). Embedded and nonembedded units need to fulfil different requirements. Furthermore, different requirements have been defined for the two TS areas of Scotland and England/Wales, which are also depending on the rated power output of the wind farm. Rather peculiarly, the PCC of a wind farm might also be defined on a remote TS bus-bar in the UK, making the compliance verification process more complicated.

7.2.5 Ireland

Even though it is a small country, Ireland is aiming for a high wind penetration and according to reference [1] the ca. 1 GW installed now should be increased to ca. 2.4 GW by 2013, reaching a penetration of over 15 %.

The relevant document is:

- 'The EirGrid Grid Code', version 3.1, 3 May 2008, elaborated by the National Grid Electricity Transmission plc [11].

The GC of Ireland is interesting as it poses technical challenges, as this country has very favourable wind potential and the political will to increase penetration but low resources for balancing wind power.

7.2.6 US

With over 25 GW by 2008 and in excess of 77 GW installed by 2013 [1], the US will be by far the country with the largest accumulation of wind energy worldwide.

The United States Order 661A published in 2004 [12] by the Federal Energy Regulatory Commission (FERC) regulates the connection of generators to the grid, but it was not specifically adapted for wind farms. In 2009, WECC (Western Electricity Coordinating Council), the largest regional entity in the US counting over 1.8 million square miles stretching from Canada to Mexico and passing through 14 western US states, published a draft of 'The Technical Basis for the New WECC LVRT Standard' [13] to replace the WECC LVRT standard from June 2005. This document was prepared by the Wind Generation Task Force (WGTF) for WECC and applies for all generation units with cumulative ratings higher than 20 MVA connected to transmission levels higher than 60 kV after 1 May 2009.

7.2.7 China

With over 12 GW installed by 2008 and close to 55 GW planned by 2013 [1] China is a major player with a huge potential for wind power integration. The CEPRI national research institute for electrical power has drafted 'Technical Rule for Connecting Wind Farm to Power System' in 2005 and due to the big growth in wind power penetration, they initiated in 2007 the three-years intergovernmental Sino–Danish Wind Energy Development (WED) project

[14]. The main idea behind this programme is to convey to China the experience of Denmark from achieving a record high wind energy penetration.

7.2.8 Summary

The grid connection requirements for wind turbines given are valid at the PCC (point of common coupling) with the TS in most of the cases defined on the high-voltage side of the wind farm step-up transformer.

The grid codes are very complex documents dealing with a lot of requirements in both stationary and transients. For the sake of relevance only the following requirements have been considered in this study:

Normal operation

- Frequency and voltage deviations.
- Active power control.
- Reactive power control.

Behaviour under grid disturbances

- Voltage ride-through LVRT.
- Reactive current injection.

7.3 Frequency and Voltage Deviation under Normal Operation

WPP are required to operate within a range around the rated voltage and frequency at PCC to avoid disconnection due to transient disturbances. Typically this requirement is described using different zones:

- continuous operation in a limited range below and above the rated point.
- time-limited operation with possible reduced output in extended ranges.
- immediate disconnection.

The voltage–frequency operational window for DK, D, E, UK, IE and CN are represented graphically in Figure 7.1.

The strictest continuous operation limits for frequency appear in the British code [10] (47.5–52 Hz) and for voltage in the Danish code [2] (90–106% nominal voltage). For lower frequencies outside the continuous operational area, some countries, like, for example, the UK, require maintaining the output power in order to support frequency control.

It is obvious that the most extreme frequency limits, 46.5 and 53.5 Hz, are for E.ON offshore [5]. In countries like Ireland, characterized by an isolated power system with weak interconnections, larger frequency ranges are allowed.

The Spanish regulation (RD 661-2007) allows the widest continuous frequency range of 48–51 Hz, with disconnection for frequencies lower than 48 Hz for more than 3 sec. The disconnection time for overfrequencies (>51 Hz) has to be agreed with the TSO.

In the German GC [3], even if permanent operation is allowed for frequencies higher than 50 Hz, power curtailment is required starting at 50.2 Hz in order to contribute to the primary

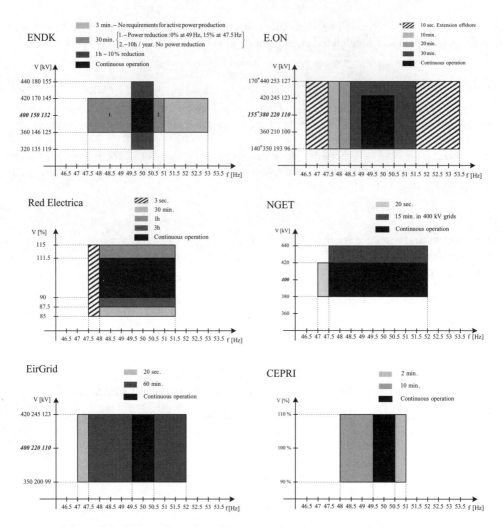

Figure 7.1 Voltage–frequency operation window for DK, D, E, UK, IE and CN

control (see Section 7.4.2). Resynchronization after disconnection can take place once the voltage increases again to about 105 kV in 110 kV networks, to 210 kV in 220 kV networks and to 370 kV in 380 kV networks.

7.4 Active Power Control in Normal Operation

This is a typical requirement aligned with the conventional power plants reflecting the ability to adjust the output power as required by the TSO in order to support balancing the load with two different goals:

- Power curtailments (participation in secondary control).
- Frequency control (participation in primary control).

Table 7.2 Power curtailments defined by the Danish GC

Type of regulation	Purpose	Primary regulation aim
Absolute production constraint	Limit the wind farm's current power production in the connection point to a maximum, specifically indicated MW value	Limit production to optional MW_{max}
Delta production constraint	Reduce the power production of the wind farm to a lower value than the available power, therefore creating regulating reserves	Limit production by MW_{delta}
Balance regulation	Adjust the power production to the power setup requirement imposed by the TSO in order to participate in the power balance of the downward/upward regulation of production that must be possible	Change current production by $\pm MW$ with the set gradient
Stop regulation	Maintain the power production at the current level (if the wind makes it possible); the function results in a stop for upward regulation and production constraints if the wind increases	Maintain current production
Power gradient limiter	Limit the maximum gradient at which the power output changes in relation to changes in wind speed	Power gradients do not exceed the maximum settings
System protection	System protection is a protective function that must be able to automatically downward regulate the power production of the wind farm to a level that is acceptable to the power system. In the case of unforeseen incidents in the power system (for instance forced outage of lines), the power grid may be overloaded at the risk of power system collapse. The system protection regulation must be able to contribute rapidly to avoid system collapse	Downward regulate power production automatically on the basis of an external system protection signal

7.4.1 Power Curtailment

The requirements for power curtailments are different across the countries.

In *Denmark* [2], a ramp rate to up to 100 % of rated power per minute with curtailments all the way down to 20 % and accuracy of 5 % (5 min. average) are required. In particular, six different profiles of curtailments or gradient limiters are defined as shown in Table 7.2.

In *Germany* [4], the active power of WPP is required to be changed with a ramp rate of at least 10 % of grid connection capacity per minute to any level required by TSO though no lower than 0.1 p.u.

In *Spain* [7], the rate is not specified but the WPP has to be able to reach any set-point sent by the TSO and additionally should transmit to the TSO the difference in the actual power and the maximum possible power for the case when it works in the derated mode.

Table 7.3 Active power gradients in the Chinese GC

WPP installation capacity (MW)	10 min. maximum ramp (MW)	1 min. maximum ramp (MW)
< 30	20	6
30–150	Installation capacity/1.5	Installation capacity/5
> 150	100	30

In *Ireland* [11], two ramp rates averaged over 1 min. and over 10 min. should be in the range of 1–30 MW per min., as required by the TSO, and should be activated in less than 10 sec.

In *China* [14], the ramp rate depends on the WPP rating, as described in Table 7.3.

7.4.2 Frequency Control

The active power output has also to be changed during frequency variation in order to ensure smooth participation of the WPP in the primary control. The ramps in this case are much higher than in the case of power curtailments. If the WPP uses a different P gradient as the other participants of the balancing act, stability issues may occur. This gradient is typically given in MW/Hz, indicating the required variation of power as a response to a frequency variation.

The participation in frequency control is different from country to country, as shown in the table in Figure 7.2. All generating units must reduce, while in operation, at a frequency of more than 50.2 Hz, the instantaneous active power with a gradient of 40 % of the generator's instantaneously available power per hertz, as shown in Figure 7.2.

According to the *German* code [4], when frequency exceeds the value 50.2 Hz wind farms must reduce their active power with a gradient of 0.4 p.u./Hz (40 % of the available power of the WPP).

The *British* code [10] requires wind farms larger than 50 MW to have a frequency control device capable of supplying primary and secondary frequency control, as well as an

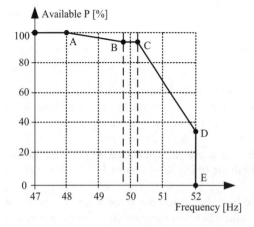

	Transmission system frequency (Hz)		Available active power [%]
F_A	47.0 – 51.0	P_A	50 – 100
F_B	49.5 – 51.0	P_B	50 – 100
F_C	49.5 – 51.0	P_C	
F_D	50.5 – 52.0	P_D	20 – 100
F_E		P_E	0

Figure 7.2 Irish power–frequency curve

overfrequency control. It is remarkable that it also prescribes tests that validate that wind farms indeed have the capability of the demanded frequency response.

The *Irish* code [11] demands a frequency response as described in the curve shown in Figure 7.2.

The values for the power and frequency of points ABCDE should be on-line modified by the TSO within the range mentioned in Figure 7.2. This is due to the fact that in order to obtain a smooth participation of WPP in the TSO frequency control, the power ramp should be imposed by the TSO in harmony with the frequency response of the other participants to the balancing act.

In the power system, the frequency is an indicator of the imbalance between production and consumption. For normal power system operation, the frequency should be close to its nominal value. In the case of an imbalance between supply and demand, the primary and secondary controls are used to reduce the imbalance of power. In a power system, conventional generating units are normally equipped with a governor control, which works as the primary load frequency control. The time span for this control is 1–30 sec. In order to restore the frequency to its nominal value and release used primary reserves, the secondary control is employed with a time span of 10–15 min. The secondary control thus results in a slower increase or decrease of generation. Some regulations require wind farms to be able to participate in secondary frequency control. During the overfrequencies, this can be achieved by shutting down some of the turbines in the wind farm or by reducing the power output using pitch control. Since wind cannot be controlled, power production at normal frequency would be intentionally kept lower than possible in order for the wind farm to be able to provide secondary control at underfrequencies. They will then be required to provide primary, secondary and high frequency responses.

7.5 Reactive Power Control in Normal Operation

The new GC aim for turn WPP is to make it behave more like conventional synchronous generator power plants in terms of Q regulation in response to the grid voltage variation, a feature known as automatic voltage regulation (AVR). The Q requirements are related to the characteristics of each grid as the voltage changing capability depends on the grid short-circuit power. This requirement can be given in three different ways:

- Q set-point.
- Power factor control.
- Voltage control.

7.5.1 Germany

The minimum requirements for reactive power generation [4] are given in the form of areas as a function of voltage at nominal active power and as a function of active power for the cases when the WPP is working at derated power for different ranges of voltages inside the normal operation range. The requirement can be given as a Q requirement or a PF requirement. As the characteristics of the grid may differ depending on location and strength, three variants are defined by the German TSOs, as depicted in Figure 7.3(a) to (c).

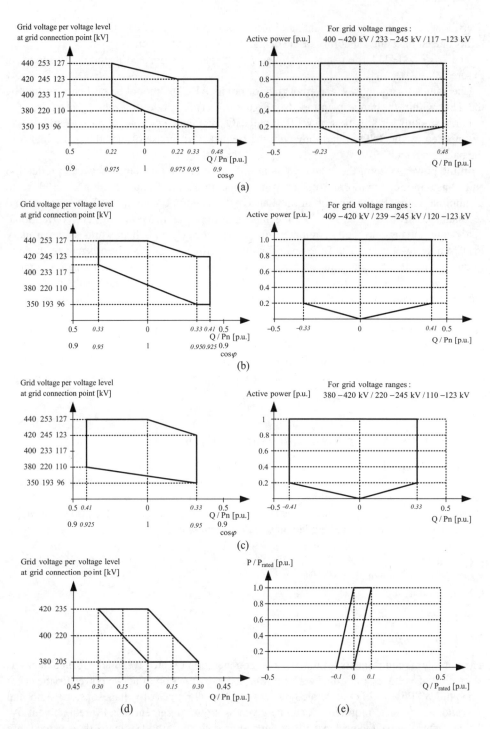

Figure 7.3 (a) to (c) Three variants of *VQ* and *PQ* dependencies defined in Germany, (d) *VQ* dependence in Spain and (e) *PQ* dependence in Denmark

The WPP should be able to cycle the entire Q range in voltage or active power planes within 4 min. The Q controller refers to the positive-sequence system components and should be slow with a settling time in the minutes range.

7.5.2 Spain

The Q requirements during normal operations as defined by the directive P.O.7.4 [15] from 2000 applies to all generation on the HV level, both conventional and renewable. The following requirements are defined as a function of active power and transmission voltages as follows:

- minimum range 0.15i–0.15c for all technical P range and nominal voltage.
- minimum range 0.30i–0.30c as a function of the voltage, as shown in Figure 7.3(d).

7.5.3 Denmark

In the Danish requirement [2], the 10 sec. average PQ diagram is given as shown in Figure 7.3(e), which applies for the whole range of voltage in normal operation. Basically it defines a control band of 0.1 p.u. In comparison with the German and Spanish GCs, the minimum required Q is lower. This should be regarded as the minimum requirement.

After agreement with the TSO, the Q set-point or voltage control mode can also be adopted by the WPP, resulting in larger amounts of Q.

7.5.4 UK

The British code[10] is specifically formulated for nonsynchronous embedded generation and requires a PF in the range 0.95i to 0.95c at 1 p.u. active power for connection to the HV system (132/275/400 kV). This requirement is equivalent to 0.33 p.u. reactive power and should be maintained for active power down to 0.2 p.u. for lagging PF and down to 0.5 p.u. for leading PF. The grey area in Figure 7.4(a) is an extension of the Q requirements in the dashed area for P lower than 0.2 p.u. and a lower band of ±0.05 p.u. of reactive power is required at low-power leading PF, which can be required after agreement with the TSO (NGET).

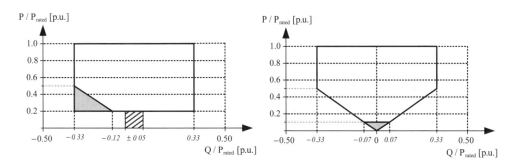

Figure 7.4 The Q requirements in the PQ plane of: (a) UK and (b) Ireland

7.5.5 Ireland

The Irish code [11] is quite similar but with 0.33 p.u. Q for both lagging and leading PF, as shown in Figure 7.4(b), and with the Q requirements decreasing linearly to 0 proportional with P for P lower than 0.5 p.u.

7.5.6 US

The US FERC 661 code [12] concludes that reactive power in the PF range 0.95i to 0.95c can be required by the TSO on a case-by-case situation and that it is not mandatory for it to be dynamic (STATCOM). This is why even now a large amount of reduced variable-speed range wind turbines with variable rotor resistance are still installed in the US. Only the modern variable-speed turbines with either DFIG or FSC technology are able to comply with dynamic reactive power requirements.

7.6 Behaviour under Grid Disturbances

Grid faults in the form of voltage sags or swells can typically lead to tripping of WPP, which can unbalance the grid and may yield blackouts. To avoid this, the GCs typically require three things: do not disconnect from the grid even if the voltage goes down to zero for up to 150 ms, support the voltage recovery by injecting reactive current and ramp up the active power after the fault clearance with a limited ramp to harmize with the 'natural' recovery of the grid after fault clearance. The following typical features are generally defined in GCs:

- Voltage ride-through (VRT) in terms of minimum (LVRT) and maximum (HVRT) voltage magnitude and recovery slope for symmetrical and asymmetrical faults the WPP should be able to withstand without disconnection and timeframe, and the circumstances under which WPP can be disconnected from the grid.
- P and Q limitation during faults and recovery.
- Reactive current injection (RCI) for voltage support during fault and recovery.
- Resuming active power with limited ramp after fault clearance.

In the following, some relevant national VRT requirements are described with a focus on the ones that pose more technical challenges for the WPP.

7.6.1 Germany

The VRT and RCI are described in Figure 7.5 [4]. The following in particular apply:

VRT

- Within the black area no interruptions are allowed. The WPP must stay connected even when the PCC voltage is zero. The 150 ms accounts for the typical operating time of the protection relays.

Grid Requirements for WT Systems

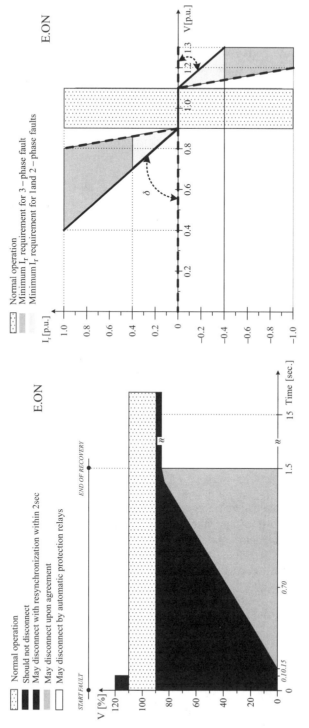

Figure 7.5 VRT (left) and I_r (right) requirements for Germany

- Within the dark grey area, if the facility is facing stability issues, short time interruptions (STI) with resynchronization for a maximum of 2 sec. are allowed. Modern WT with FSC are allowed to initiate STI at higher voltage levels providing the interruption time is limited to 2 sec. and the reactive current injection continues during the interruption period.
- Within the light grey area short disconnection with resynchronization later than 2 sec. can be allowed after agreement with the TSO.
- For faults longer than 1.5 sec., stepwise interruptions are allowed.
- The voltage value in Figure 7.5 refers to the highest value of all three-phase grid voltages measured at the low-voltage side of the transformer in each wind turbine.

P and Q limitation during faults and recovery

- During faults, the active current can be reduced in order to fulfil the reactive current requirements after the fault period. A fast return to normal active power generation is essential to ensure the power balance within the grid, and thus frequency stability.

Minimum reactive current injection

- In case of significant deviation of the voltage, proportional reactive current has to be injected/absorbed, as shown in Figure 7.5.
- The slope of the minimum reactive current injection ($K = \tan(\delta) = \Delta I_r / \Delta V$) can be variable between 0 and 10.
- A deadband of 10 % of voltage variation is used to improve stability. This deadband can be eliminated in future for an HV transmission connection.
- The response time of the reactive current controller should be a maximum of 30 ms and the control band should be between -10 and $+20$ % of the rated current.
- The reactive current requirements in Figure 7.5 apply for the highest value of the three-phase voltages in the case of faults within the black area.
- For one- and two-phase faults, the maximum reactive current can be limited to 40 % of the rated current.
- After fault clearance, the reactive current reference should not change stepwise in order to avoid stability issues.
- For voltages below 0.85 p.u., if the facility is unable to supply the reactive power required for voltage support, the so-called 'Safeguard I' implemented in PCC will trip the wind farm after 0.5 sec. 'Safeguard II' at the wind turbine level is implemented as system protection acting after 1.5 sec. and includes the stepwise tripping of wind turbines.

Resuming active power

- After fault clearance without disconnection, the active power feed-in must be continued immediately after fault clearance and increased to the original value with a gradient of at least 20 %/sec.
- In case of short disconnection, the active power feed-in must be resumed immediately after fault clearance with a gradient of at least 10 %/sec.

7.6.2 Spain [7]

VRT (see Figure 7.6)

- No disconnection is allowed within the black area for one-, two- and three-phase faults.

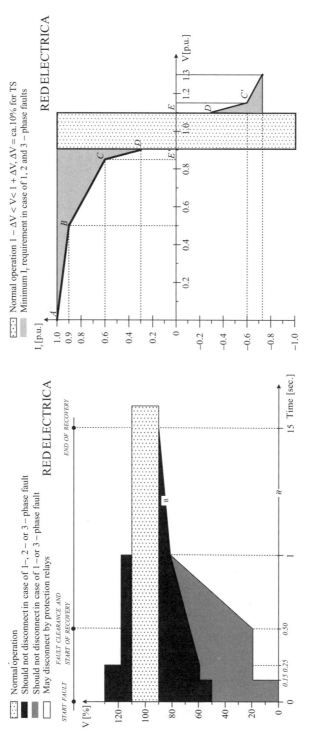

Figure 7.6 VRT (left) and I_r (right) requirements for Spain

- No disconnection is allowed within the grey area for one- and three-phase faults.
- No disconnection is allowed for 1 sec. for the 1.15 p.u. level and 250 ms for 1.3 sec.
- During the whole transient regime, the facility must be able to inject to the grid at least the nominal apparent current.

P and Q limitation during faults and recovery

- The facility might not consume active and reactive power at the grid connection point during both the fault duration and the duration of voltage recovery following fault clearance.
- Momentary active or reactive power consumption (< 0.6 p.u.) is allowed during just the first 40 ms after the start of the fault and the first 80 ms after the clearance of balanced (three-phase) faults.
- Momentary active or reactive power consumption (< 0.4 p.u.) is allowed during just the first 80 ms after the start of the fault and the first 80 ms after the clearance of unbalanced (single-phase and two-phase) faults.

Reactive current injection

The requirements of Q generation under voltage faults ($V < 0.85$ p.u.) are implemented similarly as for the case of automatic voltage regulation (AVR) in conventional synchronous generation, i.e. in the form of a PI voltage controller with reactive current reference I_r as output, as shown below:

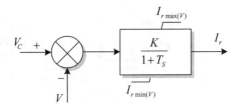

Figure 7.7 Reactive current injection requirements in Spain during FRT

where V_c is the voltage set-point (RMS), V is the PCC voltage (RMS) and I_r is the instantaneous reactive current reference. The saturation levels are voltage dependent, as explained in Figure 7.7.

The following conditions apply:

- The controller will be enabled for any voltage outside the normal operation range.
- If the WPP was working in the voltage control mode in normal operation, the voltage set-point during the fault will remain unchanged.
- If the WPP was working in the Q or PF control mode, during the disturbance the voltage set-point will be the voltage prior to the fault if the normal operation is set to reactive power or power factor allocation.
- During the fault, the facility should inject/absorb positive sequence reactive currents based on the action of the voltage controller with minimum saturation levels defined by the polygonal curve ABCDE, as shown in Figure 7.6. In case of overvoltage, the saturation levels are mirrored but for voltages higher than 1.3 p.u. disconnection is required by protection relays.

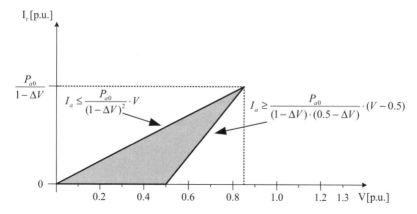

Figure 7.8 Active power limitation during FRT in Spain

- These levels should be implemented as saturation levels for the voltage controller that runs in both normal and faulty operation.
- For the range $0.85 \leq V \leq 1.15$ p.u., the injected reactive current will react according to the voltage control, possibly saturating the regulator limits.
- Once the fault is cleared, the voltage controller will remain enabled for at least 30 sec. after the voltage level reenters the normal operation range. Afterwards, the voltage controller will be disabled and the reactive power requirements for normal operation will apply.

Active current injection

- During faults, the facility should limit the active current within the grey area, as shown in Figure 7.8 (excluding the active current increments/reductions due to frequency control or, if applicable, inertia emulation).
- It can be seen that the active current limitation is a function of P_{ao}, the active power that the facility was generating prior to the disturbance and voltage level.
- For voltage levels lower than 0.5 the active current can be reduced to zero.
- Any possible violation of these active current limits must be corrected before 40 ms.
- In the case of current saturation, reactive current limitation given by voltage controller saturation has priority over active current limitation.
- For voltages higher than the normal operation, the facility will seek, if possible, to maintain the active power level prior to the disturbance.
- The gain of the active current controller should ensure a dynamic response (90 % rise) in less than 40 ms for $V < 0.85$ p.u. and 250 ms for $V > 0.85$ p.u.

Resuming active power

- The voltage-dependent active current control previously mentioned ensures that after the fault clearance without disconnection, the active power level prior to disturbance will be restored smoothly within 250 ms.

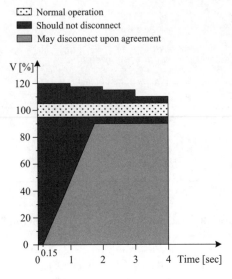

Figure 7.9 VRT in US-WECC

7.6.3 US-WECC

The recent WECC LVRT standard [13] is an effort to create compliance with the federal regulation FERC Order 661-A [12] in terms of the fault voltage level and duration (0 V for 9 cycles) and boundaries for the time of voltage recovery for both LVRT (until the voltage became higher than 90 %) and HVRT (until the voltage became lower than 110 %).

VRT (see Figure 7.9)

- All generators are required to remain in-service during three-phase faults with normal clearing (for a maximum of 9 cycles) unless clearing the fault disconnects the generator from the transmission system.
- The voltage is measured at the high-voltage side of the WPP step-up transformer.
- For single-phase faults, delayed clearing times apply unless clearing the fault disconnects the generator from the transmission system.
- The TSO should provide to the WPP owner the normal breaker clearing time for three-phase faults and the delayed clearing time for single-line-to-ground faults at the high-voltage side of the generating plant step-up transformer.
- There is no requirement for power limitation during fault or reactive power injection during fault or recovery.

A recent study by Transpower [16] summarizes the LVRT requirements in over 20 different countries.

7.7 Discussion of Harmonization of Grid Codes

From the survey presented above, it can be observed that the interconnection regulations vary considerably from country to country. It is often difficult to find a general technical justification for the existing technical regulations that are currently in use worldwide due to the different wind power penetration levels in different countries and operational methodology of power systems.

For instance, countries with a weak power system, such as Ireland, have considered the impact of wind power on network stability issues, which means that they require fault ride-through capabilities for wind turbines already at a lower wind power penetration level compared with countries that have very robust systems. It is interesting to note that inclusion of FRT regulations for DFIG increase the overall cost by 5 %. The European Wind Energy Association (EWEA) recommends that regulations for the European grid connection (or other nations) are to be developed in a more consistent and harmonized manner [17]. Harmonized technical requirements will bring maximum efficiency for all parties and should be employed wherever possible and appropriate. While this applies to all generation technologies, there is a particular urgency in the case of wind power. As wind penetration is forecasted to increase significantly in the short to medium term, it is essential that grid code harmonization should be tackled immediately. It will help manufacturers to internationalize their products/services, developers to reduce cost and TSOs to share experience, mutually, in operating power systems.

It is also important that the national GC should aim at an overall economically efficient solution; i.e. the costly technical requirements such as the 'fault ride-through' capability for wind turbines should be included only if they are technically required for reliable and stable power system operation. Hence, it can be summarized that GCs should be harmonized at least in the areas that have little impact on the overall costs of wind turbines. In other areas, GCs should take into account the specific power system robustness, the penetration level and/or the generation technology.

Moreover, interconnection standards of different countries may also vary in future.

7.8 Future Trends

The following requirements are expected to be included in future GCs.

7.8.1 Local Voltage Control

Both the Spanish and the German GCs have increased the complexity of the reactive current injection during fault and recovery and a continuous local voltage control may prove to be necessary, particularly for offshore wind farms [18].

7.8.2 Inertia Emulation (IE)

The Spanish GC [7] mentions that even if for the moment the ability to emulate inertia is not yet compulsory it is strongly recommended and it may be introduced as a requirement later.

The implementation of emulated inertia should be in the form of a PD controller acting on frequency variation as input and outputting the necessary power variation, as shown in Figure 7.10.

The following conditions apply:

- The gain K_d should be adjustable between 0 and 15 sec. and the response time should be such that in 50 ms the active power should increase at least by $\Delta P = 5\,\%$.

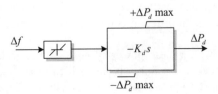

Figure 7.10 Inertia emulation recommended requirement in Spain

- In order to be able to generate the required saturation levels, $\pm\Delta P_{max}$, energy storage of any technology is required in order to inject or absorb at least 10 % active power for at least 2 sec.
- The deadband of frequency variation will be limited to ± 10 mHz.
- The IE should be disabled for voltages lower than 0.85 p.u.

7.8.3 Power Oscillation Dumping (POD)

This is another feature strongly recommended by the Spanish GC [7] where, just as in the case of the synchronous generators, the system should be able to increase or decrease the output power in such a way to reduce the power oscillations in the low-frequency range (0.15–2.0 Hz). The following specific requirements apply:

- The POD can be implemented by 'sharing' the existing power–frequency regulator.
- The POD can 'share' the energy storage used for IE.
- The deadband of frequency variation will be limited to ± 10 mHz.
- The POD should be disabled for voltages lower than 0.85 p.u.

Another important trend is to harmonize the GC worldwide by standardization of some requirements and testing procedures, say, for example, VRT. Actions in these directions are undertaken by the European Wind Energy Association (EWEA) and IEC. However, as grid systems have quite different characteristics worldwide it may be a long-term objective.

7.9 Summary

In this chapter, the grid codes for HV transmission systems were summarized for the connection of wind farms. The comparison for the recent grid codes of countries, which have high wind penetration levels achieved or announced for the future, were conducted for the common technical requirements such as active and reactive power control, frequency control, and behaviour of the WPPs during the grid disturbances. The grid codes require the wind farms to have the control and regulation capabilities similar to the conventional power plants in order to provide stable, reliable, and efficient operation of the power system. The technological development of wind turbines over the last decade has been heavily influenced by these requirements, and initiated the future requirements for the wind turbine manufacturers and WPP developers. During these studies and progress, harmonization of the common requirements has been also needed to have common language for the wind farm connection. Today, modern wind turbine technology has the capability to satisfy the grid code requirements with the use of power electronic interfaces.

References

[1] BTM Consult, 'World Market Update', March 2008, http://www.btm.dk.
[2] Regulation TF3.2.5, 'Technical Regulations for the Properties and the Regulation of Wind Turbines. Wind Turbines Connected to Grids with Voltages above 100 kV'. *Eltra and Elkraft System*, 3 December 2004, www.energinet.dk.
[3] TransmissionCode 2007, 'Networks and System Rules of the German Transmission System Operators'. *VDN-e.v. beim VDEW*, August 2997, www.vdn-berlin.de.
[4] 'Ordinance on System Services by Wind Energy Plants (System Service Ordinance – SDLWindV)'. *Draft*, 2009, www.erneuerbare-energien.de.
[5] 'Requirements for Offshore Grid Connections in the E.ON Netz Network'. *E.ON*, 2008, www.eon-netz.com.
[6] Resolution P.O.12.3, 'Response Requirements against Voltage Dips in Wind Installations'. *Red Electrica*, March 2006, www.ree.es (translated into English by Spanish Wind Association AEE in www.aeolica.es).
[7] Annex of O.P. 12.2, 'Restricted to the Technical Requirements of Wind Power and Photovoltaic Facilities (draft)' *Red Electrica*, October 2008, www.ree.es (translated into English by Spanish Wind Association AEE in www.aeolica.es).
[8] *'Spanish Verification, Validation, and Certification Procedure for the Assessment of the Response of Wind Installations against Voltage Dips According to the Requirements of P.O.12.3 (VV&CP)*, Spanish Wind Association AEE, 2007, www.aeolica.es.
[9] Morales, A., Robe, X., Sala, M., Prats, P., Aguerri, C. and Torres, E., 'Advanced Grid Requirements for the Integration of Wind Farms into the Spanish Transmission System'. *Renewable Power Generation, IET*, **2**(1), March 2008, 47–59.
[10] 'The Grid Code', Issue 4, *National Grid Electricity Transmission plc*, NGET, June 2009.
[11] 'Grid Code', version 3.1, elaborated by *The EirGrid*, April, 2008.
[12] Order 661, 'Interconnection with Wind Energy', issued by Federal Energy Regulatory Commission (FERC) of United States, 2 June 2005.
[13] WECC-0060 – PRC-024-WECC-1-CR, 'Generator Low Voltage Ride-Through Criterion – Regional Criterion (draft), 2009, www.wecc.biz.
[14] 'Revised National Grid Code (draft)', Report WED-QR-C01-E-06, Elaborated by CEPRI as part of the Sino-Danish Wind Energy Development Programm (WED), July 2009, www.dwed.org.
[15] Resolution P.O.7.4, 'Servicio Complementario de Dontrol de Tension de la Red de Transporte'. *Red Electrica*, March 2000, www.ree.es.
[16] 'Generator Fault Ride Through (FRT) Investigation'. *Transpower Ltd*, February 2009.
[17] *Wind Power Interconnection into the Power System: A Review of Grid Code Requirements – Singh*. Elsevier, 2009.
[18] Erlich, I., Feltes, C., Shewarega, F. and Wilch, M., 'Interaction of large offshore wind parks with the electrical grid'. In *Third International Conference on Electric Utility Deregulation and Restructuring and Power Technologies, DRPT 2008*, **6–9** April 2008, pp. 2658–2663.

8

Grid Synchronization in Three-Phase Power Converters

8.1 Introduction

One of the lessons learned from the intensive research conducted on distributed power systems during last few years is that the electricity networks of the future will be based to a large extent on new power electronics and ICT applications, some of which have already been in use in other sectors of industry for decades [1]. This implies that grid-connected power converters applied in distributed power generation systems should be carefully designed and controlled in order to achieve even better performance than the conventional power plants they replace. One of the most important aspects to consider in the control of power converters connected to electrical grids is the proper synchronization with the three-phase utility voltages. This three-phase synchronization is not just a matter of multiplying by three the synchronization system used in single-phase applications, since the three phases of a three-phase system do not work autonomously but do it in a coordinated way, keeping particular relationships in terms of phase shifting and phase sequencing. Therefore, the three-phase voltage should be understood as a vector consisting of three voltage components, which provides the capability of generating and consuming power in a three-phase system.

The module and the rotation speed of the three-phase grid voltage vector keep constant when balanced sinusoidal waveforms are present in the three phases of the system – with equal amplitude, frequency and relative phase shifting. As shown in Figure 8.1, under such ideal operating conditions, the voltage vector describes a circular locus on a Cartesian plane, generally known as the $\alpha\beta$ plane.

In power systems, this rotating voltage vector is mainly supplied by big synchronous generators, and the electrical equipments located at the transmission, distribution and utilization levels are designed assuming that such a voltage vector has both a constant module and a constant positive rotation speed. In practice, however, there are multiple nonidealities in power systems that originate disturbances on the three-phase voltage vector. These voltage disturbances can be classified according to their harmonic spectrum, duration and amplitude

---

Grid Converters for Photovoltaic and Wind Power Systems   Remus Teodorescu, Marco Liserre, and Pedro Rodríguez
© 2011 John Wiley & Sons, Ltd

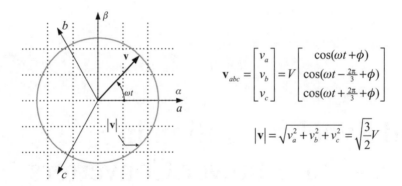

Figure 8.1 Ideal three-phase voltage vector

and give rise to undesirable effects on electrical equipments, such as resonances, increasing power losses or premature ageing [2].

A grid-connected power converter is particularly sensitive to voltage disturbances since its control system might lose controllability on the power signals under such distorted operating conditions, which could trip any of its protection systems or might even destroy the power converter. Moreover, a power converter can interact with the grid at the point of common coupling in order to attenuate the voltage disturbances and reduce their undesirable effects. For these reasons, the voltage vector disturbances should be properly detected by the synchronization system, and the control system of the power converter should react to both ride-through such operating conditions and provide some support to the grid.

In the case where the voltage vector at the connection point of the power converter is distorted by high-order harmonics with reasonable amplitude, the detection system bandwidth can be reduced in order to cancel out the effect of these harmonics on the output. Despite this bandwidth reduction, the detection system should still operate satisfactorily in the presence of slow voltage fluctuations. In the case where the voltage vector is unbalanced, the bandwidth reduction is not an acceptable solution since the overall dynamic performance of the detection system would become unsatisfactorily deficient. In such a case, the sequence components of the unbalanced voltage vector should be identified by using specific detection techniques and passed as inputs to the control system to react accordingly to such voltage disturbance.

Therefore, grid synchronization of three-phase power converters entails the usage of advanced detection systems, specially designed to both reject high-order harmonics and identify the sequence components of the voltage vector in a fast and precise way. Particularly, the real-time detection of the sequence components of the voltage vector in three-phase networks is an essential issue in the control of distributed generation and storage systems, flexible AC transmission systems (FACTS), power line conditioners and uninterruptible power supplies (UPS) [3, 4]. In such systems, the magnitude and phase angle of the positive- and negative-sequence voltage components are generally used for the synchronization of the converter output variables, calculation of the power flow or transformation of stationary variables into rotating reference frames [5–7].

This chapter presents some three-phase synchronization systems suitable to be applied under unbalanced and distorted grid operating conditions.

8.2 The Three-Phase Voltage Vector under Grid Faults

Three-phase voltages can become unbalanced and distorted because of the effect of nonlinear loads and transient grid faults. Ideally, power converters used in distributed generation should be properly synchronized with the grid under such adverse operating conditions to stay actively connected, supporting the grid services (voltage/frequency) and keeping up generation. The faulty three-phase voltages can be generically understood as a summation of unbalanced harmonic components. Therefore, in a general way, the three-phase voltage vector can be written as

$$\boldsymbol{v}_{abc} = \begin{bmatrix} v_a \\ v_b \\ v_c \end{bmatrix} = \sum_{n=1}^{\infty} \left(\boldsymbol{v}_{abc}^{+n} + \boldsymbol{v}_{abc}^{-n} + \boldsymbol{v}_{abc}^{0n} \right), \qquad (8.1)$$

where

$$\boldsymbol{v}_{abc}^{+n} = V^{+n} \begin{bmatrix} \cos(n\omega t + \phi^{+n}) \\ \cos(n\omega t - \frac{2\pi}{3} + \phi^{+n}) \\ \cos(n\omega t + \frac{2\pi}{3} + \phi^{+n}) \end{bmatrix} \qquad (8.2.a)$$

$$\boldsymbol{v}_{abc}^{-n} = V^{-n} \begin{bmatrix} \cos(n\omega t + \phi^{-n}) \\ \cos(n\omega t + \frac{2\pi}{3} + \phi^{-n}) \\ \cos(n\omega t - \frac{2\pi}{3} + \phi^{-n}) \end{bmatrix} \qquad (8.2.b)$$

$$\boldsymbol{v}_{abc}^{0n} = V^{0n} \begin{bmatrix} \cos(n\omega t + \phi^{0n}) \\ \cos(n\omega t + \phi^{0n}) \\ \cos(n\omega t + \phi^{0n}) \end{bmatrix} \qquad (8.2.c)$$

In (8.1) and (8.2), superscripts $+n$, $-n$ and $0n$ respectively represent the positive-, negative- and zero-sequence components of the nth harmonic of the voltage vector v.

Distributed generators are usually linked to three-phase networks by using a three-wire connection and hence they do not inject zero-sequence current into the grid. Thus, the zero-sequence component of the voltage vector will be intentionally ignored in the equations describing the synchronization systems presented in this chapter since it is not necessary to synchronize any current with such a zero-sequence voltage component. Nonetheless, if necessary, the zero-sequence component could be easily extracted from the voltage vector by applying the *Clarke* transformation, defined by (A.14) in Appendix A, and its characteristic module and phase angle can be determined by using any of the single-phase synchronization systems presented in Chapter 4. Moreover, three-phase power converters used in WT and PV systems generally inject positive-sequence currents at the fundamental frequency into the grid and only intentionally inject negative-sequence and harmonics currents in unusual cases, i.e. either avoiding power oscillations to protect the power converter or injecting unbalanced reactive currents to compensate the unbalanced grid voltage at the point of common coupling. Therefore, the correct detection of the positive-sequence component at the fundamental frequency of the three-phase grid voltage can be considered as the main task of the synchronization system of a grid-connected three-phase power converter.

In a general form, a positive-sequence voltage vector at the fundamental frequency interacting with either a positive- or negative-sequence nth-order component can be expressed by

$$\boldsymbol{v}_{abc} = \boldsymbol{v}_{abc}^{+1} + \boldsymbol{v}_{abc}^{n} = V^{+1} \begin{bmatrix} \cos(\omega t) \\ \cos(\omega t - \frac{2\pi}{3}) \\ \cos(\omega t + \frac{2\pi}{3}) \end{bmatrix} + V^{n} \begin{bmatrix} \cos(n\omega t) \\ \cos(n\omega t - \frac{2\pi}{3}) \\ \cos(n\omega t + \frac{2\pi}{3}) \end{bmatrix} \quad (8.3)$$

where $n > 0$ means a positive-sequence component and $n < 0$ a negative-sequence one. The voltage vector of (8.3) can be expressed on the Cartesian $\alpha\beta$ stationary reference frame by using a reduced version of the *Clarke* transformation, resulting in

$$\boldsymbol{v}_{\alpha\beta} = \begin{bmatrix} v_\alpha \\ v_\beta \end{bmatrix} = [T_{\alpha\beta}] \cdot \boldsymbol{v}_{abc} = \sqrt{\frac{3}{2}} V^{+1} \begin{bmatrix} \cos(\omega t) \\ \sin(\omega t) \end{bmatrix} + \sqrt{\frac{3}{2}} V^{n} \begin{bmatrix} \cos(n\omega t) \\ \sin(n\omega t) \end{bmatrix} \quad (8.4)$$

where

$$[T_{\alpha\beta}] = \sqrt{\frac{2}{3}} \begin{bmatrix} 1 & -\frac{1}{2} & -\frac{1}{2} \\ 0 & \frac{\sqrt{3}}{2} & -\frac{\sqrt{3}}{2} \end{bmatrix} \quad (8.5)$$

The voltage vector of (8.3) can also be expressed on a Cartesian dq rotating reference frame by using the *Park* transformation as

$$\boldsymbol{v}_{dq} = \begin{bmatrix} v_d \\ v_q \end{bmatrix} = [T_{dq}] \cdot \boldsymbol{v}_{\alpha\beta} = \sqrt{\frac{3}{2}} V^{+1} \begin{bmatrix} \cos(\omega t - \theta') \\ \sin(\omega t - \theta') \end{bmatrix} + \sqrt{\frac{3}{2}} V^{n} \begin{bmatrix} \cos(n\omega t - \theta') \\ \sin(n\omega t - \theta') \end{bmatrix} \quad (8.6)$$

where

$$[T_{dq}] = \begin{bmatrix} \cos(\theta') & \sin(\theta') \\ -\sin(\theta') & \cos(\theta') \end{bmatrix} \quad (8.7)$$

with θ' the angular position of the dq rotating reference frame.

As an illustrative example, Figure 8.2 shows the evolution of the three-phase voltage vector of (8.3) in two different cases. Figure 8.2(a) shows the interaction of the fundamental frequency positive-sequence component with a fundamental frequency negative-sequence component ($n = -1$), whereas Figure 8.2(b) shows the interaction of the fundamental frequency positive-sequence component with a fifth harmonic negative-sequence component ($n = -5$).

Assuming that the dq reference frame rotates synchronously with the positive-sequence voltage vector, with the d axis in the same direction as the positive-sequence voltage vector v^{+1}, i.e. with $\theta' = \omega t$, the expression of (8.6) gives rise to

$$\boldsymbol{v}_{dq} = \sqrt{\frac{3}{2}} V^{+1} \begin{bmatrix} 1 \\ 0 \end{bmatrix} + \sqrt{\frac{3}{2}} V^{n} \begin{bmatrix} \cos((n-1)\omega t) \\ \sin((n-1)\omega t) \end{bmatrix} \quad (8.8)$$

Grid Synchronization in Three-Phase Power Converters

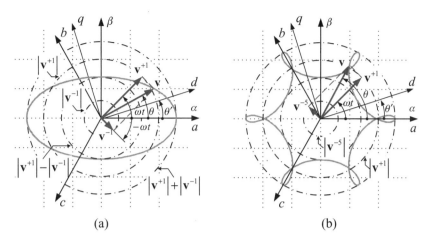

Figure 8.2 Locus of (a) an unbalanced and (b) a distorted voltage vector

From (8.8), the module and angular position of the three-phase voltage vector, $|v|$ and θ respectively are given by

$$|v| = \sqrt{v_\alpha^2 + v_\beta^2} = \sqrt{\frac{3}{2}\left[\left(V^{+1}\right)^2 + \left(V^n\right)^2 + 2V^{+1}V^n \cos\left((n-1)\omega t\right)\right]} \qquad (8.9)$$

$$\theta = \tan^{-1}\frac{v_\beta}{v_\alpha} = \omega t + \tan^{-1}\left(\frac{v_q}{v_d}\right) = \omega t + \tan^{-1}\left[\frac{V^n \sin\left((n-1)\omega t\right)}{V^{+1} + V^n \cos\left((n-1)\omega t\right)}\right] \qquad (8.10)$$

Equations (8.9) and (8.10) are evidence that the compound voltage vector v has neither constant module nor rotational frequency. Moreover, these equations show that both the amplitude and the angular position of the positive-sequence component cannot be extracted by just filtering the detected module and phase angle of the compound voltage vector v.

As an example, Figure 8.3(a) shows the phase voltage waveforms of a three-phase system affected by a phase-to-phase grid fault. Just the positive- and negative-sequence components are considered in this example, being $V^{+1} = 0.75$ p.u. and $V^{-1} = 0.25$ p.u. (it is assumed here that the pre-fault voltage amplitude is equal to 1 p.u.). Therefore, the module and angular position of the voltage vector during the grid fault are given by

$$|v| = \sqrt{0.9375 + 0.5625 \cos\left(2\omega t\right)} \qquad (8.11)$$

$$\theta = \omega t + \tan^{-1}\left[\frac{0.25 \sin\left(-2\omega t\right)}{0.75 + 0.25 \cos\left(2\omega t\right)}\right] \qquad (8.12)$$

Figure 8.3(b) shows the locus described by the voltage vector normalized with respect to the pre-fault vector module. In this figure, the instantaneous value of the angular frequency of the voltage vector, $d\theta/dt$, has been represented on a vertical axis, orthogonal to the $\alpha\beta$ plane. It can

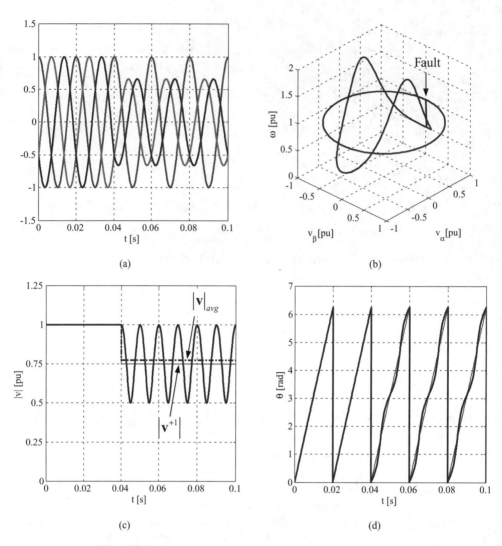

Figure 8.3 Space vector evolution in a phase-to-phase grid fault

be appreciated in this figure how the voltage vector locus changes from a circle to a completely different shape once the fault happens. However, the projection of this three-dimensional shape on the $\alpha\beta$ plane matches the ellipse plotted in Figure 8.2(a). This figure highlights that the instantaneous angular frequency of the voltage vector is not a constant during the grid fault, which should be taken into account when a three-phase synchronization system is designed.

Figure 8.3(c) shows the evolution of the voltage vector module $|v|$, normalized with respect to the pre-fault vector module. The average value of the voltage vector module during the grid fault, $|v|_{avg}$, is plotted by a dashed line, whereas the module of the positive-sequence voltage vector, $|v^{+1}|$, is plotted by a thin continuous line. Figure 8.3(d) shows the evolution of the voltage vector phase angle θ. Both figures show that the information about the positive-sequence voltage vector v^{+1} cannot be properly obtained by just applying conventional filtering techniques to the module and phase-angle signals of the compound vector v.

8.2.1 Unbalanced Grid Voltages during a Grid Fault

Before presenting some solutions for detecting the sequence components of an unbalanced voltage vector, an example of how such unbalanced voltages are generated during a grid fault will be presented here. The analysis procedure presented in the following can be applied to any kind of grid fault to obtain its characteristic parameters (three-phase, three-phase to ground, phase to ground, etc.). Nevertheless, a more detailed explanation about this topic can be found in the literature [8–10].

The phase-to-phase grid fault that originated the unbalanced waveforms of Figure 8.3(a) can be represented by the circuit of Figure 8.4, where it is assumed that the line impedance is equal for all three phases and the voltage supplied by the three-phase generator is sinusoidal, balanced with positive-sequence components, at the fundamental frequency.

The phase voltages and current in the faulty lines of Figure 8.4 verify that

$$v_{b'} = v_{c'}; \quad i_{b'} = -i_{c'}; \quad i_{a'} = 0 \quad (8.13)$$

From (8.13) and using phasors, the positive-, negative- and zero-sequence voltage components at the fault point, $\vec{V}_{a'}^+$, $\vec{V}_{a'}^-$, $\vec{V}_{a'}^0$, can be calculated by equation (A.1) of Appendix A, resulting in

$$\mathbf{V}_{+-0(a')} = \begin{bmatrix} \vec{V}_{a'}^+ \\ \vec{V}_{a'}^- \\ \vec{V}_{a'}^0 \end{bmatrix} = \frac{1}{3} \begin{bmatrix} 1 & \alpha & \alpha^2 \\ 1 & \alpha^2 & \alpha \\ 1 & 1 & 1 \end{bmatrix} \begin{bmatrix} \vec{V}_{a'} \\ \vec{V}_{b'} \\ \vec{V}_{c'} \end{bmatrix} = \frac{1}{3} \begin{bmatrix} \vec{V}_{a'} - \vec{V}_{b'} \\ \vec{V}_{a'} - \vec{V}_{b'} \\ \vec{V}_{a'} + 2\vec{V}_{b'} \end{bmatrix} \quad (8.14)$$

where $\alpha = e^{j2\pi/3} = 1\angle 120°$ is the *Fortescue* operator [11]. Voltage phasors of (8.14) indicate that the positive- and negative-sequence voltage components at the fault point are equal, i.e. $\vec{V}_{a'}^+ = \vec{V}_{a'}^-$.

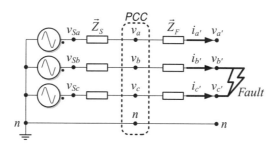

Figure 8.4 Three-phase circuit of a phase-to-phase grid fault

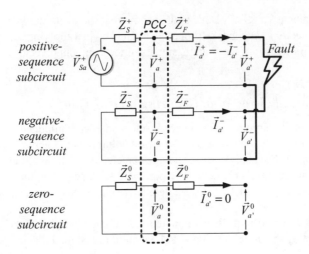

Figure 8.5 Sequence components based on the equivalent circuit of a phase-to-phase grid fault

Likewise, from (8.13), the positive-, negative- and zero-sequence components of the line currents during the grid fault are given by

$$\mathbf{I}_{+-0(a')} = \begin{bmatrix} \vec{I}_{a'}^+ \\ \vec{I}_{a'}^- \\ \vec{I}_{a'}^0 \end{bmatrix} = \frac{1}{3} \begin{bmatrix} 1 & \alpha & \alpha^2 \\ 1 & \alpha^2 & \alpha \\ 1 & 1 & 1 \end{bmatrix} \begin{bmatrix} \vec{I}_{a'} \\ \vec{I}_{b'} \\ \vec{I}_{c'} \end{bmatrix} = \frac{1}{\sqrt{3}} \begin{bmatrix} j\vec{I}_{b'} \\ -j\vec{I}_{b'} \\ 0 \end{bmatrix} \quad (8.15)$$

Current phasors of (8.15) indicate that the zero-sequence current component is equal to zero and, consequently, the addition of the positive- and negative-sequence current components are equal to zero as well, namely $\vec{I}_{a'}^0 = 0$ and $\vec{I}_{a'}^+ + \vec{I}_{a'}^- = 0$.

Once the relationships between the sequence components of voltages and currents during the grid fault are defined, the actual three-phase circuit of Figure 8.4 can be transformed into the equivalent circuits of Figure 8.5 based on the sequence components. In this figure, $\vec{V}_{Sa}^+$ represents the pre-fault voltage vector of the phase a, i.e. one of the three phase voltages of the positive-sequence balanced pre-fault voltage vector.

Assuming that the positive- and negative-sequence line impedances at the source side are equal, i.e. $\vec{Z}_S^+ = \vec{Z}_S^- = \vec{Z}_S$, which is true in most cases, the circuit of Figure 8.5 gives rise to the following sequence voltages at the point of common coupling (PCC):

$$\vec{V}_a^+ = \frac{\vec{Z}_S + \left(\vec{Z}_F^+ + \vec{Z}_F^-\right)}{2\vec{Z}_S + \left(\vec{Z}_F^+ + \vec{Z}_F^-\right)} \vec{V}_{Sa}^+ \quad (8.16a)$$

$$\vec{V}_a^- = \frac{\vec{Z}_S}{2\vec{Z}_S + \left(\vec{Z}_F^+ + \vec{Z}_F^-\right)} \vec{V}_{Sa}^+ \quad (8.16b)$$

$$\vec{V}_a^0 = 0 \quad (8.16c)$$

Grid Synchronization in Three-Phase Power Converters

The severity of the grid fault seen from the PCC can be assessed by the dip parameter $\vec{D}$, which defines the relationship between the line impedances at the fault side and the source side, i.e.

$$\vec{D} = D\angle\rho_D = \frac{\left(\vec{Z}_F^+ + \vec{Z}_F^-\right)}{2\vec{Z}_S + \left(\vec{Z}_F^+ + \vec{Z}_F^-\right)} \tag{8.17}$$

It is worth highlighting here that the magnitude of the faulty voltage depends on the distance from the PCC to the fault point, namely it mainly depends on the module of $\vec{D}$. The difference in the phase angle between the pre-fault and the faulty voltage depends on the phase angle of $\vec{D}$. If the X/R ratio of the impedances at both sides of the PCC remains constant, i.e. if the phase angle of $\vec{Z}_S$ is equal to that of $\vec{Z}_F^+ + \vec{Z}_F^-$, there is no phase-angle jump between the pre-fault and the faulty voltage.

Substituting (8.17) in (8.16), the sequence voltages at the PCC can be written as

$$\mathbf{V}_{+-0(pcc)} = \begin{bmatrix} \vec{V}_a^+ \\ \vec{V}_a^- \\ \vec{V}_a^0 \end{bmatrix} = \frac{1}{2} \vec{V}_{Sa}^+ \begin{bmatrix} 1 + \vec{D} \\ 1 - \vec{D} \\ 0 \end{bmatrix} \tag{8.18}$$

From (8.18), the phase voltages at the PCC can be calculated by using equation (A.4) of Appendix A, resulting in

$$\mathbf{V}_{abc(pcc)} = \begin{bmatrix} \vec{V}_a \\ \vec{V}_b \\ \vec{V}_c \end{bmatrix} = \begin{bmatrix} 1 & 1 & 1 \\ \alpha^2 & \alpha & 1 \\ \alpha & \alpha^2 & 1 \end{bmatrix} \begin{bmatrix} \vec{V}_a^+ \\ \vec{V}_a^- \\ \vec{V}_a^0 \end{bmatrix} = \vec{V}_{Sa}^+ \begin{bmatrix} 1 \\ -\frac{1}{2} - \frac{\sqrt{3}}{2}\vec{D} \\ -\frac{1}{2} + \frac{\sqrt{3}}{2}\vec{D} \end{bmatrix} \tag{8.19}$$

In this example, the voltage phasors of (8.19) describe the unbalanced voltage waveforms of Figure 8.3(a), which are related to a phase-to-phase grid fault.

8.2.2 Transient Grid Faults, the Voltage Sags (Dips)

A voltage sag, also called a voltage dip, is a sudden reduction of the grid voltage at the PCC, generally between 10 and 90 % of the rated value, during a period lasting from half a cycle to a few seconds. Voltage sags usually happen as a consequence of short-circuits, faults to ground, transformers energizing and connection of large induction motors. Depending on both the type of grid fault and the transformer connections along the power lines, it is possible to distinguish between different types of voltage sags.

The definition of guidelines to classify voltage sags is a matter that remains under discussion still today [12]. The product $\vec{D}\vec{V}_{Sa}^+$ is known as the 'characteristic voltage' of the voltage sag and represents either the phase voltage in phase-to-ground faults or the line-to-line voltage in phase-to-phase faults. Likewise, the phase angle of $\vec{D}$ is known as the 'characteristic phase angle jump' of the voltage sag. Figure 8.6 shows four types of voltage sag resulting from

Sag type **A**. Three-phase fault and three-phase-to-ground fault

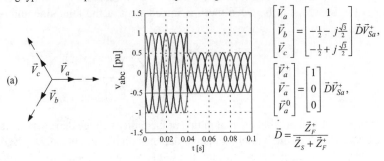

$$\begin{bmatrix} \vec{V}_a \\ \vec{V}_b \\ \vec{V}_c \end{bmatrix} = \begin{bmatrix} 1 \\ -\frac{1}{2}-j\frac{\sqrt{3}}{2} \\ -\frac{1}{2}+j\frac{\sqrt{3}}{2} \end{bmatrix} \vec{D}\vec{V}_{Sa}^+,$$

$$\begin{bmatrix} \vec{V}_a^+ \\ \vec{V}_a^- \\ \vec{V}_a^0 \end{bmatrix} = \begin{bmatrix} 1 \\ 0 \\ 0 \end{bmatrix} \vec{D}\vec{V}_{Sa}^+,$$

$$\vec{D} = \frac{\vec{Z}_F^+}{\vec{Z}_S + \vec{Z}_F^+}$$

Sag type **B**. Single-phase-to-ground fault

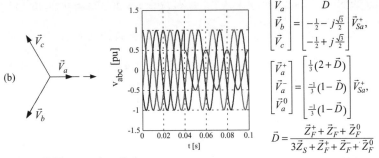

$$\begin{bmatrix} \vec{V}_a \\ \vec{V}_b \\ \vec{V}_c \end{bmatrix} = \begin{bmatrix} \vec{D} \\ -\frac{1}{2}-j\frac{\sqrt{3}}{2} \\ -\frac{1}{2}+j\frac{\sqrt{3}}{2} \end{bmatrix} \vec{V}_{Sa}^+,$$

$$\begin{bmatrix} \vec{V}_a^+ \\ \vec{V}_a^- \\ \vec{V}_a^0 \end{bmatrix} = \begin{bmatrix} \frac{1}{3}(2+\vec{D}) \\ \frac{-1}{3}(1-\vec{D}) \\ \frac{-1}{3}(1-\vec{D}) \end{bmatrix} \vec{V}_{Sa}^+,$$

$$\vec{D} = \frac{\vec{Z}_F^+ + \vec{Z}_F^- + \vec{Z}_F^0}{3\vec{Z}_S + \vec{Z}_F^+ + \vec{Z}_F^- + \vec{Z}_F^0}$$

Sag type **C**. Phase-to-phase fault

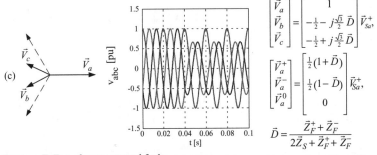

$$\begin{bmatrix} \vec{V}_a \\ \vec{V}_b \\ \vec{V}_c \end{bmatrix} = \begin{bmatrix} 1 \\ -\frac{1}{2}-j\frac{\sqrt{3}}{2}\vec{D} \\ -\frac{1}{2}+j\frac{\sqrt{3}}{2}\vec{D} \end{bmatrix} \vec{V}_{Sa}^+,$$

$$\begin{bmatrix} \vec{V}_a^+ \\ \vec{V}_a^- \\ \vec{V}_a^0 \end{bmatrix} = \begin{bmatrix} \frac{1}{2}(1+\vec{D}) \\ \frac{1}{2}(1-\vec{D}) \\ 0 \end{bmatrix} \vec{V}_{Sa}^+,$$

$$\vec{D} = \frac{\vec{Z}_F^+ + \vec{Z}_F^-}{2\vec{Z}_S + \vec{Z}_F^+ + \vec{Z}_F^-}$$

Sag type **E**. Two-phase to ground fault

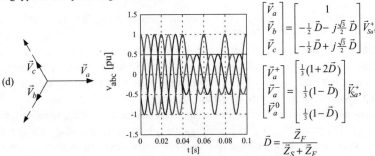

$$\begin{bmatrix} \vec{V}_a \\ \vec{V}_b \\ \vec{V}_c \end{bmatrix} = \begin{bmatrix} 1 \\ -\frac{1}{2}\vec{D}-j\frac{\sqrt{3}}{2}\vec{D} \\ -\frac{1}{2}\vec{D}+j\frac{\sqrt{3}}{2}\vec{D} \end{bmatrix} \vec{V}_{Sa}^+,$$

$$\begin{bmatrix} \vec{V}_a^+ \\ \vec{V}_a^- \\ \vec{V}_a^0 \end{bmatrix} = \begin{bmatrix} \frac{1}{3}(1+2\vec{D}) \\ \frac{1}{3}(1-\vec{D}) \\ \frac{1}{3}(1-\vec{D}) \end{bmatrix} \vec{V}_{Sa}^+,$$

$$\vec{D} = \frac{\vec{Z}_F}{\vec{Z}_S + \vec{Z}_F}$$

Figure 8.6 Voltage sags due to grid faults in three-phase systems with $\vec{D} = 0.5\angle 0°$

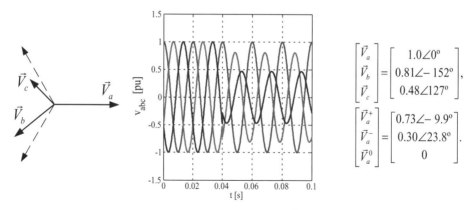

Figure 8.7 Voltage sag type C with $\vec{D} = 0.5\angle -30°$

different grid faults in which the characteristic phase angle jump was assumed equal to zero. In this figure, the voltage sags have been typified according to the nomenclature proposed in reference [8]. The characteristic parameters of the voltage sags of Figure 8.6 can be calculated by following a similar procedure to the one described in the previous example for a phase-to-phase fault. A more detailed explanation about this mathematical procedure can be found in reference [9].

There are many practical cases in which the X/R ratio of the impedances at both sides of the PCC of Figure 8.4 does not keep constant during a fault, which implies a phase-angle jump is different to zero. This is particularly true when the fault affects to power lines consisting of sections with different impedances, or when big induction motors are connected to the grid. In such a case, the voltage phasors during the grid fault lose the symmetry, shown by the sags of Figure 8.6. As an example, Figure 8.7 shows a voltage sag type C with $\vec{D} = 0.5\angle -30°$.

8.2.3 Propagation of Voltage Sags

The type of sag experienced by a system connected to a given AC bus not only depends on the number of phases affected by the grid fault but it is also influenced by the transformers located in between the AC bus and the fault point. The amplitude and phase angle of the unbalanced voltage resulting from a given grid fault will be modified when propagated through regular three-phase transformers used in power systems, which will give rise to new types of voltage sags different to the ones shown in Figure 8.6. Moreover, the zero-sequence component, generally present in phase-to-ground faults, will be removed. As an example, Figure 8.8 shows how the line-to-line voltages of a sag type C applied to the primary of a Dy transformer are propagated to the secondary winding with different voltage amplitudes and phase angles, which results in a new type of voltage sag (type D).

To identify the different types of voltage sag existing in a generic distribution system, voltage on the buses of a power line like the one shown in Figure 8.9 are analysed in this section. In this analysis, three possible points of common coupling (PCC$_1$, PCC$_2$ and PCC$_3$) are considered, which result from the cascade connection of two Dy transformers. An analysis

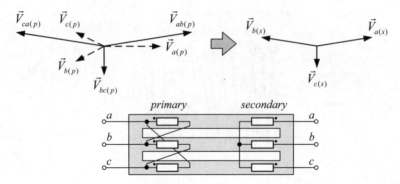

Figure 8.8 Propagation of a voltage sag type C ($\vec{D} = 0.5\angle - 0°$) through a Dy transformer

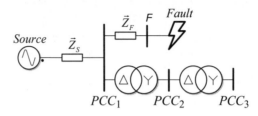

Figure 8.9 Voltage sag propagation along three points of common coupling (PCC$_1$, PCC$_2$ and PCC$_3$) in a power line with two cascade-connected Dy transformers

of the voltages measured on PCC$_2$ and PCC$_3$ allows three new types of voltage sags (types D, F and G) to be identified from the original voltage sags (types A, B, C and E) existing on bus PCC$_1$ as a consequence of different types of faults occurred at bus F.

The relationship between the different types of voltage sag is summarized in Table 8.1. The sequence components and phase-voltages of the voltage sag types D, F and G are shown in Figure 8.10.

Table 8.1 Propagation of voltage sags through Dy transformers

	Point of common coupling		
Fault type	PCC$_1$	PCC$_2$	PCC$_3$
Three-phase/three-phase to ground	A	A	A
Single-phase to ground	B	C	D
Two-phase	C	D	C
Two-phase to ground	E	F	G

Sag type **D**. Propagation of a sag type **C**

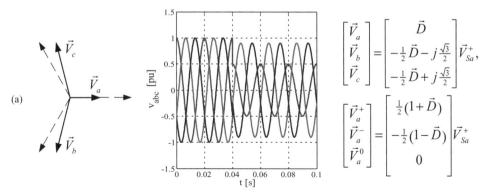

Sag type **F**. Propagation of a sag type **E**

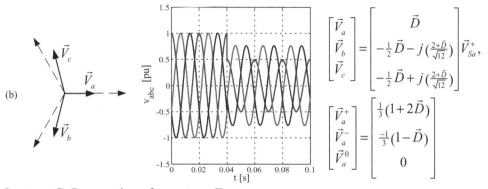

Sag type **G**. Propagation of a sag type **F**

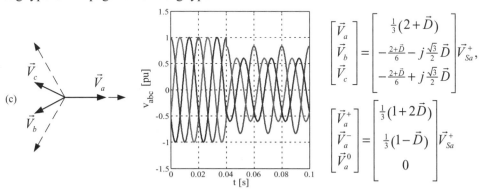

Figure 8.10 Voltage sags due to the propagation of grid faults in three-phase systems with $\vec{D} = 0.5 \angle 0°$

8.3 The Synchronous Reference Frame PLL under Unbalanced and Distorted Grid Conditions

The most extended technique used for frequency-insensitive grid synchronization in three-phase systems is the PLL based on the synchronous reference frame (SRF-PLL) [13]. The conventional SRF-PLL translates the three-phase voltage vector from the *abc* natural reference frame to the *dq* rotating reference frame by using Park's transformation $[T_\theta]$, as shown in Figure 8.11. The angular position of this *dq* reference frame is controlled by a feedback loop that regulates the *q* component to zero. As shown in (8.20), the $[T_\theta]$ transformation in this PLL has been rescaled by using a 2/3 factor in order to detect the amplitude of the sinusoidal input signal instead of the module of the input voltage vector. Therefore, in the steady state, the *d* component depicts the amplitude of the sinusoidal positive-sequence input voltage (V^{+1}) and its phase angle is determined by the output of the feedback loop (θ').

$$\begin{bmatrix} v_d \\ v_q \end{bmatrix} = [T_\theta] \begin{bmatrix} v_a \\ v_b \\ v_c \end{bmatrix}, [T_\theta] = \frac{2}{3} \begin{bmatrix} \cos(\theta') & \cos(\theta' - \frac{2\pi}{3}) & \cos(\theta' + \frac{2\pi}{3}) \\ -\sin(\theta') & -\sin(\theta' - \frac{2\pi}{3}) & -\sin(\theta' + \frac{2\pi}{3}) \end{bmatrix} \quad (8.20)$$

$$[T_\theta] = [T_{dq}] \cdot [T_{\alpha\beta}], [T_{dq}] = \begin{bmatrix} \cos(\theta') & \sin(\theta') \\ -\sin(\theta') & \cos(\theta') \end{bmatrix}, [T_{\alpha\beta}] = \frac{2}{3} \begin{bmatrix} 1 & -\frac{1}{2} & -\frac{1}{2} \\ 0 & \frac{\sqrt{3}}{2} & -\frac{\sqrt{3}}{2} \end{bmatrix} \quad (8.21)$$

Under ideal grid conditions, i.e. when the grid voltage is not affected by either harmonic distortion or unbalances, setting a high bandwidth for the SRF-PLL feedback loop yields a fast and precise detection of the phase angle and amplitude of the grid voltage. Column (a) of Figure 8.12 shows some waveforms illustrating the response of an SRF-PLL, tuned with a high gain, i.e. a high bandwidth, in the presence of a voltage sag type A. As shown in this figure, the SRF-PLL almost instantaneously detects the amplitude and phase angle of the balanced input voltage vector by making $v_q = 0$. Column (b) of Figure 8.12 shows the response of the SRF-PLL when the voltage sag type A is polluted by a fifth-order harmonic ($V^{-5} = 0.1V^{+1}$). In this case, the SRF-PLL makes a small error in tracking the instantaneous position of the input voltage vector and consequently $v_q \neq 0$. This is in fact an advantage, since the PLL automatically will reduce the effect of the fifth-order harmonic on the angular position of the *dq* reference frame. Hence, the average value of the voltage on the *d* axis will match the amplitude of the positive-sequence fundamental voltage, i.e. $\bar{v}_d = V^{+1}$. Therefore, it can be concluded that a slight reduction in the PLL bandwidth improves its response, almost completely rejecting the effect of high-order harmonics on the PLL output signals. Column (c) of Figure 8.12 shows the response of an SRF-PLL when the grid voltage experienced a sag type C with $\vec{V}_a^{+1} = 0.75\angle 0°$ and $\vec{V}_a^{-1} = 0.25\angle 0°$. In this SRF-PLL, the control loop bandwidth was high enough to make $v_q \approx 0$, which means that the SRF-PLL was able to instantaneously track the evolution of the unbalanced voltage vector applied to its input. Therefore, the detected phase angle, shown in the second plot of Figure 8.12(c), matches the one calculated by (8.12) and presents oscillations at twice the input frequency. On the other hand, the voltage v_d, shown in the third plot in Figure 8.12(c), matches the value calculated by the expression (8.11) – multiplied by $\sqrt{2/3}$ since the *Park* transformation of (8.20) was rescaled. Consequently, as commented in Section 8.2, the amplitude of

Grid Synchronization in Three-Phase Power Converters

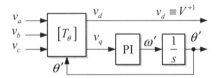

Figure 8.11 Basic block diagram of the SRF-PLL

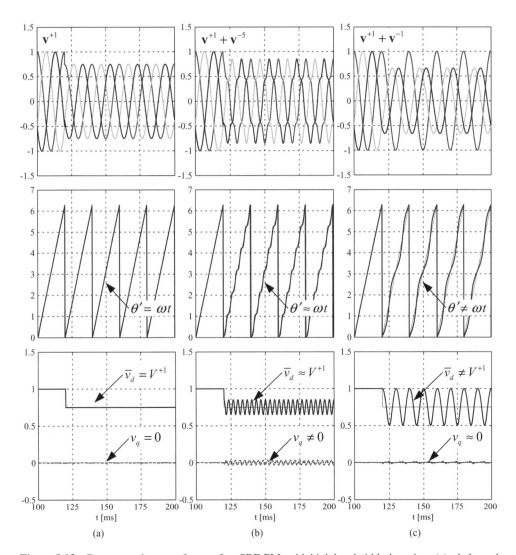

Figure 8.12 Representative waveforms of an SRF-PLL with high bandwidth detecting: (a) a balanced voltage sag, (b) a balanced and distorted voltage sag and (c) an unbalanced voltage sag. From top to bottom, each column shows the three-phase input voltage (p.u.), the detected phase angle (rad) and the detected voltages on the dq axis (p.u.).

the positive-sequence component cannot be properly evaluated by just using conventional filtering techniques to extract the average value of v_d.

As previously mentioned, a reduction of the PLL bandwidth can help to attenuate the effect of the distorting components on the SRF-PLL output signals. Figure 8.13 shows some representative plots from an SRF-PLL, tuned with a low bandwidth, when the grid voltage is affected by the same amount of fifth harmonic as in the case of Figure 8.12(b). It can be appreciated in Figure 8.13(c) how the PLL is not able to instantaneously track the angular position of the fifth-order component and thus it gives rise to an oscillating error signal on both axes of the dq reference frame. Therefore, a simple low-pass filter can be used to obtain the average value of the voltage on the d axis, $\bar{v}_d$. Figure 8.13(c) shows the result of applying a second-order low-pass filter with a cut-off frequency of 20 Hz to extract $\bar{v}_d$. It is possible to appreciate in this figure how $\bar{v}_d$ almost perfectly matches the positive-sequence voltage amplitude V^{+1}. Figure 8.13(d) and (e) shows the reconstruction of the detected positive-sequence voltage and its spectrum respectively. These figures confirm that reduction of the PLL bandwidth is an effective measure to obtain high-quality signals at the SRF-PLL output when synchronizing with three-phase voltages polluted by high-order harmonics [14].

However, as evidenced in the following, limitation of the PLL bandwidth is not the most effective solution to extract the positive-sequence component from the unbalanced three-phase voltages resulting from an asymmetrical grid fault.

After applying the rescaled $[T_{\alpha\beta}]$ transformation of (8.21), the unbalanced grid voltage can be expressed on the $\alpha\beta$ reference frame as

$$\boldsymbol{v}_{\alpha\beta} = \begin{bmatrix} v_\alpha \\ v_\beta \end{bmatrix} = V^{+1} \begin{bmatrix} \cos(\omega t) \\ \sin(\omega t) \end{bmatrix} + V^{-1} \begin{bmatrix} \cos(-\omega t) \\ \sin(-\omega t) \end{bmatrix} \tag{8.22}$$

Therefore, if it is assumed that the PLL bandwidth is low enough only to allow tracking the evolution of the positive-sequence component of the input voltage, which means that the dq reference frame rotates at the positive-sequence fundamental frequency, the voltage on the dq axes, resulting from applying the $[T_{dq}]$ transformation of (8.21) on the vector of (8.22), will be given by

$$\boldsymbol{v}_{dq} = V^{+1} \begin{bmatrix} 1 \\ 0 \end{bmatrix} + V^n \begin{bmatrix} \cos(-2\omega t) \\ \sin(-2\omega t) \end{bmatrix} \tag{8.23}$$

In (8.23), it has been additionally assumed that the d axis of the SRF perfectly matches the angular position of the positive-sequence component of the input voltage vector. Therefore, (8.23) indicates that the amplitude of the positive-sequence component might be easily obtained by just using any filtering technique to cancel out the oscillation at 2ω present on the d axis signal.

Figure 8.13 allows the performance of the SRF-PLL to be evaluated when a second-order low-pass filter with a cut-off frequency of 20 Hz is used to extract $\bar{v}_d$. Figure 8.13(a) shows the unbalanced input voltage (sag type C with $\vec{V}_a^{+1} = 0.75\angle 0°$, $\vec{V}_a^{-1} = 0.25\angle 0°$). Figure 8.13(b) shows the phase angle detected by the PLL, confirming that the low bandwidth set to the PLL only allows tracking of the positive-sequence component of the unbalanced input voltage vector. Figure 8.13(c) shows in thin lines the voltages on the dq axes of the SRF and in thick

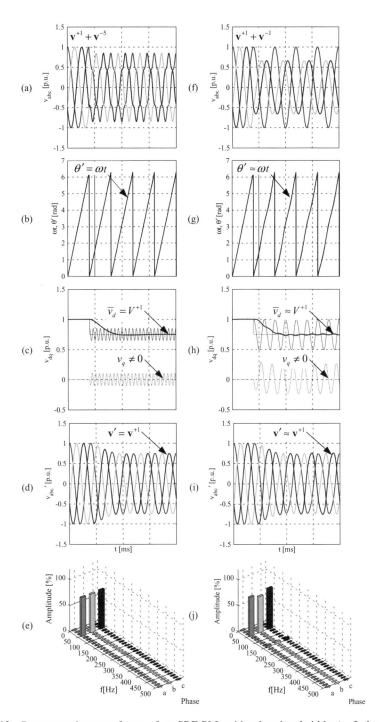

Figure 8.13 Representative waveforms of an SRF-PLL with a low bandwidth: (a, f) three-phase input voltage, (b, g) detected phase angle, (c, h) detected amplitude for the positive-sequence voltage component, (d, i) positive-sequence detected voltage and (e, j) spectrum of the three-phase detected voltage

line the signal $\bar{v}_d$ extracted by the second-order low-pass filter. Using an estimation of the amplitude and phase angle, the positive-sequence voltage waveforms can be reconstructed as shown in Figure 8.13(d). The spectrum of these waveforms is shown in Figure 8.13(e). As appreciated in these last two plots, the detection of the positive-sequence component of the input voltage vector is not accurately detected since the detection of both the amplitude and phase angle is just based on an approximation, i.e. on the attenuation of the oscillation at 2ω generated by the negative-sequence component, and not in the accurate cancellation of such oscillation. Of course, the better the filtering technique applied to cancel out oscillations at 2ω on the dq voltages the better is the synchronization system that will be obtained [15], which is necessary to guarantee that the selected filtering technique presents a frequency adaptive response [16]. The next section of this chapter presents an enhanced synchronization technique based on decoupling the effects of the positive- and negative-sequence components of the input voltage vector.

8.4 The Decoupled Double Synchronous Reference Frame PLL (DDSRF-PLL)

This section presents an improved three-phase synchronous PLL based on using two synchronous reference frames, rotating with positive and negative synchronous speeds, respectively. The usage of this double synchronous reference frame allows decoupling of the effect of the negative-sequence voltage component on the dq signals detected by the synchronous reference frame rotating with positive angular speed, and vice versa, which makes possible accurate grid synchronization even under unbalanced grid faults [17].

8.4.1 The Double Synchronous Reference Frame

Figure 8.14 shows the positive- and negative-sequence components of the unbalanced voltage vector together with a double synchronous reference frame (DSRF) consisting of two rotating reference frames: dq^{+1}, rotating with the positive speed ω' and whose angular position is θ', and dq^{-1}, rotating with the negative speed $-\omega'$ and whose angular position is $-\theta'$.

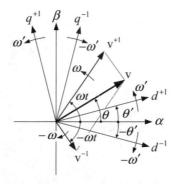

Figure 8.14 Voltage vectors and axes of the DSRF

Grid Synchronization in Three-Phase Power Converters

If it is assumed that the angular position of the positive reference frame dq^{+1} matches the angular position of the positive-sequence voltage vector v^{+1}, i.e. if $\theta' = \omega t$, the unbalanced input voltage vector v can be expressed on the DSRF, yielding

$$v_{dq^{+1}} = \begin{bmatrix} v_{d^{+1}} \\ v_{q^{+1}} \end{bmatrix} = [T_{dq^{+1}}] \cdot v_{\alpha\beta} = V^{+1} \begin{bmatrix} 1 \\ 0 \end{bmatrix} + V^{-1} \begin{bmatrix} \cos(-2\omega t) \\ \sin(-2\omega t) \end{bmatrix} \quad (8.24)$$

$$v_{dq^{-1}} = \begin{bmatrix} v_{d^{-1}} \\ v_{q^{-1}} \end{bmatrix} = [T_{dq^{-1}}] \cdot v_{\alpha\beta} = V^{+1} \begin{bmatrix} \cos(2\omega t) \\ \sin(2\omega t) \end{bmatrix} + V^{-1} \begin{bmatrix} 1 \\ 0 \end{bmatrix} \quad (8.25)$$

where

$$[T_{dq^{+1}}] = [T_{dq^{-1}}]^T = \begin{bmatrix} \cos(\theta') & \sin(\theta') \\ -\sin(\theta') & \cos(\theta') \end{bmatrix} \quad (8.26)$$

Expressions of (8.24) and (8.25) are evidence that the DC values on the dq^{+1} and the dq^{-1} frames correspond to the amplitude of the sinusoidal signals of v^{+1} and v^{-1}, while the oscillations at 2ω correspond to the coupling between axes appearing as a consequence of the voltage vectors rotating in opposite directions. Therefore, instead of using any filtering technique for attenuating oscillations at 2ω, a decoupling network is presented in the following to completely cancel out the effect of such oscillations on the synchronous reference frame voltages of the PLL.

8.4.2 The Decoupling Network

To generalize the explanation of the decoupling network used in the DSRF, one supposes a voltage vector consisting of two generic components rotating with $n\omega$ and $m\omega$ frequencies respectively, where n and m can be either positive or negative. Therefore, this generic voltage vector is given by

$$v_{\alpha\beta} = \begin{bmatrix} v_\alpha \\ v_\beta \end{bmatrix} = v_{\alpha\beta}^n + v_{\alpha\beta}^m = V^n \begin{bmatrix} \cos(n\omega t + \phi^n) \\ \sin(n\omega t + \phi^n) \end{bmatrix} + V^m \begin{bmatrix} \cos(m\omega t + \phi^m) \\ \sin(m\omega t + \phi^m) \end{bmatrix} \quad (8.27)$$

Additionally, two rotating reference frames are considered, dq^n and dq^m, whose angular positions are $n\theta'$ and $m\theta'$ respectively, where θ' is the phase angle detected by the PLL. If a perfect synchronization of the PLL is possible, i.e. if $\theta' = \omega t$, with ω the fundamental grid frequency, the voltage vector in (8.27) can be expressed on the dq^n and dq^m reference frames as follows:

$$\boldsymbol{v}_{dq^n} = \begin{bmatrix} v_{d^n} \\ v_{q^n} \end{bmatrix} = \underbrace{\begin{bmatrix} \bar{v}_{d^n} \\ \bar{v}_{q^n} \end{bmatrix} + \begin{bmatrix} \tilde{v}_{d^n} \\ \tilde{v}_{q^n} \end{bmatrix} = V^n \begin{bmatrix} \cos(\phi^n) \\ \sin(\phi^n) \end{bmatrix}}_{DC\ terms}$$

$$+ \underbrace{V^m \cos(\phi^m) \begin{bmatrix} \cos((n-m)\omega t) \\ -\sin((n-m)\omega t) \end{bmatrix} + V^m \sin(\phi^m) \begin{bmatrix} \sin((n-m)\omega t) \\ \cos((n-m)\omega t) \end{bmatrix}}_{AC\ terms} \quad (8.28)$$

$$\boldsymbol{v}_{dq^m} = \begin{bmatrix} v_{d^m} \\ v_{q^m} \end{bmatrix} = \underbrace{\begin{bmatrix} \bar{v}_{d^m} \\ \bar{v}_{q^m} \end{bmatrix} + \begin{bmatrix} \tilde{v}_{d^m} \\ \tilde{v}_{q^m} \end{bmatrix} = V^m \begin{bmatrix} \cos(\phi^m) \\ \sin(\phi^m) \end{bmatrix}}_{DC\ terms}$$

$$+ \underbrace{V^n \cos(\phi^n) \begin{bmatrix} \cos((n-m)\omega t) \\ \sin((n-m)\omega t) \end{bmatrix} + V^n \sin(\phi^n) \begin{bmatrix} -\sin((n-m)\omega t) \\ \cos((n-m)\omega t) \end{bmatrix}}_{AC\ terms} \quad (8.29)$$

As shown by (8.28) and (8.29), the amplitude of the AC terms in the dp^n axes depends on the DC terms of the signals on the dq^m axes, and vice versa. Therefore, once the coupling terms between both reference frames are identified, a decoupling cell, such as the one shown in Figure 8.15, can be designed to cancel out the oscillations generated by the voltage vector v^m on the dq^n axes signals. To cancel out the oscillations in the dq^m axes signals, the same structure may be used, but with swapping of the m and n indexes in it. In Figure 8.15, the DC terms on the dq^m axes are represented as $\bar{v}_{d^m}$ and $\bar{v}_{q^m}$. As shown in Figure 8.16, a

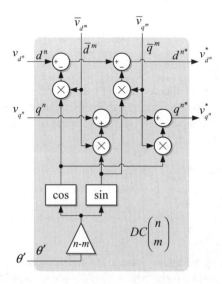

Figure 8.15 Decoupling cell for cancelling the effect of v^m on the dq^n frame signals

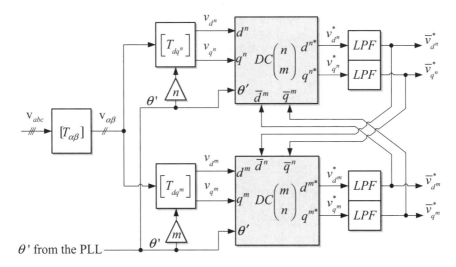

Figure 8.16 Decoupled double synchronous reference frame (DDSRF)

cross-feedback decoupling network is used to estimate the value of these DC terms on the positive and negative reference frames. In this decoupling network, the estimated DC terms are named as $\bar{v}^*_{d^m}$, $\bar{v}^*_{q^m}$, $\bar{v}^*_{d^n}$ and $\bar{v}^*_{q^n}$ and the LPF block is a low-pass filter such as

$$LPF(s) = \frac{\omega_f}{s + \omega_f} \qquad (8.30)$$

The decoupled double synchronous reference frame (DDSRF) of Figure 8.16 allows free-oscillation signals to be obtained on the dq^m and dq^n reference frames. By setting $n = +1$ and $m = -1$, this network decouples information about the positive- and negative-sequence components of either voltage or current in unbalanced three-phase systems, which makes it a useful tool for synchronous controllers during unbalanced grid faults. This network can be also used to decouple other frequency/sequence components simply by setting the proper values for the m and n coefficients.

8.4.3 Analysis of the DDSRF

The state-space model of the DDSRF and some relevant expressions showing its performance have already been presented in reference [17], so this section will present a more intuitive analysis on the complex-frequency domain to improve understanding about the DDSRF performance during unbalanced grid faults. From (8.27), the unbalanced voltage during a grid fault, consisting of positive- and negative-sequence components at the fundamental frequency, can be generically described as

$$\boldsymbol{v}_{\alpha\beta} = \begin{bmatrix} v_\alpha \\ v_\beta \end{bmatrix} = \boldsymbol{v}^{+1}_{\alpha\beta} + \boldsymbol{v}^{-1}_{\alpha\beta} = V^{+1}\begin{bmatrix} \cos(\omega t + \phi^{+1}) \\ \sin(\omega t + \phi^{+1}) \end{bmatrix} + V^{-1}\begin{bmatrix} \cos(-\omega t + \phi^{-1}) \\ \sin(-\omega t + \phi^{-1}) \end{bmatrix} \qquad (8.31)$$

The projection of this voltage vector on the dq^{+1} and dq^{-1} reference frames can be easily obtained simply by setting $n = +1$ and $m = -1$ in (8.28) and (8.29). After rearranging equations, the dq signals on the positive and negative reference frames are given by

$$\mathbf{v}_{dq^{+1}} = \begin{bmatrix} v_{d+1} \\ v_{q+1} \end{bmatrix} = V^{+1} \begin{bmatrix} \cos(\phi^{+1}) \\ \sin(\phi^{+1}) \end{bmatrix} + V^{-1} \begin{bmatrix} \cos(2\omega t) & \sin(2\omega t) \\ -\sin(2\omega t) & \cos(2\omega t) \end{bmatrix} \begin{bmatrix} \cos(\phi^{-1}) \\ \sin(\phi^{-1}) \end{bmatrix} \quad (8.32)$$

$$\mathbf{v}_{dq^{-1}} = \begin{bmatrix} v_{d-1} \\ v_{q-1} \end{bmatrix} = V^{-1} \begin{bmatrix} \cos(\phi^{-1}) \\ \sin(\phi^{-1}) \end{bmatrix} + V^{+1} \begin{bmatrix} \cos(2\omega t) & -\sin(2\omega t) \\ \sin(2\omega t) & \cos(2\omega t) \end{bmatrix} \begin{bmatrix} \cos(\phi^{+1}) \\ \sin(\phi^{+1}) \end{bmatrix} \quad (8.33)$$

These expressions clearly give evidence that the AC terms in the dq^{+1} axes result from the DC terms in the dq^{-1} axes being affected by a rotating transformation matrix at 2ω frequency. A similar conclusion can be obtained for AC signals on the dq^{-1} reference frame. These rotating transformation matrices are given by

$$[T_{dq^{+2}}] = [T_{dq^{-2}}]^T = \begin{bmatrix} \cos(2\omega t) & \sin(2\omega t) \\ -\sin(2\omega t) & \cos(2\omega t) \end{bmatrix} \quad (8.34)$$

Therefore, (8.32) and (8.33) can be rewritten as

$$\mathbf{v}_{dq^{+1}} = \begin{bmatrix} v_{d+1} \\ v_{q+1} \end{bmatrix} = \overline{\mathbf{v}}_{dq^{+1}} + [T_{dq^{+2}}] \overline{\mathbf{v}}_{dq^{-1}} \quad (8.35)$$

$$\mathbf{v}_{dq^{-1}} = \begin{bmatrix} v_{d-1} \\ v_{q-1} \end{bmatrix} = \overline{\mathbf{v}}_{dq^{-1}} + [T_{dq^{-2}}] \overline{\mathbf{v}}_{dq^{+1}} \quad (8.36)$$

where

$$\overline{\mathbf{v}}_{dq^{+1}} = \begin{bmatrix} \overline{v}_{d+1} \\ \overline{v}_{q+1} \end{bmatrix} = V^{+1} \begin{bmatrix} \cos(\phi^{+1}) \\ \sin(\phi^{+1}) \end{bmatrix} \quad \text{and} \quad \overline{\mathbf{v}}_{dq^{-1}} = \begin{bmatrix} \overline{v}_{d-1} \\ \overline{v}_{q-1} \end{bmatrix} = V^{-1} \begin{bmatrix} \cos(\phi^{-1}) \\ \sin(\phi^{-1}) \end{bmatrix}$$

represent the amplitude of sequence components applied to the input of the DDSRF. Thus, (8.35) and (8.36) give evidence that the relationship between the signals on the positive and negative reference frames are given by

$$\mathbf{v}_{dq^{+1}} = [T_{dq^{+2}}] \mathbf{v}_{dq^{-1}} \quad \text{and} \quad \mathbf{v}_{dq^{-1}} = [T_{dq^{-2}}] \mathbf{v}_{dq^{+1}} \quad (8.37)$$

As a result, the estimated values at the output of the DDSRF can be written as:

$$\overline{\mathbf{v}}^*_{dq^{+1}} = \begin{bmatrix} \overline{v}^*_{d+1} \\ \overline{v}^*_{q+1} \end{bmatrix} = [F] \left\{ \mathbf{v}_{dq^{+1}} - [T_{dq^{+2}}] \overline{\mathbf{v}}^*_{dq^{-1}} \right\} \quad (8.38)$$

$$\overline{\mathbf{v}}^*_{dq^{-1}} = \begin{bmatrix} \overline{v}^*_{d-1} \\ \overline{v}^*_{q-1} \end{bmatrix} = [F] \left\{ \mathbf{v}_{dq^{-1}} - [T_{dq^{-2}}] \overline{\mathbf{v}}^*_{dq^{+1}} \right\} \quad (8.39)$$

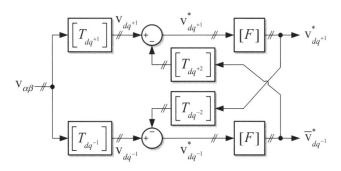

Figure 8.17 Block diagram of the DDSRF with n = +1 and m = −1

where

$$[F] = \begin{bmatrix} LPF(s) & 0 \\ 0 & LPF(s) \end{bmatrix}$$

Therefore, the DDSRF of Figure 8.16 can be represented for the particular case of the positive- and negative-sequence components at the fundamental frequency, as Figure 8.17 shows.

In order not to extend excessively the analysis of the DDSRF, only an estimation of the positive-sequence component will be considered in the following. Therefore, substituting (8.39) in (8.38) we obtain

$$\overline{v}^*_{dq^{+1}} = [F] \left\{ v_{dq^{+1}} - [T_{dq^{+2}}][F] \left(v_{dq^{-1}} - [T_{dq^{-2}}] \overline{v}^*_{dq^{+1}} \right) \right\} \quad (8.40)$$

and using the relationships of (8.37), we arrive at

$$\overline{v}^*_{dq^{+1}} = [F] \left\{ v_{dq^{+1}} - [T_{dq^{+2}}][F] \left([T_{dq^{-2}}] v_{dq^{+1}} - [T_{dq^{-2}}] \overline{v}^*_{dq^{+1}} \right) \right\} \quad (8.41)$$

$$\overline{v}^*_{dq^{+1}} = [F] \left\{ v_{dq^{+1}} - [T_{dq^{+2}}][F][T_{dq^{-2}}] \left(v_{dq^{+1}} - \overline{v}^*_{dq^{+1}} \right) \right\} \quad (8.42)$$

A very amusing academic exercise is to determine the matrix resulting from $[T_{dq^{+2}}][F][T_{dq^{-2}}]$. Some useful directions regarding how these matrices should be operated can be found in references [18] and [19]. As a result of such operations it can be concluded that

$$[F_{-2}] = [T_{dq^{+2}}][F][T_{dq^{-2}}]$$

$$= \frac{1}{2} \begin{bmatrix} (LPF(s+j2\omega) + LPF(s-j2\omega)) & j(-LPF(s+j2\omega) + LPF(s-j2\omega)) \\ j(LPF(s+j2\omega) - LPF(s-j2\omega)) & (LPF(s+j2\omega) + LPF(s-j2\omega)) \end{bmatrix}$$

$$[F_{-2}] = [F_{+2}]^T = \frac{1}{2} \begin{bmatrix} \dfrac{\omega_f(s+\omega_f)}{s^2+2s\omega_f+\omega_f^2+(2\omega)^2} & -\dfrac{\omega_f \omega}{s^2+2s\omega_f+\omega_f^2+(2\omega)^2} \\ \dfrac{\omega_f \omega}{s^2+2s\omega_f+\omega_f^2+(2\omega)^2} & \dfrac{\omega_f(s+\omega_F)}{s^2+2s\omega_f+\omega_f^2+(2\omega)^2} \end{bmatrix} \quad (8.43)$$

Therefore, (8.42) can be simplified as

$$\overline{v}^*_{dq+1} = [F]\left\{v_{dq+1} - [F_{-2}]\left(v_{dq+1} - \overline{v}^*_{dq+1}\right)\right\} \tag{8.44}$$

and regrouping terms yields

$$\{[I] - [F][F_{-2}]\}\overline{v}^*_{dq+1} = [F]\{[I] - [F_{-2}]\} v_{dq+1} \tag{8.45}$$

where $[I]$ is the identity matrix. As a result, the following expression describes the relationship between the signals on the positive-sequence frame and the estimated value for the positive-sequence component at the output of the DDSRF:

$$\overline{v}^*_{dq+1} = \{[I] - [F][F_{-2}]\}^{-1}[F]\{[I] - [F_{-2}]\} v_{dq+1} \tag{8.46}$$

After operating, this relationship is given by

$$\frac{\overline{v}^*_{dq+1}}{v_{dq+1}} = \begin{bmatrix} H_{11} & H_{12} \\ H_{21} & H_{22} \end{bmatrix} ; \quad \begin{cases} H_{11} = H_{22} = H\left(s^3 + 2\omega_f s^2 + 4\omega^2 s + 4\omega_f \omega^2\right) \\ H_{12} = -H_{21} = -H\left(2\omega_f \omega s\right) \end{cases} \tag{8.47}$$

where

$$H = \frac{\omega_f}{s^4 + 4\omega_f s^3 + 4\left(\omega_f^2 + \omega^2\right)s^2 + 8\omega_f \omega^2 s + 4\omega_f^2 \omega^2} \tag{8.48}$$

The transfer function for the negative-sequence output of the DDSRF is given simply by transposing the matrix shown in (8.47), and is obtained by following the same steps as in the positive-sequence case.

The DDSRF is a very useful tool when dealing with three-phase systems, since it is a sequence separator that allows independent control of the positive- and negative-sequence components of voltage and/or current during unbalanced grid faults. For this reason, the transfer functions shown in (8.47) are very important for correct implementation of the low-voltage ride-through capability in power converters under unbalanced grid fault conditions.

In the transfer functions of (8.47), it is supposed that the frequency ω is given by a PLL and matches the fundamental frequency of the grid, while the cut-off frequency of the low-pass filer, ω_f, is properly set in design time to obtain the required performance of the system. Figure 8.18 shows the evolution of the signal on the d^{+1} axis of the positive-sequence reference frame of the DDSRF, $\overline{v}^*_{d+1}$, when v_{dq+1} is suddenly applied to its input in the form of a unitary step for different values of ω_f. From this figure, it can be concluded that a reasonable trade-off between the time response and oscillation damping can be achieved by setting $\omega_f = \omega/\sqrt{2}$ rad/s [17].

8.4.4 Structure and Response of the DDSRF-PLL

The block diagram of the DDSRF-PLL is shown in Figure 8.19. As shown in the figure, this PLL is an extension of the conventional three-phase SRF-PLL structure. In this PLL, in order to

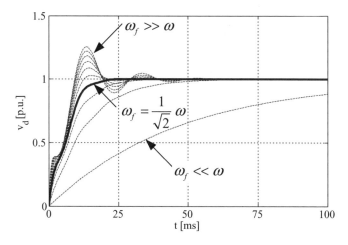

Figure 8.18 Step response of the signal $\bar{v}^*_{d+1}$ at the output of the DDSRF

obtain a similar dynamic response for different grid voltage amplitudes, the phase-angle error signal v_q^{+1} is adaptively normalized to the amplitude of the positive-sequence input vector. Moreover, the rated grid frequency is added as a feed-forward parameter, ω_{ff}, to accelerate the pulling process of the PLL. However, the most significant performance improvement in this PLL comes from the decoupling network added to the DSRF.

The decoupling network of the DDSRF-PLL completely cancels out the oscillations at 2ω on the dq^{+1} and dq^{-1} reference frame signals. Therefore, there is no need to reduce the bandwidth of the PLL to attenuate such oscillations and the real amplitude of the unbalanced input voltage sequence components are indeed exactly detected.

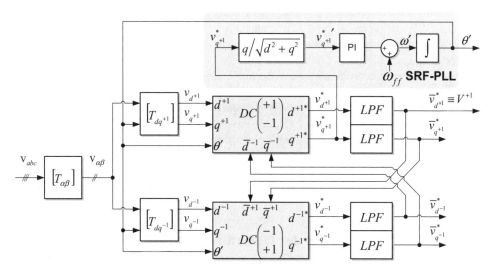

Figure 8.19 Structure of the DDSRF-PLL

Figure 8.20 shows the response of the DDSRF-PLL when used in the estimation of the sequence components of the grid voltage during a sag type C with $\vec{V}_a^{+1} = 0.5\angle -30°$ and $\vec{V}_a^{-1} = 0.25\angle +60°$. In this study case, the fundamental grid frequency is $\omega = 314.1$ rad/s. Therefore, the cut-off frequency of the low-pass filter was set to $\omega_f = \omega/\sqrt{2} = 222.1$ rad/s and the parameters of the PI controller were set to $k_p = 222.1$ and $k_i = 9 \times 10^{-3}$, which results in a settling time around 40 ms according to the guidelines given in [20] and in Section 4.2.2 of Chapter 4.

The plot of Figure 8.20(a) shows the three-phase unbalanced voltages, which experience a jump in the phase angle, as evidenced by the vector locus shown in Figure 8.20(b). The actual and the detected phase angle are plotted in Figure 8.20(c). This figure shows how the PLL is able to lock the phase-angle jump after a transient period, which roughly matches the settling time of the PLL calculated by the expression (4.37) in Chapter 4. This settling time, approximately 40 ms, can be clearly observed in Figure 8.20(d), which represents the frequency detected by the DDSRF-PLL. From this figure, it is worth highlighting the fact that the large amplitude of the transient oscillation in the detected frequency as a consequence of the phase-angle jump occurred in the grid voltage. The existence of such significant oscillations in the detected frequency, which is indeed one of the most stable magnitudes in power systems, can be considered as a drawback of the DDSRF-PLL in certain applications. Figure 8.20(e) shows the dq^{+1} signals on the axes of the positive reference frame (thin traces) and the resultant signals at the output of the DDSRF (thick traces). Likewise, the dq^{-1} signals on the negative-sequence reference frame are shown Figure 8.20(f). In this last figure, the d^{-1} component is equal to zero because of a geometrical coincidence. Since the PLL is controlling the position of the dq^{+1} reference frame, only the q^{+1} signal at the output of the DDSRF is forced to be equal to zero. Both d^{-1} and q^{-1} signals at the output of the DDSRF can take any arbitrary value depending on the relative angular position between the positive- and negative-sequence voltage vectors. Figure 8.20(e) and (f) shows how the DDSRF perfectly decouples the positive- and negative-sequence voltage components and obtains free oscillation signals describing the amplitude of the positive- and negative-sequence voltage vectors applied to its input. From the detected phase angle and amplitudes, the positive- and negative-sequence three-phase voltages can be readily reconstructed as shown in Figure 8.20(g) and (h).

The DDSRF-PLL is an effective synchronization solution for the implementation of synchronous controllers for three-phase power converters, mainly if they provide low-voltage ride-through capabilities under unbalanced grid faults. However, as presented in Chapter 9, the power converter controllers can also be implemented on the stationary reference frame by using resonant controllers. In such a case, the grid voltage phase angle is not the most important synchronization variable – the grid frequency is. Since the grid frequency is a more stable variable than the grid phase angle, it is intuitive to think that controllers based on grid frequency detection will present a more robust performance than those based on phase-angle detection during grid faults. In the next section, a synchronization system based on adaptive filters working on the stationary reference frame is presented as a suitable technique to be applied in the implementation of resonant controllers for three-phase converters.

8.5 The Double Second-Order Generalized Integrator FLL (DSOGI-FLL)

The DSOGI-FLL exploits the instantaneous symmetrical components method by using adaptive filters based on the second-order generalized integrator [21, 22]. As previously presented

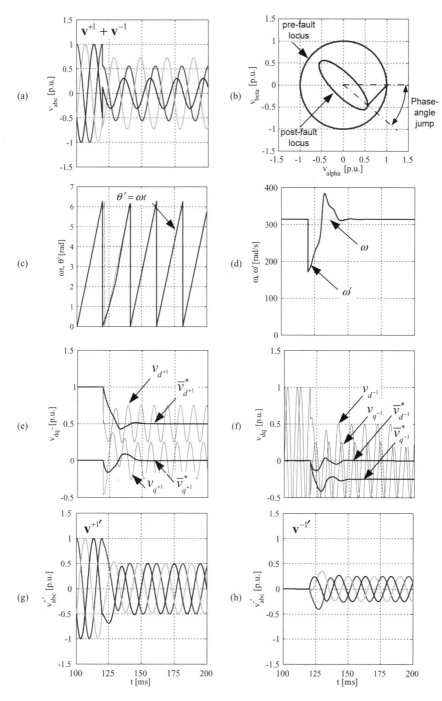

Figure 8.20 Representative waveforms of a DDSRF-PLL: (a) three-phase input voltage, (b) voltage vector locus, (c) detected phase angle, (d) detected frequency, (e) detected *dq* signals for the positive-sequence component, (f) detected *dq* signals for the negative-sequence component, (g) detected positive-sequence three-phase voltages and (h) detected negative-sequence three-phase voltages

in this chapter, an unbalanced three-phase system can be systematically analysed by transforming its unbalanced phasors into a set of symmetrical components according to the *Fortescue* transformation matrix shown in (8.14) [23]. The symmetrical components method can also be applied in the time domain analysis by using the *Lyon* transformation [24]. According to this method, a voltage vector v_{abc} consisting of three unbalanced sinusoidal waveforms can be split up into its instantaneous positive-, negative- and zero-sequence components, $v_{abc} = v_{abc}^+ + v_{abc}^- + v_{abc}^0$, by applying the following transformations:

$$v_{abc}^+ = [T_+]\, v_{abc}; \quad \begin{bmatrix} v_a^+ \\ v_b^+ \\ v_c^+ \end{bmatrix} = \frac{1}{3} \begin{bmatrix} 1 & a & a^2 \\ a^2 & 1 & a \\ a & a^2 & 1 \end{bmatrix} \begin{bmatrix} v_a \\ v_b \\ v_c \end{bmatrix} \tag{8.49}$$

$$v_{abc}^- = [T_-]\, v_{abc}; \quad \begin{bmatrix} v_a^- \\ v_b^- \\ v_c^- \end{bmatrix} = \frac{1}{3} \begin{bmatrix} 1 & a^2 & a \\ a & 1 & a^2 \\ a^2 & a & 1 \end{bmatrix} \begin{bmatrix} v_a \\ v_b \\ v_c \end{bmatrix} \tag{8.50}$$

$$v_{abc}^0 = [T_0]\, v_{abc}; \quad \begin{bmatrix} v_a^0 \\ v_b^0 \\ v_c^0 \end{bmatrix} = \frac{1}{3} \begin{bmatrix} 1 & 1 & 1 \\ 1 & 1 & 1 \\ 1 & 1 & 1 \end{bmatrix} \begin{bmatrix} v_a \\ v_b \\ v_c \end{bmatrix} \tag{8.51}$$

where a is a particular version of the *Fortescue* operator and represents a kind of time-shifting over the instantaneous sinusoidal input signals at the fundamental grid frequency, equivalent to a 120° phase-shifting.

In three-phase three-wire grid-connected power converters, the main interest lies in controlling the positive- and negative-sequence components of the injected current. In turn, the grid synchronization system should be focused on perfectly tracking the positive- and negative-sequence components of the grid voltage at the point of common coupling.

The sequence components of v_{abc} can be expressed on the $\alpha\beta$ reference frame by using either the transformation matrix of (8.5) or its rescaled version of (8.21), yielding

$$\begin{aligned} v_{\alpha\beta}^+ &= [T_{\alpha\beta}]\, v_{abc}^+ \\ v_{\alpha\beta}^- &= [T_{\alpha\beta}]\, v_{abc}^- \end{aligned} \tag{8.52}$$

Substituting (8.49) and (8.50) we obtain

$$\begin{aligned} v_{\alpha\beta}^+ &= [T_{\alpha\beta}]\, [T_+]\, v_{abc} \\ v_{\alpha\beta}^- &= [T_{\alpha\beta}]\, [T_-]\, v_{abc} \end{aligned} \tag{8.53}$$

and applying the inverse transformation $[T_{\alpha\beta}]^{-1}$ we have

$$\begin{aligned} v_{\alpha\beta}^+ &= [T_{\alpha\beta}]\, [T_+]\, [T_{\alpha\beta}]^{-1}\, v_{\alpha\beta} \\ v_{\alpha\beta}^- &= [T_{\alpha\beta}]\, [T_-]\, [T_{\alpha\beta}]^{-1}\, v_{\alpha\beta} \end{aligned} \tag{8.54}$$

Finally, after operating these transformation matrixes we arrive at the following expressions:

$$v_{\alpha\beta}^{+} = [T_{\alpha\beta+}] \, v_{\alpha\beta}; \quad [T_{\alpha\beta+}] = \frac{1}{2}\begin{bmatrix} 1 & -q \\ q & 1 \end{bmatrix} \quad (8.55)$$

$$v_{\alpha\beta}^{-} = [T_{\alpha\beta-}] \, v_{\alpha\beta}; \quad [T_{\alpha\beta-}] = \frac{1}{2}\begin{bmatrix} 1 & q \\ -q & 1 \end{bmatrix} \quad (8.56)$$

where $q = \mathrm{e}^{-\mathrm{j}\pi/2}$ is a 90°-lagging phase-shifting operator applied on the time domain to obtain an in-quadrature version of the input waveforms.

8.5.1 Structure of the DSOGI

Different techniques to implement a quadrature signal generator (QSG) were presented in Chapter 4. In the DSOGI, the operator q of (8.55) and (8.56) is implemented by using the second-order AF based on a SOGI (SOGI-QSG), which was presented in Section 4.5.3 of Chapter 4 as an effective method to obtain a set of two in-quadrature output signals from a given sinusoidal input signal. Moreover, the filtering characteristic of the SOGI-QSG attenuates the effect of the distorting high-order harmonics from the input to the output.

The structure of the DSOGI is presented in Figure 8.21. As observed in this figure, two SOGI-QSGs are in charge of generating the direct and in-quadrature signals for the α and β components of the input vector, i.e. v'_α, v'_β, qv'_α and qv'_β respectively. These signals are

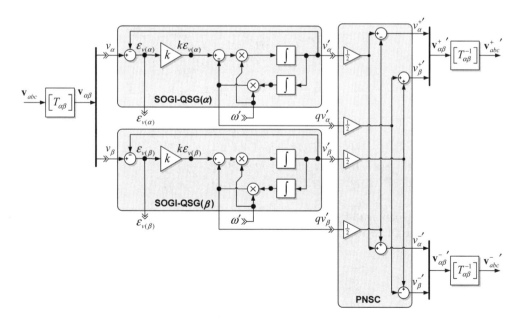

Figure 8.21 Structure of the DSOGI

provided as inputs to a positive-/negative-sequence calculation block (PNSC), which computes the sequence components on the $\alpha\beta$ reference frame according to (8.55) and (8.56).

8.5.2 Relationship between the DSOGI and the DDSRF

According to (8.55), the transfer function from the unbalance input voltage vector to the positive-sequence component detected by the DSOGI is given by

$$v_{\alpha\beta}^+ = \begin{bmatrix} T_{\alpha\beta+} \end{bmatrix} v_{\alpha\beta} = \frac{1}{2}\begin{bmatrix} D(s) & -Q(s) \\ Q(s) & D(s) \end{bmatrix} v_{\alpha\beta} = \frac{1}{2}\frac{k\omega'}{s^2 + k\omega's + \omega'^2}\begin{bmatrix} s & -\omega' \\ \omega' & s \end{bmatrix} v_{\alpha\beta} \tag{8.57}$$

where $D(s)$ and $Q(s)$ are the characteristic transfer functions of the SOGI-QSG and were already presented in Section 4.5.3 of Chapter 4. The negative-sequence component at the output of the DSOGI can be calculated by simply transposing the matrix of (8.57).

The DDSRF was analysed in Section 8.4.3 and its transfer function on the synchronous reference frame was presented in (8.47). To translate the transfer function of the DDSRF from the synchronous reference frame to the stationary one it is necessary to operate the following transformation:

$$v_{\alpha\beta}^+ = \begin{bmatrix} T_{dq+} \end{bmatrix} \begin{bmatrix} H_{11} & H_{12} \\ H_{21} & H_{22} \end{bmatrix} \begin{bmatrix} T_{dq+} \end{bmatrix}^{-1} v_{\alpha\beta} \tag{8.58}$$

Taking into account that $H_{11} = H_{22}$ and $H_{12} = -H_{21}$, we arrive at

$$v_{\alpha\beta}^+ = \frac{1}{2}\begin{bmatrix} H_a - jH_b & jH_c + H_d \\ -jH_c - H_d & H_a - jH_b \end{bmatrix} v_{\alpha\beta} \tag{8.59}$$

where

$$\begin{aligned} H_a &= (H_{11}(s + j\omega') + H_{11}(s - j\omega')) \\ H_b &= (H_{12}(s + j\omega') - H_{12}(s - j\omega')) \\ H_c &= (H_{11}(s + j\omega') - H_{11}(s - j\omega')) \\ H_d &= (H_{12}(s + j\omega') + H_{12}(s - j\omega')) \end{aligned} \tag{8.60}$$

Expanding and regrouping (8.59), the following transfer functions are obtained to describe the performance of the DDSRF on the stationary $\alpha\beta$ reference frame:

$$v_{\alpha\beta}^+ = \frac{\omega_f}{s^2 + 2\omega_f s + \omega'^2}\begin{bmatrix} s & -\omega' \\ \omega' & s \end{bmatrix} v_{\alpha\beta} \tag{8.61}$$

where ω_f is the cuf-off frequency of the first-order low-pass filter and ω' is the frequency detected by the PLL.

Expressions (8.57) and (8.61) show that the DSOGI and the DDSRF are two equivalent systems, which perform the same function – sequence separation – on two different reference frames. In principle, the DSOGI and the DDSRF would have the same dynamic response when $k = 2\omega_f/\omega'$ in (8.57). However, it is worth remarking that the DSOGI performance depends on the frequency detected by the FLL, while the DDSRF depends on the phase angle detected by the PLL. Therefore, the response of the DSOGI-FLL and the DDSRF-PLL will not exactly be equal in practice since the FLL and the PLL are two completely different systems with a different dynamic response.

To analyse the frequency response of the DSOGI, the expression of (8.57) can be written in the frequency domain ($s = j\omega$) as follows:

$$\begin{bmatrix} v_\alpha^+ \\ v_\beta^+ \end{bmatrix} = \frac{1}{2}\begin{bmatrix} D(j\omega) & -Q(j\omega) \\ Q(j\omega) & D(j\omega) \end{bmatrix}\begin{bmatrix} v_\alpha \\ v_\beta \end{bmatrix} = \frac{1}{2}\frac{k\omega'}{(\omega'^2 - \omega^2) + jk\omega'\omega}\begin{bmatrix} j\omega & -\omega' \\ -\omega' & j\omega \end{bmatrix}\begin{bmatrix} v_\alpha \\ v_\beta \end{bmatrix} \quad (8.62)$$

Considering that the $\alpha\beta$ components of a balanced positive-sequence voltage vector at frequency ω keep the following steady-state relationship on the frequency domain:

$$v_\beta(j\omega) = -jv_\alpha(j\omega) \quad (8.63)$$

Therefore, the steady-state transfer function of the DSOGI on the frequency domain can be written as

$$\begin{bmatrix} v_\alpha^+ \\ v_\beta^+ \end{bmatrix} = \frac{1}{2}\frac{k\omega'(\omega + \omega')}{k\omega'\omega + j(\omega^2 - \omega'^2)}\begin{bmatrix} v_\alpha \\ v_\beta \end{bmatrix} \quad (8.64)$$

This transfer function describes the relationship between the amplitude of the positive-sequence component detected by the DSOGI and the actual amplitude of a given positive-sequence voltage vector applied to its input. This transfer function is plotted in the Bode diagram of Figure 8.22 as $P(j\omega) = |v^+_{\alpha\beta}{}'|/|v^+_{\alpha\beta}|$. By simply substituting ω by $-\omega$ in (8.64), another transfer function $N(j\omega) = |v^+_{\alpha\beta}{}'|/|v^-_{\alpha\beta}|$ can be defined. This second transfer function, also plotted in the Bode diagram of Figure 8.21, describes the relationship between the

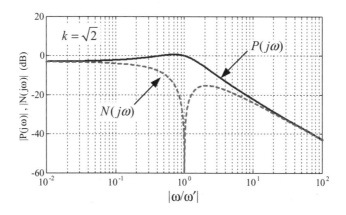

Figure 8.22 Frequency response of the DSOGI

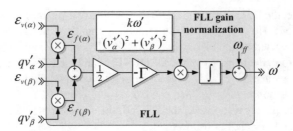

Figure 8.23 Structure of the FLL for the DSOGI.

detected amplitude for the positive-sequence component and the actual amplitude of a given negative-sequence voltage vector applied to the input of the DSOGI. As this *Bode* diagram shows, the DSOGI acts either as a low-pass filter or as a notch filter in the detection of the positive-sequence component, depending on whether the input voltage shows either a positive or a negative sequence respectively.

8.5.3 The FLL for the DSOGI

As presented in Section 4.6 of Chapter 4, a SOGI-QSG needs an FLL to become frequency adaptive. Moreover, the gain of the FLL has to be normalized in runtime according to the amplitude of the input signal in order to linearize the response of the frequency adaptation loop. Although mathematically correct, the use of two independent FLLs in the DSOGI-FLL might, however, seem conceptually odd since its two input signals, v_α and v_β, have the same frequency. For this reason, the DSOGI uses a single FLL (see Figure 8.23) in which the frequency error signals generated by the QSGs of the α and β signals have been combined by calculating an average error signal, i.e.

$$\varepsilon_f = \frac{\varepsilon_{f(\alpha)} + \varepsilon_{f(\beta)}}{2} = \frac{1}{2}\left(\varepsilon_\alpha q v'_\alpha + \varepsilon_\beta q v'_\beta\right) \tag{8.65}$$

The gain of this two-dimensional FLL is normalized by using the square of the amplitude of the positive-sequence component, i.e. $(v_\alpha^+)^2 + (v_\beta^+)^2$, which results in a first-order exponential linearized response with a settle time that still matches very well that one calculated by (4.109) in Chapter 4. In this manner, the DSOGI-FLL permits a decoupled estimation to be carried out of the symmetrical components of the input three-phase voltage on the $\alpha\beta$ reference frame, as well as the value of the grid frequency, something that is essential to implement power converter controllers on the stationary reference frame by using generalized integrators.

8.5.4 Response of the DSOGI-FLL

To evaluate the response of DSOGI-FLL the same unbalanced grid voltage as in the case of the DDSRF-PLL is applied to its input, i.e. a sag type C with $\vec{V}_a^{+1} = 0.5\angle -30°$ and $\vec{V}_a^{-1} = 0.25\angle +60°$. In this study case, the gain of the SOGI-QSGs was set to $k = \sqrt{2}$ to have the same tuning conditions as in the case of the evaluation of the DDSRF response ($\omega_f = \omega/\sqrt{2}$). The gain of the FLL was set to $\Gamma = 100$, which results in a settling time of around 45 ms according to the guidelines given in Section 4.6.1 of Chapter 4.

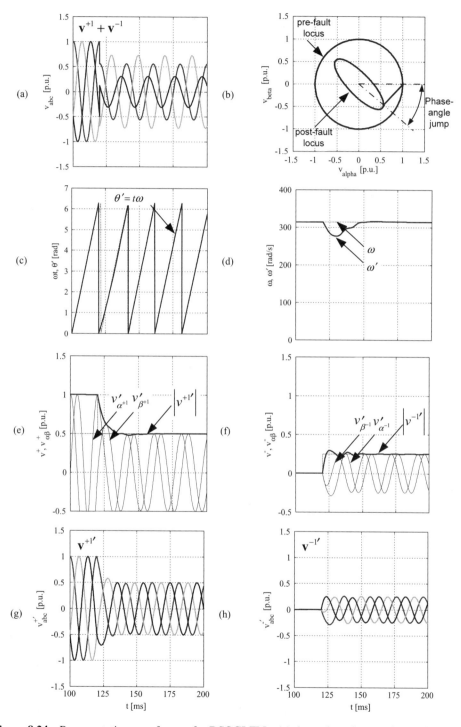

Figure 8.24 Representative waveforms of a DSOGI-FLL: (a) three-phase input voltage, (b) voltage vector locus, (c) detected phase angle, (d) detected frequency, (e) detected positive-sequence amplitude and $\alpha\beta$ signals, (f) detected negative-sequence amplitude and $\alpha\beta$ signals, (g) detected positive-sequence three-phase voltages and (h) detected negative-sequence three-phase voltages

The plot of Figure 8.24(a) shows the three-phase unbalanced voltages, which experience a jump in the phase angle as evidenced by the vector locus shown in Figure 8.24(b). The frequency detected by the FLL is shown in Figure 8.24(d). It is possible to appreciate in this figure how the detected frequency does not present high oscillations as in the case of the DDSRF-PLL. Moreover, the settling time in frequency adaptation matches the theoretical calculations. The amplitude and the phase angle of the sequence components detected by the DSOGI-FLL can be calculated by

$$|v'| = \sqrt{\left(v'_\alpha\right)^2 + \left(v'_\beta\right)^2}; \quad \theta' = \tan^{-1} \frac{v'_\beta}{v'_\alpha} \qquad (8.66)$$

The actual and the detected phase angle of the positive-sequence component of the unbalanced input voltage are plotted in Figure 8.24(c). This figure shows that the DSOGI-FLL completely cancels the steady-state error in the detected phase angle. Figure 8.24(e) shows the amplitude of the positive-sequence component together with the $\alpha\beta^{+1}$ signals. The amplitude of the negative-sequence component, together with the $\alpha\beta^{-1}$ signals are shown in Figure 8.24(f). The positive- and negative-sequence three-phase voltages can be reconstructed from the detected phase angle and amplitudes, and are shown in Figure 8.24(g) and (h).

After comparing the plots shown in Figures 8.20 and 8.24, it is possible to highlight the fact that the waveforms of the DSOGI-FLL are smoother than those of the DDSRF-PLL when the same unbalanced voltage is applied to their inputs and an equivalent set of parameters are used in both systems. This difference in the response of both synchronization systems gives rise to a significant divergence between the performances of power converter controllers working on the synchronous reference frame and on the stationary one, mainly when they operate under unbalanced grid faults.

8.6 Summary

This chapter has studied the characteristics of the three-phase voltage vector under unbalanced grid faults and presented expressions to determine its sequence components as a function of both the type of fault and the grid impedances.

The conventional SRF-PLL, although commonly used as a essential building block in the implementation of controllers for grid-connected converters, has demonstrated that it is not a suitable solution when a fast and precise grid synchronization is required during unbalanced grid faults, as is the case of the controllers for wind turbines and photovoltaics generators implementing the low-voltage ride-through functionality.

The DDSRF-PLL and the DSOGI-FLL, two advanced grid synchronization systems, have been presented in this chapter as suitable solutions to be used in the implementation of synchronous and stationary controllers for power converters respectively. The fundamental variable estimated by the DDSRF-PLL is the grid phase-angle, whereas the grid frequency is the one for the DSOGI-FLL. Since the grid frequency is a more stable variable than the grid phase-angle, the DSOGI-FLL use to present a smoother response than the DDSRF-PLL during transient faults.

References

[1] 'Towards Smart Power Networks – Lessons Learned from European Research FP5 Projects'. Luxembourg: European Commission – Office for Official Publications of the European Communities, Ref. EUR 21970, 2005.

[2] Dugan, R.C., McGranaghan, M.F., Santoso, S. and Beaty, H.W., *Electrical Power Systems Quality*, 2nd edition, New York: McGraw-Hill, 2002.

[3] Cichowlas, M., Malinowski, M., Sobczuk, D.L., Kazmierkowski, M. P., Rodríguez, P. and Pou, J., 'Active Filtering Function of Three-Phase PWM Boost Rectifier under Different Line Voltage Conditions'. *IEEE Transactions on Industrial Electronics*, **52**, April 2005, 410–419.

[4] Teodorescu, R. and Blaabjerg, F., 'Flexible Control of Small Wind Turbines with Grid Failure Detection Operating in Stand-alone and Grid-Connected Mode'. *IEEE Transactions on Power Electronics*, **19**, September 2004, 1323–1332.

[5] Nielsen, J. G., Newman, M., Nielsen, H. and Blaabjerg, F., 'Control and Testing of a Dynamic Voltage Restorer (DVR) at Medium Voltage Level'. *IEEE Transactions on Power Electronics*, **19**, May 2004, 806–813.

[6] Haque, M.H., 'Power Flow Control and Voltage Stability Limit: Regulating Transformer versus UPFC'. In *Proceedings of the IEE on Generation, Transmission and Distribution*, Vol. **151**, May 2004, pp. 299–304.

[7] Mattavelli, P., 'A Closed-Loop Selective Harmonic Compensation for Active Filters'. *IEEE Transactions on Industry Applications*, **37**, January/February 2001, 81–89.

[8] Bollen, M.H.J., *Understanding Power Quality Problems*, New York: IEEE Press, 2000.

[9] Bollen, M.H.J. and Gu, I., *Signal Processing of Power Quality Disturbances*, Wiley–IEEE Press, 2006. ISBN: 978-0-471-73168-9.

[10] Bollen, M.H.J. and Zhang, L. D., 'Different Methods for Classification of Three-Phase Unbalanced Voltage Dips Due to Faults'. *Electric Power Systems Research*, **66**(1), July 2003, 59–69.

[11] Fortescue, C.L., 'Method of Symmetrical Coordinates Applied to the Solution of Polyphase Networks'. *Transactions of AIEE*, Part II, **37**, 1918, 1027–1140.

[12] Joint Working Group JWG C4.110, 'Voltage Dip Immunity of Equipment Used in Installations', CIGRE/CIRED/UIE, http://www.jwgc4-110.org.

[13] Kaura, V. and Blasco, V., 'Operation of a Phase Locked Loop System under Distorted Utility Conditions'. *IEEE Transactions on Industry Applications*, **33**, January/February 1997, 58–63.

[14] Chung, S., 'A Phase Tracking System for Three Phase Utility Interface Inverters'. *IEEE Transactions on Power Electronics*, **15**, May 2000, 431–438.

[15] Timbus, A.V., Teodorescu, R., Blaabjerg, F., Liserre, M. and Rodriguez, P., 'PLL Algorithm for Power Generation Systems Robust to Grid Voltage Faults'. In *Proceedings of the Power Electronics Specialists Conference*, PESC '06, June 2006, pp. 1–7.

[16] McGrath, B. P., Holmes, D. G. and Galloway, J.J.H., 'Power Converter Line Synchronization Using a Discrete Fourier Transform (DFT) Based on a Variable Sample Rate'. *IEEE Transactions on Power Electronics*, **20**, July 2005, 877–884.

[17] Rodriguez, P., Pou, J., Bergas, J., Candela, J. I., Burgos, R.P. and Boroyevich, D., 'Decoupled Double Synchronous Reference Frame PLL for Power Converters Control'. *IEEE Transactions on Power Electronics*, **22**, March 2007, 584–592.

[18] Teodorescu, R., Blaabjerg, F., Liserre, M. and Loh, P., 'Proportional-Resonant Controllers and Filters for Grid-Connected Voltage-Source Converters'. *IEE Proceedings of Electrical Power Applications*, **153**, September 2006, 750–762.

[19] Zmood, D. and Holmes, D., 'Stationary Frame Current Regulation of PWM Inverters with Zero Steady-State Error'. *IEEE Transactions on Power Electronics*, **18**, May 2003, 814–822.

[20] Rodríguez, P., Bergas, J. and Gallardo, J.A., 'A New Positive Sequence Voltage Detector for Unbalanced Power Systems'. In *Proceedings of the European Conference on Power Electronics and Applications*, September 2002, CD Ref. T6-015.

[21] Rodriguez, P., Luna, A., Ciobotaru, M., Teodorescu, R. and Blaabjerg, F., 'Advanced Grid Synchronization System for Power Converters under Unbalanced and Distorted Operating Conditions'. In *IEEE Industrial Electronics (IECON 2006)*, 6–10 November 2006, pp. 5173–5178.

[22] Rodriguez, P., Luna, A., Candela, J. I., Rosas, R., Teodorescu, R. and Blaabjerg, F., 'Multi-Resonant Frequency-Locked Loop for Grid Synchronization of Power Converters under Distorted Grid Conditions'. *IEEE Transactions on Industrial Electronics*, **PP**(99), April 2010, 1.

[23] Anderson, P., 'Analysis of Faulted Power Systems'. In *IEEE Power, Energy, and Industry Applications*, 2009. ISBN 9780470544129.

[24] Lyon, W.V., *Application of the Method of Symmetrical Components*, New York: McGraw-Hill, 1937.

9

Grid Converter Control for WTS

9.1 Introduction

This chapter discusses the control issues related to the use of a voltage source converter in wind turbine (WT) systems. Before going into detail on the control of the converter, its model is presented. Then different control structures are discussed. In all of them directly or indirectly the two main issues are the control of the DC link voltage (if there is not a DC/DC converter in charge of it) and the control of the AC power. The AC power can be controlled with the aim either of feeding the main grid or of feeding stand-alone loads or a micro-grid. In the first case the WT system may also offer support to the grid. Once the operation of the grid converter is decided and the strategy to select the reference power is also selected, the power control results, using instantaneous power theory (outlined in Appendix B) in a current and/or voltage control (as shown in Figure 9.1). In fact, the converter can be operated as a controlled current source (typically adopted if the converter is grid connected and does not offer any support to the grid) or as a controlled voltage source (typically adopted if the converter is in stand-alone, micro-grid or grid-supporting mode). In the second case, if an LC filter is also employed the current is controlled (see Figure 9.2). The control of currents and voltages can be done in state variables or in a cascade structure. In the case of operation in a micro-grid and in a grid-supporting mode the droop control (or an equivalent one) is needed.

Hence AC voltage and DC link voltage controls are briefly outlined. However, the chapter gives details only on the DC link voltage control because the presence of the converter between the DC voltage state variable and the AC current state variable makes the analysis more complex and a deep analysis is needed. The AC voltage control design is a straightforward consequence of the AC current control described in Chapter 12.

Then the structures are adopted for implementation of the control ($\alpha\beta$, abc, dq) [1]. Direct power control is also discussed in a separate section. In fact, in this case the current control loop is not present and the control is linear with respect to the decoupling between AC and DC dynamics but it is nonlinear with respect to the AC dynamics that are expressed directly in terms of power. The direct power control can be implemented with or without a separate PWM modulator. If this is present the scheme is quite similar to those

---

Grid Converters for Photovoltaic and Wind Power Systems   Remus Teodorescu, Marco Liserre, and Pedro Rodríguez
© 2011 John Wiley & Sons, Ltd

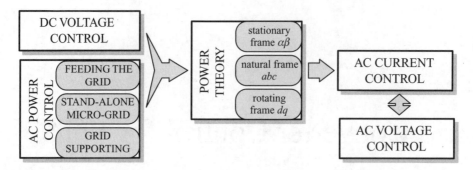

Figure 9.1 The main control issues of grid converters are transformed through the instantaneous power theory, with reference to AC current and voltage controllers

adopting current control, but it remains nonlinear with respect to the fact that the power is controlled directly.

Finally, it is shown how the converter can be controlled in stand-alone operation or to manage a micro-grid in case more DG units are connected in parallel. Typically in this case the frequency and voltage droop methods are adopted. A brief outline of their principle of operation closes the chapter.

9.2 Model of the Converter

In the following the mathematical models of the L-filter-based and LCL-filter-based inverters are presented. These models are relevant for study of the AC voltage control treated in this chapter as well as the AC current control treated in Chapter 12.

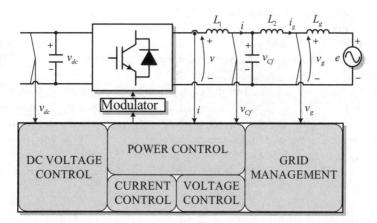

Figure 9.2 Overall scheme of the LCL filter grid converter control showing all the functions

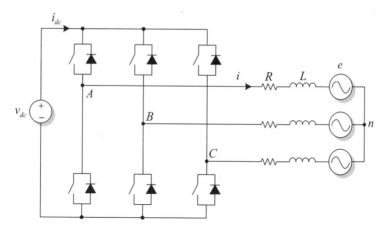

Figure 9.3 L-filter inverter connected to the grid

9.2.1 Mathematical Model of the L-Filter Inverter

The state of the three-phase inverter is modelled by means of a switching space vector defined with the switching functions $p_j(t)$ ($j = a, b, c$) equal to 1 when the upper switch is closed and 0 when the lower switch is closed:

$$\bar{p}(t) = \frac{2}{3}\left(p_a(t) + \alpha p_b(t) + \alpha^2 p_c(t)\right) \tag{9.1}$$

where $\alpha = e^{j2\pi/3}$. Hence the inverter produces on the AC side the following voltage:

$$\bar{v}(t) = \bar{p}(t)v_{dc}(t) \tag{9.2}$$

If the inverter is connected to the grid through an L filter (Figure 9.3), the equation that describes the evolution of the grid current is

$$\bar{v}(t) = \bar{e}(t) + R\bar{i}(t) + L\frac{d\bar{i}(t)}{dt} \tag{9.3}$$

Obviously $\bar{v}(t)$ is the space vector of the inverter voltages, $\bar{i}(t)$ is the space vector of the inverter input currents and $\bar{e}(t)$ is the space vector of the input line voltages. Each of these vectors can be obtained by substituting in (9.1) the phase converter voltages, the phase currents and the phase grid voltages respectively.

The mathematical model of the system written in the state-space form is

$$\frac{d\bar{i}(t)}{dt} = \frac{1}{L}\left[-R\bar{i}(t) - \bar{e}(t) + \bar{p}(t)v_{dc}(t)\right] \tag{9.4}$$

A commonly adopted approach in analysing three-phase systems is to use a stationary or rotating frame (Figure 9.4) [1]. In the first case the frame will be denoted as $\alpha\beta$ and in

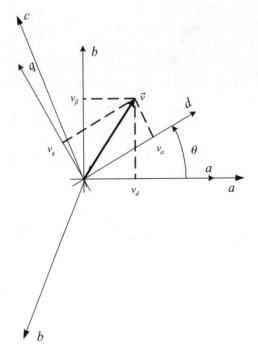

Figure 9.4 Stationary $\alpha\beta$ frame and rotating dq frame

the second as dq and also called synchronous. In fact the dq frame is synchronized with the angular speed ω (where $\omega = 2\pi f$ and f is the fundamental frequency of the power grid voltage waveform). The space vectors that express the inverter electrical quantities are projected on the α axis and β axis or on the d axis and q axis.

The following transformations can be used to obtain the switching functions in the $\alpha\beta$ frame and dq frame knowing the switching functions of each inverter leg:

$$\begin{bmatrix} p_\alpha \\ p_\beta \end{bmatrix} = \frac{2}{3} \begin{bmatrix} 1 & -1/2 & -1/2 \\ 0 & -\sqrt{3}/2 & \sqrt{3}/2 \end{bmatrix} \begin{bmatrix} p_a \\ p_b \\ p_c \end{bmatrix} \tag{9.5}$$

$$\begin{bmatrix} p_d \\ p_q \end{bmatrix} = \begin{bmatrix} \cos\theta & \cos\left(\theta - \dfrac{2\pi}{3}\right) & \cos\left(\theta + \dfrac{2\pi}{3}\right) \\ \sin\theta & \sin\left(\theta - \dfrac{2\pi}{3}\right) & \sin\left(\theta + \dfrac{2\pi}{3}\right) \end{bmatrix} \begin{bmatrix} p_a \\ p_b \\ p_c \end{bmatrix} \tag{9.6}$$

Obviously (9.5) can be obtained from (9.6) by assuming that $\theta = 0$. The same transformations can be adopted for all the electrical quantities involved in (9.4).

The mathematical model in the $\alpha\beta$ frame is

$$\begin{cases} \dfrac{di_\alpha(t)}{dt} = \dfrac{1}{L}\left[-Ri_\alpha(t) - e_\alpha(t) + p_\alpha(t)v_{dc}(t)\right] \\ \dfrac{di_\beta(t)}{dt} = \dfrac{1}{L}\left[-Ri_\beta(t) - e_\beta(t) + p_\beta(t)v_{dc}(t)\right] \end{cases} \quad (9.7)$$

It should be noted that the particular feature of the dq frame is that if a space vector with constant magnitude rotates at the same speed of the frame, it has constant d and q components while if it rotates at a different speed or it has a time-variable magnitude it has pulsating components. Thus in a dq frame rotating at the angular speed ω (9.7) becomes

$$\begin{cases} \dfrac{di_d(t)}{dt} - \omega i_q(t) = \dfrac{1}{L}\left[-Ri_d(t) - e_d(t) + p_d(t)v_{dc}(t)\right] \\ \dfrac{di_q(t)}{dt} + \omega i_d(t) = \dfrac{1}{L}\left[-Ri_q(t) - e_q(t) + p_q(t)v_{dc}(t)\right] \end{cases} \quad (9.8)$$

In the dq frame, the d and q differential equations for the current are dependent due to the cross-coupling terms $\omega i_q(t)$ and $\omega i_d(t)$.

The mathematical model of a single-phase voltage source inverter (H-bridge) in the case where an L filter is connected on the grid side can be obtained from (9.7) simply by considering one of the two equations.

9.2.2 Mathematical Model of the LCL-Filter Inverter

In the case where an LCL filter (Figure 9.5) is adopted to connect the inverter to the grid the mathematical formulation becomes more complex:

$$\frac{d}{dt}\begin{bmatrix} i_d \\ i_q \\ v_{C_f d} \\ v_{C_f q} \\ i_{gd} \\ i_{gq} \end{bmatrix} = \begin{bmatrix} -\dfrac{R_1}{L_1} & \omega & -\dfrac{1}{L_1} & 0 & 0 & 0 \\ -\omega & -\dfrac{R_1}{L_1} & 0 & -\dfrac{1}{L_1} & 0 & 0 \\ \dfrac{1}{C_f} & 0 & 0 & \omega & -\dfrac{1}{C_f} & 0 \\ 0 & \dfrac{1}{C_f} & -\omega & 0 & 0 & -\dfrac{1}{C_f} \\ 0 & 0 & \dfrac{1}{(L_2+L_g)} & 0 & -\dfrac{(R_2+R_g)}{(L_2+L_g)} & \omega \\ 0 & 0 & 0 & \dfrac{1}{(L_2+L_g)} & -\omega & -\dfrac{(R_2+R_g)}{(L_2+L_g)} \end{bmatrix} \begin{bmatrix} i_d \\ i_q \\ v_{C_f d} \\ v_{C_f q} \\ i_{gd} \\ i_{gq} \end{bmatrix}$$

$$+ \begin{bmatrix} 0 & 0 \\ 0 & 0 \\ 0 & 0 \\ 0 & 0 \\ -\dfrac{1}{L_1} & 0 \\ 0 & -\dfrac{1}{L_1} \end{bmatrix} \begin{bmatrix} e_d \\ e_q \end{bmatrix} + \begin{bmatrix} \dfrac{v_{dc}}{(L_2+L_g)} & 0 \\ 0 & \dfrac{v_{dc}}{(L_2+L_g)} \\ 0 & 0 \\ 0 & 0 \\ 0 & 0 \\ 0 & 0 \end{bmatrix} \begin{bmatrix} p_d \\ p_q \end{bmatrix} \quad (9.9)$$

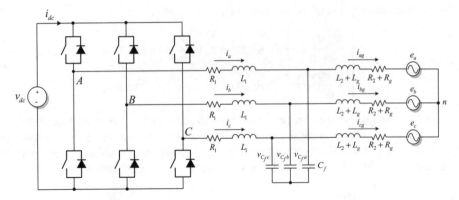

Figure 9.5 LCL-filter inverter connected to the grid

This kind of mathematical formulation can be obtained by writing the KVL and KCL equations and then reporting them in the dq frame, paying attention to the derivative of the vectors that generate the cross-coupling terms already highlighted in (9.7). The $\alpha\beta$ model will not be reported since it can be obtained from (9.9) by substituting d with α and q with β and eliminating the cross-coupling terms.

The presence of an LCL filter can cause resonance problems, as will be discussed in Chapter 11, in other words instability of the current loop that might not be evident in a normal situation but can arise due to disturbances from other sources connected to the grid. It is always recommended to damp the possible resonance with resistors or with a specific control algorithm called 'active damping'. These solutions will be discussed in the Chapter 11.

9.3 AC Voltage and DC Voltage Control

The control of the AC voltage across the capacitor could be needed because the system should operate in a stand-alone mode or in a micro-grid. However, in the grid-connected mode the AC voltage control can also be useful for supporting local loads, the local electrical power system or even the power grid. Of course this depends on the power level of the wind turbine. In any case a multiloop control should be adopted: the AC capacitor voltage is controlled through the AC converter current (see Figure 9.6). In fact, the current-controlled converter is operated as a current source used to charge/discharge the capacitor.

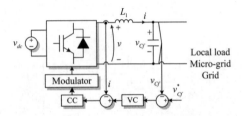

Figure 9.6 Multiloop control strategy (CC stands for current control, VC stands for voltage control)

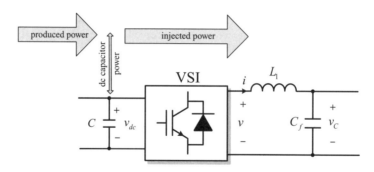

Figure 9.7 Power flow in a WT system

The inner loop feedback variable can also be the capacitor current. The different choice does not change the reference tracking stability but does change the load current rejection capability [2, 3]. For both feedback alternatives, the measured output voltage can also be fed forward to reduce the inner loop control action (see Chapter 12).

Once an AC voltage control loop is adopted, it is then in charge of controlling the power exchange with a load, with the micro-grid or with the main grid [4, 5].

9.3.1 Management of the DC Link Voltage

The DC voltage can be subjected to transient conditions due to the change of the power produced by the generator (see Figure 9.7). The increase of the produced power results in voltage overshoot while its decrease results in voltage undershoot. From the point of view of the DC voltage control, power changes result in voltage variations that should be compensated by charge or discharge processes.

The DC voltage control is achieved through the control of the power exchanged by the converter with the grid or through the control of a DC/DC converter. In the first case the decrease or increase of the DC voltage level is obtained by injecting more or less power to the grid with respect to that produced by the WTS, thus changing the value of the reference for the AC current control loop or the phase displacement of the AC voltage across the capacitor of the LCL-filter. In the second case the grid converter does not play a role in the management of the DC link. Hence this second case will not be considered in this section.

From a control perspective the DC voltage control can be achieved only indirectly through the grid current/voltage control. This indirect control is motivated by the fact that the zero dynamics of the DC voltage, if the average switching functions of the converter are taken as control input, are not stable. If the zero dynamics diverge this means that it is not possible to stabilize the system using that control input [6].

In the following the variations in the DC link voltage will be discussed from an energy perspective.

In a grid-connected converter there are two possible variations of the DC link voltage: the DC type (caused by a change of the average power exchanged by the DC link or by a change in its set-point) or the AC type (caused by an oscillation in the instantaneous power due to grid unbalance conditions) [7,8]. These two variations will be discussed in the following separately.

In the first case, the fact that the DC voltage is different from its reference value v_{dc}^* implies that the amount of energy that the capacitor must receive to come back at the set-point is

$$\Delta E = \frac{(v_{dc}^*)^2 - v_{dc}^2}{2} \cdot C \approx v_{dc}^* \Delta v_{dc} C \qquad (9.10)$$

The power that the converter should exchange corresponding to this energy is

$$\Delta P = \frac{2\Delta E}{(3+n)T_s} \qquad (9.11)$$

The controller plays a role reacting after $(3+n)T_s$, where $3T_s$ is the current control delay (if designed with the technical optimum) and nT_s is the DC link filtering delay [9]. Hence by substituting (9.10) in (9.11), the estimated DC link voltage error is

$$\Delta v_{dc} = \frac{\Delta P(3+n)T_s}{2Cv_{dc}^*} \qquad (9.12)$$

In the case of oscillatory instantaneous power and assuming that unbalance causes mainly a 2nd harmonic steady-state input–output power mismatch (50 Hz grid frequency):

$$p_{in} - p_{out} = \frac{\Delta p_{pk-pk}}{2} \sin(2\pi \times 100 t) \qquad (9.13)$$

The amount of energy associated with this instantaneous oscillation is

$$\Delta E = \int_0^{T/2} (p_{in} - p_{out}) dt = \int_0^{0.005} \frac{\Delta p_{pk-pk}}{2} \sin(2\pi \times 100 t) dt \qquad (9.14)$$

$$\Delta E = \frac{\Delta p_{pk-pk}}{4\pi \times 100} [-\cos(2\pi \times 100 \times 0.005) + \cos(0)] = \frac{\Delta p_{pk-pk}}{2\pi \times 100} \qquad (9.15)$$

On the other hand, by substituting (9.15) in (9.10), the DC link voltage ripple amplitude is

$$\Delta v_{dc} = \frac{\Delta p_{pk-pk}}{2\pi \times 100 C v_{dc}^*} \qquad (9.16)$$

In conclusion, the control of the DC voltage passes through the control of the power exchanged by the converter with the grid. This can be done either by controlling the current or controlling the AC voltage across the capacitor. The former will be discussed in the following while the latter can be achieved using the droop control, the theory of which is treated in Section 9.5.3.

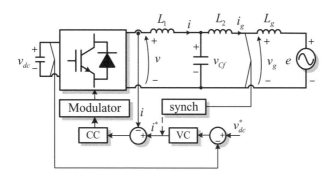

Figure 9.8 Cascaded control of the DC voltage control (VC) through the AC current control (CC)

9.3.2 Cascaded Control of the DC Voltage through the AC Current

The control of the DC voltage through the AC current can result in the identification of two loops, an outer DC voltage loop and an internal current loop (see Figure 9.8). The internal loop is designed to achieve short settling times. On the other hand, the outer loop main goals are optimum regulation and stability; thus the voltage loop could be designed to be somewhat slower. Therefore, the internal and the external loops can be considered decoupled, and thereby the actual grid current components can be considered equal to their references when designing the outer DC controller. If this assumption is used in the design of the controller the control problem is linearized.

However, during the grid converter startup and under unbalance conditions (that lead to a second harmonic oscillation in the DC voltage, as previously discussed) this kind of dynamics decoupling is not valid and the inner and outer loops interact. As a consequence the controller designed on the basis of the linearization can no longer guarantee the expected performances. In the following the DC links dynamics will be analysed in order to derive a proper tuning procedure for the PI controller.

Considering the instantaneous input–output power balance for the grid converter in a synchronous rotating dq frame under a no-loss condition, the following equation holds:

$$\frac{3}{2}\{e_d i_d + e_q i_q\} = -v_{dc} C \frac{dv_{dc}}{dt} + v_{dc} i_o \qquad (9.17)$$

assuming that the previous stage is injecting a current $i_o(t)$ (Figure 9.9).

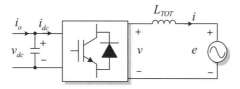

Figure 9.9 Grid converter with highlighted DC link quantities

The small-signal linearization leads to

$$\frac{3}{2}\{(E_d + \hat{e}_d)(I_d + \hat{i}_d) + (E_q + \hat{e}_q)(I_q + \hat{i}_q)\} = -(V_{dc} + \hat{v}_{dc})C\frac{d(V_{dc} + \hat{v}_{dc})}{dt}$$
$$+ (V_{dc} + \hat{v}_{dc})(I_o + \hat{i}_o) \quad (9.18)$$

If the purpose is to control the DC voltage v_{dc} through the i_d current component, the transfer function $\hat{v}_{dc}/\hat{i}_d$ has to be found. Thus the other perturbations have to be considered null, resulting in

$$\frac{3}{2}\{E_d I_d + E_d \hat{i}_d + E_q I_q\} = -V_{dc}C\frac{d\hat{v}_{dc}}{dt} + \hat{v}_{dc}I_o + V_{dc}I_o \quad (9.19)$$

where the second-order signal perturbations have also been assumed to be zero. Once it is considered that

$$\frac{3}{2}\{E_d I_d + E_q I_q\} = V_{dc}I_o \quad (9.20)$$

then

$$\frac{3}{2}\{E_d \hat{i}_d\} = -V_{dc}C\frac{d\hat{v}_{dc}}{dt} + \hat{v}_{dc}I_o \quad (9.21)$$

In the Laplace domain (indicating with s the Laplace operator), once the steady-state equivalent resistance $R_o = V_{dc}/I_o$ and $V_{dc} \simeq \sqrt{3}E_d$ have been indicated (the DC link voltage cannot be lower than this value in order to allow current controllability, but it is not much higher in order not to increase the IGBT losses too much), then

$$\frac{\hat{v}_{dc}}{\hat{i}_d} = \frac{\sqrt{3}}{2}\frac{R_o}{(1 - R_o Cs)} \quad (9.22)$$

and the PI controller for the DC voltage loop can be designed using the 'symmetrical optimum' principle but with a sufficiently lower bandwidth with respect to that of the current loop in order to ensure a proper decoupling [10].

On the contrary, the influence of a small-signal perturbation originated by the power stage connected to the source has to be calculated through $\hat{v}_{dc}/\hat{i}_o$ in (9.18). This leads to

$$\frac{\hat{v}_{dc}}{\hat{i}_o} = \frac{-R_o}{(1 - R_o Cs)} \quad (9.23)$$

Finally, if the influence of the grid voltage perturbation on the DC link voltage needs to be investigated, the transfer function $\hat{v}_{dc}/\hat{e}_d$ has to be computed in (9.18), assuming $Iq = 0$, as

$$\frac{\hat{v}_{dc}}{\hat{e}_d} = \frac{\sqrt{3}}{(1 - R_o Cs)} \quad (9.24)$$

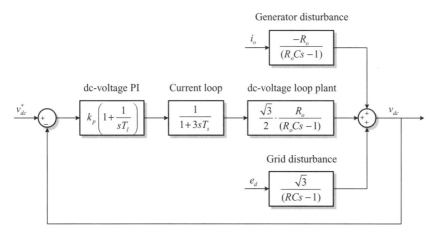

Figure 9.10 Small-signal model of the voltage control loop, generator and grid disturbances

This can be interpreted by considering that the grid perturbation influence on the DC voltage is filtered by a time constant that depends on the DC capacitor and on the equivalent resistance of the DC bus. Hence, the lower the DC storage and the higher the DC load the less immune the DC voltage will be to the grid disturbances. In Figure 9.10 the DC voltage loop is reported with all the disturbances highlighted, where T_s is the sampling period. The plant (9.22), (9.23) and (9.24) has a pole in the right side of the S plane and may be unstable.

However, what it is interesting to note is that as C tends to infinity (large DC link capacitance) the pole tends to zero and makes the system always stable again. It is important to stress that the DC link storage is designed not only in view of DC link filtering but also to offer a power buffer in view of the maximum known variation of the power on the DC bus and of the desired load ride-through protection during utility voltage sag events. These issues suggest that the DC link storage should not be limited too much.

The equivalent resistance seen by the grid converter at its DC terminal depends on the kind of upstream converter to which it is connected. In the case of a PVS the PV array can be connected through a DC/DC converter that works in the constant power mode. In fact, an MPPT algorithm is usually controlling the DC/DC converter in order to adapt the DC voltage across the PV array so that maximum power transfer is obtained. The same situation is obtained in the case of a WTS when a diode bridge plus a DC/DC converter or an inverter is adopted on the generator side. In other words, if the source (a WT generator or PV array) is controlled by a dedicated converter (i.e. situated upstream with respect to the grid converter, considering the power flow from the source to the grid) aiming to optimize the power extraction, in the hypothesis of unchanged atmospherical conditions, the injected power can be assumed constant during transients. In this case the small-signal linearization leads to consider the upstream system as a negative resistance (as shown in Figure 9.11). In fact, if the load is in the order of a few kW and it is assumed that the DC voltage variation is in the range 540–730 V then the error of this approximation at the border will be less than 1 %. Conversely, if the same DC voltage variation is assumed for a converter of hundreds of kW then the error could be 30 %. Therefore, the DC link fed by a constant power source has positive effects for the stability because the plant pole will be in the left side of the S plane.

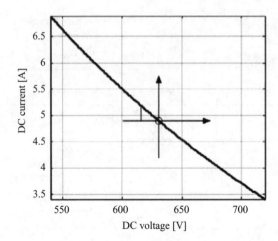

Figure 9.11 Constant power characteristic that can be approximated by a negative resistance (4 kW)

Once the plant has been derived the tuning procedure of the DC voltage PI can be described.

9.3.3 Tuning Procedure of the PI Controller

The design of the PI controllers is done using the zero/pole placement in the z plane, which aims to obtain a better compromise between the high dynamic performance of the DC output voltage and reduction of the AC current overshoot.

The method of 'symmetrical optimum' is a standard design procedure for transfer functions containing a double integration when the controller is included, such as the case of the DC voltage control open-loop transfer function:

$$H_{ov}(s) = \frac{\sqrt{3}k_P (1 + T_I s)}{2T_I s (1 + 3T_s s)(Cs)} \tag{9.25}$$

obtained by considering $R_D C \gg 3T_s$

The main idea is to choose the crossover frequency at the geometric mean of the two corner frequencies, in order to obtain the maximum phase margin ψ, which in turn will result in optimum damping of the DC voltage loop. Thus the Bode diagram shows symmetry with respect to the crossover frequency ω_c. The crossover frequency and the phase margin ψ are related as follows:

$$\omega_c = \frac{1}{3aT_s}$$
$$a = \frac{1 + \cos \psi}{\sin \psi} \tag{9.26}$$
$$a = \sqrt{\frac{T_I}{3T_s}}$$

The gain of the PI regulator at the crossover frequency ω_c is given as

$$k_P = \frac{C}{2\sqrt{3}aT_s} \qquad (9.27)$$

Thus, given the phase margin ψ or the constant a the parameters of the PI controller are determined. The closed-loop transfer function of the system is

$$H_v(s) = \frac{T_I s + 1}{12T_I T_s^2 a s^3 + 4T_I T_s a s^2 + T_I s + 1} \qquad (9.28)$$

The pair of complex poles of $H_v(s)$ result in a slightly underdamped response $\xi = 0.707$ and $45°$ phase margin for $a = 2.4$. Thus:

$$\begin{aligned} k_P &= 0.12 \cdot \frac{C}{T_s} \\ T_I &= 17 \cdot T_s \end{aligned} \qquad (9.29)$$

The bandwidth is expected to be

$$f_{bv} = \frac{\omega_c}{2\pi} = \frac{1}{6\pi a T_s} \approx \frac{1}{50 T_s} = \frac{f_s}{50} \qquad (9.30)$$

Due to the double integrating term in the open-loop transfer function, the closed-loop DC voltage loop exhibits a zero control area. This means that the step response is characterized by considerable overshoot, even though the transients are well damped. In order to eliminate this effect, which is caused by the lead term of the PI controller, a corresponding lag term could be added to the reference signal:

$$H_v(s) = \frac{1}{1 + T_I s} \qquad (9.31)$$

9.3.4 PI-Based Voltage Control Design Example

Consider a system with $C = 500\ \mu F$ on the DC side and a sampling frequency equal to 5 kHz, and the current control loop assumed to have been designed to be critically damped.

If the design of the PI controller is made following the rules expressed in the previous section then $k_P = 0.3$ and $T_I = 0.003$. With these values the crossover frequency should be 100 Hz, the phase margin $40°$ and the open-loop Bode plot should be symmetric, as shown in Figure 9.12.

The system step response is shown in Figure 9.13(a). Usually the DC voltage is kept constant so the overshoot can be a problem only at startup. However, introducing the lag network of (9.31), the overshoot reduces to 5%, which is how it appears in Figure 9.13(b).

The closed-loop Bode plot (Figure 9.14) shows that the bandwidth is 100 Hz, as could be predicted by (9.30).

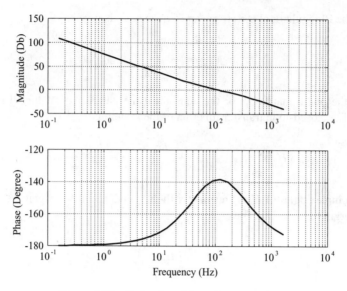

Figure 9.12 Bode plot of the voltage open loop

If a filter on the feedback voltage with a cut-off frequency of 2.5 kHz is adopted without considering it in the controller design, the system experiences a higher overshoot and an oscillatory transient, which is how it is shown in Figure 9.15(a). If the six sample period delays introduced by the 2.5 kHz filter are considered in the design this leads to $T_I = 0.01$, modifying (9.26). The result is in Figure 9.15(b) with the overshoot that is again the 40 % but with a slow down of the system.

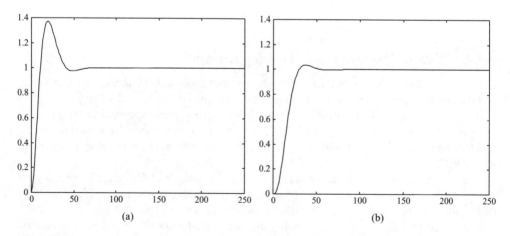

Figure 9.13 Step response of the system (a) without and (b) with a lag network on the reference signal

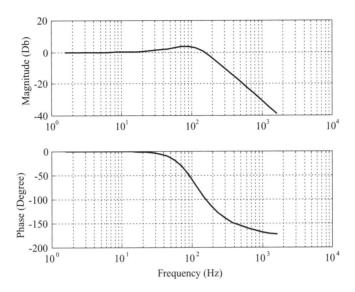

Figure 9.14 Bode plot of the voltage closed loop

9.4 Voltage Oriented Control and Direct Power Control

The power control of the grid converter is based on the instantaneous power theory and as a consequence on the definition of the power in a reference frame, as siscussed in Appendix B.

Typically the voltage oriented control is based on the use of a *dq* frame rotating at ω speed and oriented such that the *d* axis is aligned on the grid voltage vector (Figure 9.16). The space

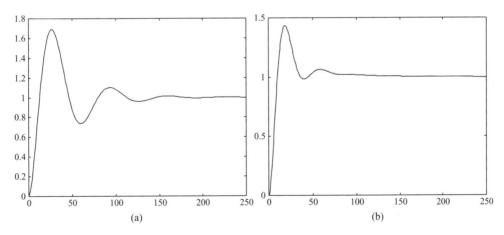

Figure 9.15 Step response of the system with: (a) 80 % overshoot and oscillations due to the filter (not considered in the design) on the feedback signal and (b) 40 % overshoot if the filter is considered in the design

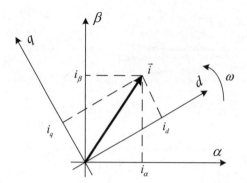

Figure 9.16 Stationary $\alpha\beta$ frame and rotating dq frame: the vector current is projected in both frames

vector of the fundamental harmonic has constant components in the dq frame while the other harmonics space vectors have pulsating components. The main purpose of the grid inverter is to generate or to absorb sinusoidal currents; thus the reference current's components in the dq frame are DC quantities.

The reference current d component i_d^* is controlled to manage active power exchange and typically to perform the DC voltage regulation while the reference current q component i_q^* is controlled to manage the reactive power exchange and typically to obtain a unity power factor. In fact, to have the grid current vector in phase with the grid voltage vector i_q^* should be zero. Thus the active and reactive power produced by the grid converter is

$$P = \frac{3}{2}\left(e_d i_d + e_q i_q\right) \tag{9.32}$$

$$Q = \frac{3}{2}\left(e_q i_d - e_d i_q\right) \tag{9.33}$$

Assuming that the d axis is perfectly aligned with the grid voltage $e_q = 0$, the active power and the reactive power will therefore be proportional to i_d and i_q respectively:

$$P = \frac{3}{2} e_d i_d \tag{9.34}$$

$$Q = -\frac{3}{2} e_d i_q \tag{9.35}$$

Similar results can be achieved in a stationary $\alpha\beta$ frame, but the relation between active/reactive power and the vector current components are more complex. In fact, the active and reactive power produced by the grid converter are

$$P = \frac{3}{2}\left(e_\alpha i_\alpha + e_\beta i_\beta\right) \tag{9.36}$$

$$Q = \frac{3}{2}\left(e_\beta i_\alpha - e_\alpha i_\beta\right) \tag{9.37}$$

However, in this case both the components of the grid voltage vector are nonzero and it is not possible to establish a direct relation between the $\alpha\beta$ components of the current vector and active/reactive power.

It is worth noting that in all the following schemes the voltage and current components in the synchronous dq frame or stationary $\alpha\beta$ frame should be properly filtered before they can be manipulated in the previous active and reactive power formulas. In fact, the previous formulas are used to control the first harmonic active and reactive power and extra controllers are needed to manage pulsating components due to the unbalance operation caused by a fault on the grid, as will be discussed in the chapter 10.

A different approach is the so-called direct power control, which is based on the direct control of the grid inverter switch states in order to obtain the desired active and reactive powers. This approach can be modified with the introduction of the modulator as will be shown in the following.

9.4.1 Synchronous Frame VOC: PQ Open-Loop Control

The most straightforward implementation of the voltage oriented control can be done using a current controller implemented in a dq frame (Figure 9.17) and active and reactive power feed-forward control. The control of the DC voltage modifies the active power reference. Then the active and reactive power command signals are translated into d and q components of the reference current, using the following matrix:

$$\begin{bmatrix} i_d^* \\ i_q^* \end{bmatrix} = \frac{1}{v_{gd}^2 + v_{gq}^2} \begin{bmatrix} v_{gd} & -v_{gq} \\ v_{gq} & v_{gd} \end{bmatrix} \begin{bmatrix} P^* \\ Q^* \end{bmatrix} \tag{9.38}$$

where v_g is the measured grid voltage. Figure 9.18 shows the resulting control scheme.

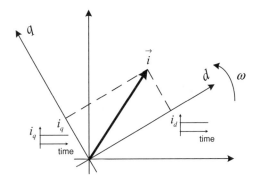

Figure 9.17 The dq frame rotating at synchronous speed with highlighted d and q components of the current

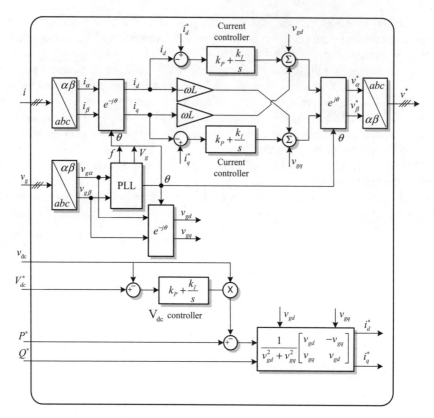

Figure 9.18 PQ open-loop voltage oriented control based on the synchronous dq frame

9.4.2 Synchronous Frame VOC: PQ Closed-Loop Control

An alternative solution to the scheme shown in Figure 9.18 is the closed-loop control of active and reactive powers. In the scheme shown in Figure 9.19 the active and reactive powers are calculated using measurements at the PCC and their values are compared with their set-points. Then PI-based controllers decide the reference d and q components of the reference current while the control of the DC voltage acts directly on the reference current i_d^*. The closed-loop control allows the dynamics of active/reactive power control to be decided as a consequence of a variation of the grid voltage change; hence substantial differences between the Figure 9.18 and Figure 9.19 scheme performances can be observed.

9.4.3 Stationary Frame VOC: PQ Open-Loop Control

The active/reactive power control can also be implemented in a stationary $\alpha\beta$ frame, leading to an indirect voltage oriented control (Figure 9.20). In the case reported in Figure 9.21 there

Grid Converter Control for WTS

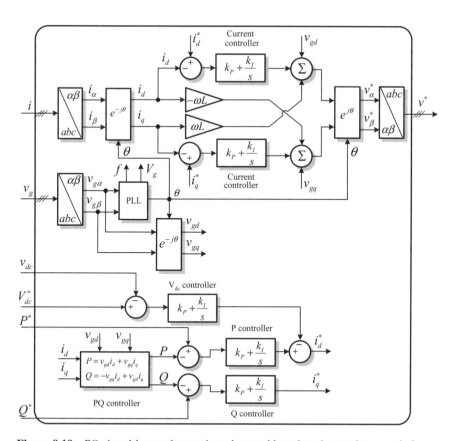

Figure 9.19 PQ closed-loop voltage oriented control based on the synchronous dq frame

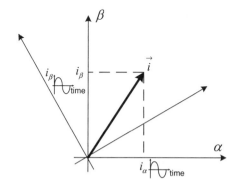

Figure 9.20 Stationary $\alpha\beta$ frame with the components of the current highlighted

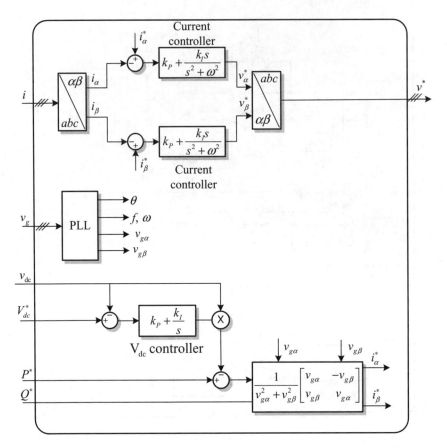

Figure 9.21 PQ open-loop voltage oriented control based on the stationary $\alpha\beta$ frame

is active and reactive power feed-forward control and the DC voltage control acts on the power reference. The PLL is still used for adapting the frequency of the resonant controllers and extracting the first harmonic of the grid voltages used for calculating the reference current.

9.4.4 Stationary Frame VOC: PQ Closed-Loop Control

In the case of the implementation of the power control in the $\alpha\beta$ frame it is also possible to have a closed-loop version. In the scheme shown in Figure 9.22 the active and reactive powers are calculated using measurements at the PCC and their values are compared with their set-points. Then PI-based controllers decide the amplitude and phase of the grid current reference. The control of the DC voltage acts directly on the amplitude value. The PLL is indispensable for providing the grid voltage reference phase with the capacity to calculate the phase displacement of the current in view of the desired reactive power injection. Also in this

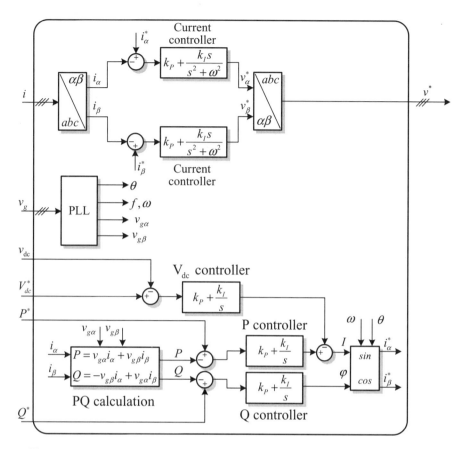

Figure 9.22 PQ closed-loop voltage oriented control based on the stationary $\alpha\beta$ frame

case, as in the case of the scheme of Figure 9.19, the closed-loop control allows the dynamics of active/reactive power control to be decided.

9.4.5 Virtual-Flux-Based Control

The virtual-flux-based approach has been proposed in reference [11] to improve the direct power control operation, but it can also be used for VOC. The idea is to model the grid as an electrical machine and estimate the equivalent air-gap flux for control purposes. The estimation obtained while integrating the measured grid voltage can be used for synchronization purposes, for replacing the PLL and/or for estimating the power injected into the grid for controlling power fluxes. In the first case the obtained results cannot outperform the PLL ones and in the second case sensible improvements with respect to the use of the measured grid voltage are obtained only in the case of voltage sensorless operation, where the grid voltage is

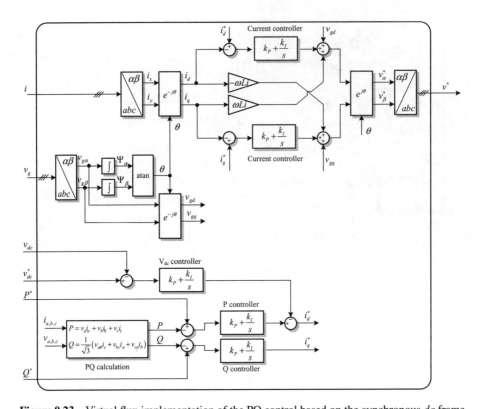

Figure 9.23 Virtual flux implementation of the PQ control based on the synchronous dq frame

approximated to the grid converter voltage that is particular noisy. The core of this technique is in the following set of equations:

$$\begin{cases} \Psi_\alpha = \int v_{g\alpha} dt \\ \Psi_\beta = \int v_{g\beta} dt \\ \sin(\vartheta) = \dfrac{\Psi_\beta}{\sqrt{\Psi_\alpha^2 + \Psi_\beta^2}} \\ \cos(\vartheta) = \dfrac{\Psi_\alpha}{\sqrt{\Psi_\alpha^2 + \Psi_\beta^2}} \end{cases} \quad (9.39)$$

The scheme of the VOC implemented using the virtual flux is reported in Figure 9.23.

9.4.6 Direct Power Control

The direct power control has been developed in analogy to the well-known direct torque control used for drives. In DPC there are no internal current loops and no PWM modulator block because the converter switching states are appropriately selected by a switching table based on the instantaneous errors between the commanded and estimated values of active and reactive power [12] (see Figure 9.24(a)). The main advantage of the DPC is in its simple

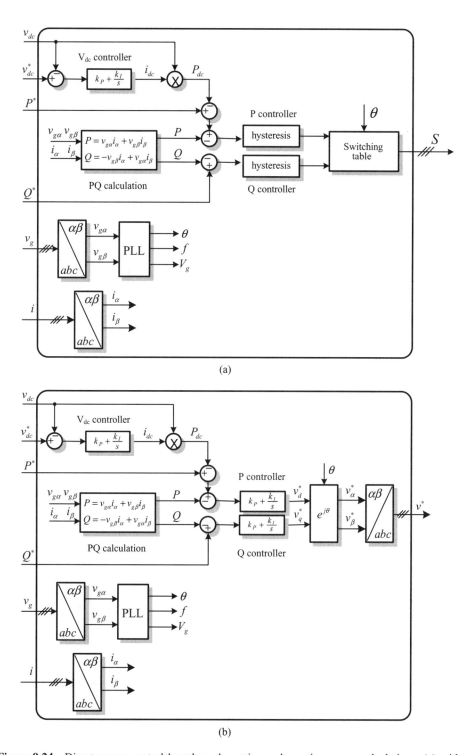

Figure 9.24 Direct power control based on the active and reactive power calculations: (a) without modulator and (b) with modulator

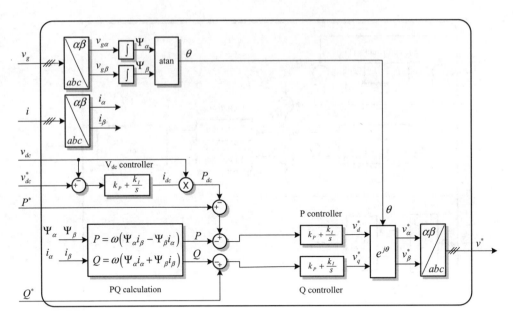

Figure 9.25 Virtual flux implementation of the DPC

algorithm while the main disadvantage is the need for a high sampling frequency to obtain satisfactory performance. A modified version proposed in reference [13] consists in the use of a modulator to synthesize the desired voltage (Figure 9.24(b)). However, if the grid is stiff the active and reactive power loops behave like classical d and q current loops. In the case where the system has the capability of influencing the grid voltage substantial differences may arise. Figure 9.25 shows the implementation of the DPC based on the use of virtual flux.

9.5 Stand-alone, Micro-grid, Droop Control and Grid Supporting

In this section the WT systems not connected to a main grid are discussed. These systems can be autonomous, isolated or forming a micro-grid. The size can be variable, from a few kW to many MW. The main reasons that lead these systems to be isolated from the main grid are:

- Far from grid.
- Difficult terrain.
- Size of load.
- Distance = high losses, poor quality of supply.

Moreover, it should be considered that 2 billion people, one-third of the world's population, do not have access to a reliable electricity supply and 300 000 houses in Europe have no access to the grid.

Hence the topic of ensuring electricity to remote locations is particularly important and the use of renewable energy sources supported by storage solutions is attractive because the widely used alternative (diesel generator) is polluting and expensive. The stand-alone wind system cost can be comparable and in the case of a hybrid wind–diesel system even cheaper than the stand-alone diesel system [14].

The main technical difficulties are related to the frequency and voltage control, the fluctuating nature of the generation/load and to the characteristic of the needed supply (short term/long term/seasonal). Possible solutions are in the use of energy storage or load control (match load to generation, define low priority loads), which usually needs the cooperation of the local community.

This micro-grid can be managed by shunt converters provided that the controllers are properly modified, as will be discussed in the following sections.

9.5.1 Grid-Connected/Stand-Alone Operation without Load Sharing

The wind turbine system developed under the Gaia project offers one good example of a back-to-back converter and control strategy for an operation in both stand-alone and grid-connected modes [15]. In order to test this kind of system a test setup with flexible control implementation features has been built at Aalborg University, as reported in Figure 9.26. The data of the small wind turbine system are reported in Table 9.1.

The developed system was equipped with a standard wind turbine controller capable of controlling a system directly connected to the grid or to a diesel grid (weak grid). Figure 9.27 shows the controller adopted for the Gaia project. The converter voltage is controlled directly; there is not a current control but a current limiter adds a voltage contribution in order to limit the current if it is too high. The DC voltage controller gives a contribution only if the DC voltage is below the natural DC link voltage in order to avoid the PWM saturation. If the DC voltage is above the DC natural voltage the chopper dissipates the power in excess. In short, these are the main features:

- All the available power that can be extracted from the wind turbine is transferred to the grid.
- Additionally, the static reactive power compensation is possible by adjusting the reactive current i_q reference.
- Standard decoupled dq PI control of the currents is used together with voltage feed-forward.
- The PI DC voltage controller provides the d axis current reference.
- The consumed power is decided by the load. Speed adjustment is used to balance the power to some extent. Eventual excess power will be quickly damped in the damping resistor by starting the chopper.
- The output voltage controller aims to control the output voltage with minimal influence on the shape of the nonlinear load currents or load transients.
- The standard PI DC voltage controller and current limiter are also part of the control.

9.5.2 Micro-Grid Operation with Controlled Storage

In the case where the system has been designed to operate in a micro-grid, load sharing is not enough to guarantee stability of the system and storage is an indispensable component. In reference [16] a controlled storage unit that adopts a flyweel is investigated (Figure 9.28).

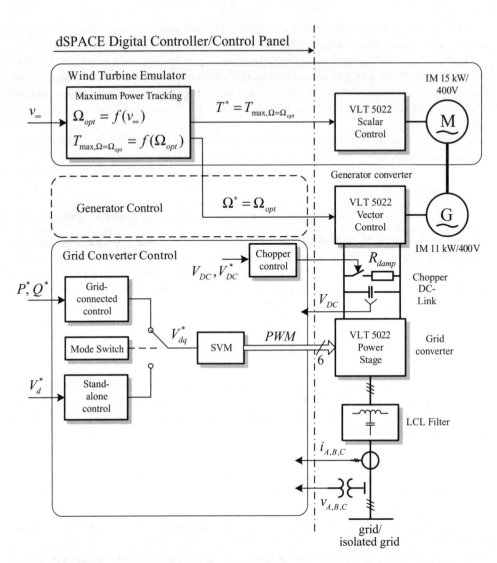

Figure 9.26 Back-to-back control setup developed for testing stand-alone/grid-connected operation

Table 9.1 Data of the small wind turbine designed for the Gaia project

Rated power	11 kW
Hub height	18.2 m
Rotor diameter	13 m
Rotor speed	56 rpm
Total weight	2400 kg
Nacelle weight	900 kg

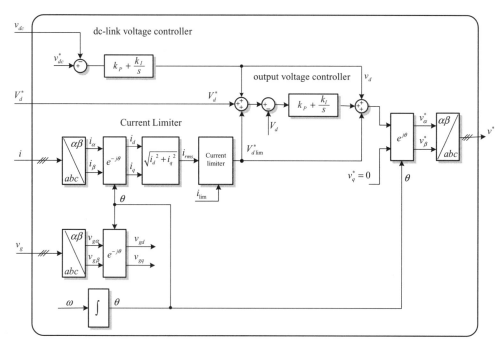

Figure 9.27 Wind turbine controller for the Gaia project

9.5.3 Droop Control

The power sharing among different sources that feed a group of loads through a grid or microgrid can be better managed using the so-called 'droop control' and all its derivatives [17–20]. The basic idea is to reproduce the characteristic of the synchronous generators connected to a steam/water turbine regulated through a speed governor, which are controlled such that the

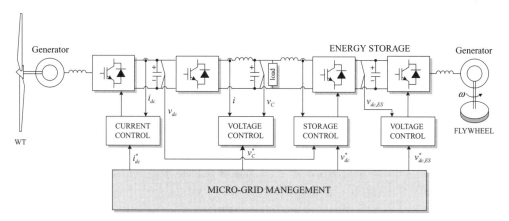

Figure 9.28 A multiloop, again with the AC voltage external and AC current internal. The interesting thing is that the management of the DC voltage is in charge of the controlled storage unit ESS

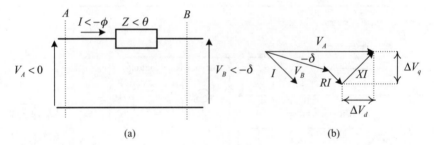

Figure 9.29 (a) Power flow through a line and (b) a phasor diagram

frequency decreases as the fed active power increases and the voltage amplitude decreases as the fed reactive power increases [21].

This principle can be explained by looking at the power transfer between two sections of the line connecting a DG converter to the grid. This can be derived using a short-line model and complex phasors. The analysis below is valid for both single-phase and balanced three-phase systems. Referring to Figure 9.29, the active and reactive power flowing into the line at section A are

$$P_A = \frac{V_A}{Z}\cos\theta - \frac{V_A V_B}{Z}\cos(\theta + \delta) \qquad (9.40)$$

$$Q_A = \frac{V_A^2}{Z}\sin\theta - \frac{V_A V_B}{Z}\sin(\theta + \delta) \qquad (9.41)$$

where δ is the power angle and θ is the power factor angle at section A. As $Z\cos\theta = R$ and $Z\sin\theta = X$ equations (9.40) and (9.41) are rewritten as

$$P_A = \frac{V_A}{R^2 + X^2}[R(V_A - V_B\cos\delta) + XV_B\sin\delta] \qquad (9.42)$$

$$Q_A = \frac{V_A}{R^2 + X^2}[-RV_B\sin\delta + X(V_A - V_B\cos\delta)] \qquad (9.43)$$

Hence

$$\Delta V_d = V_A - V_B\cos\delta = \frac{RP_A + XQ_A}{V_A} \qquad (9.44)$$

$$\Delta V_q = V_B\sin\delta = \frac{XP_A - RQ_A}{V_A} \qquad (9.45)$$

When the DG inverter is connected to the grid through a mainly inductive line (transmission and three-phase distribution line), $X \gg R$, R may be neglected. If also the power angle δ is small, then $\sin\delta \cong \delta$ and $\cos\delta \cong 1$. Equations (9.44) and (9.45) then become

$$\delta \cong \frac{XP_A}{V_A V_B} \qquad (9.46)$$

$$V_A - V_B \cong \frac{XQ_A}{V_A} \qquad (9.47)$$

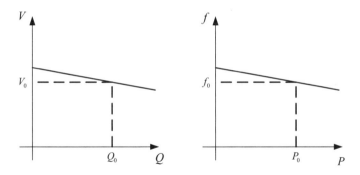

Figure 9.30 Q/V and P/f droop control characteristics

For $X \gg R$, a small power angle δ and a small difference $V_A - V_B$, equations (9.44) and (9.45) show that the power angle depends predominantly on the active power, whereas the voltage difference depends predominantly on the reactive power. In other words, the angle δ or the frequency f can be controlled by regulating the active power P whereas the inverter voltage V_A is controllable through the reactive power Q.

Assuming to have a load or a micro-grid fed by several DG sources, since it is not possible to have DG sources with exactly the same frequency and amplitude, the choice of paralleling several units will lead to an uncontrolled active/reactive power flow depending on the small differences in their frequencies and amplitudes. On the contrary, if a choice is made to control them as a function respectively of the produced active and reactive power, the overall system will find an equilibrium point that will guarantee proper power sharing as a function of the control characteristic.

In fact, assuming the use of these control laws, depicted in Figure 9.30,

$$f - f_0 = -k_P (P - P_0) \qquad (9.48)$$
$$V - V_0 = -k_Q (Q - Q_0) \qquad (9.49)$$

where f_0 and V_0 are the rated frequency and voltage respectively, while P_0 and Q_0 are the set-points for active and reactive DG powers. The micro-grid will find a working point characterized by a V, f working point that will force the DG units to feed P, Q depending on the adopted coefficients k_P and k_Q.

However, low-voltage distribution lines have a mainly resistive nature. Hence, when a DG converter, like a PV inverter, is connected to a low-voltage grid the resistance R can no longer be neglected. On the contrary, often X may be neglected instead of R and the droop regulation defined by (9.46) and (9.47) is no longer effective since adjusting the active power P influences the voltage amplitude while adjusting the reactive power Q influences the frequency. In the general case both X and R have to be considered to regulate the voltage and the frequency droop optimally.

The droop control can be implemented directly, using (9.48) and (9.49), or indirectly by measuring frequency and voltage and imposing the active/reactive power set-point to each of the DG units. Moreover, the droop control can be adopted without a current control loop, with a current control loop and with a multiloop approach (Figures 9.31). Further improvements can

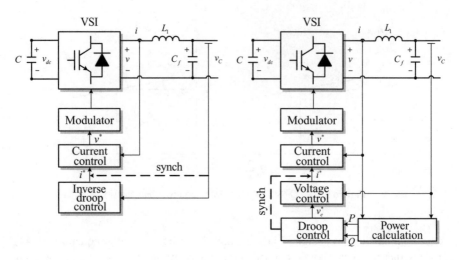

Figure 9.31 Inverse and direct droop control with a current loop and multiloop respectively

be obtained using the 'impedance emulation' approach, which allows a hot-swap of different sources to the grid or micro-grid, thus reducing current overshoot (Figure 9.32) [19].

9.6 Summary

This chapter has investigated the main control structures used for the grid converter adopted in wind turbine systems. The control of a grid converter in the WTS is characterized by three levels: the first is the inner current and/or voltage control loop, the second is the control of active and reactive power that is the subject of this chapter and the third is the supervisory control that decides the active and reactive power set-points. The second level could be enhanced in order to deal with unbalance and to provide a voltage sag ride-through capability, but this goes behind the scope of this chapter and will be treated in Chapter 10. The focus of this chapter has been on the control of the first harmonic active and reactive power that is the main task of the second-level control. The control structures have been classified as those based on the voltage

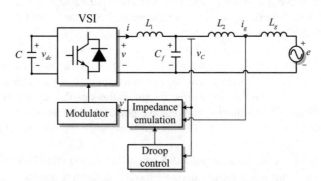

Figure 9.32 Droop method via the impedance emulation

oriented control approach and those based on the direct power control approach. Moreover, it has been shown that the active and reactive power control can be performed in open or closed loops. Finally, it has been discussed how the control structures can be modified in order to allow stand-alone or micro-grid operation.

References

[1] Kazmierkowski, M., Krishnan, R. and Blaabjerg, F., *Control in Power Electronics – Selected Problems*, Academic Press, 2002.
[2] Loh, P. C. and Holmes, D. G., 'Analysis of Multiloop Control Strategies for LC/CL/LCL-Filtered Voltage-Source and Current-Source Inverters'. *IEEE Transactions on Industry Applications*, **41**(2), March/April 2005, 644–654
[3] Loh, P. C., Newman, M. J., Zmood, D. N. and Holmes, D. G., 'A Comparative Anlysis of Multiloop Voltage Regulation Strategies for Single and Three-Phase UPS Systems'. *IEEE Transactions on Power Electronics*, **18**(5), September 2003, 1176–1185.
[4] Vasquez, J. C., Mastromauro, R. A., Guerrero, J. M., Liserre, M., 'Voltage Support Provided by a Droop-Controlled Multifunctional Inverter'. *IEEE Transactions on Industrial Electronics*, **56**(11), November 2009, 4510–4519.
[5] Mastromauro, R. A., Liserre, M., Kerekes, T., Dell'Aquila, A., 'A Single-Phase Voltage Controlled Grid Connected Photovoltaic System with Power Quality Conditioner Functionality', *IEEE Transactions on Industrial Electronics*, **56**(11), November 2009, 4436–4444.
[6] Ortega, R., Loria, A., Nicklasson, P. J. and Sira-Ramirez, H., *Passivity-Based Control of Euler–Lagrange Systems*, New York: Springer-Verlag, 1998. ISBN 1852330163.
[7] Klumpner, C., Liserre, M. and Blaabjerg, F., 'Improved Control of an Active-Front-End Adjustable Speed Drive with a Small DC-Link Capacitor under Real Grid Conditions'. In *PESC 04*, Vol. 2, 20–25 June 2004, pp. 1156–1162.
[8] Liserre, M., Klumpner, C., Blaabjerg, F., Monopoli, V. G. and Dell'Aquila, A., 'Evaluation of Ride-Through Capability of an Active-Front-End Adjustable Speed Drive under Real Grid Conditions'. In *IECON 2004*, Busan, Korea, 2–6 November 2004.
[9] Malesani, L., Rossetto, L., Tenti, P. and Tomasin, P., 'AC/DC/AC PWM Converter with Reduced Energy Storage in the DC link'. *IEEE Transactions on Industry Applications*, **31**(2), 1995, 287–292.
[10] Espinoza, J. R., Joos, G., Perez, M. and Moran, T. L. A., 'Stability Issues in Three-Phase PWM Current/Voltage Source Rectifiers in the Regeneration Mode'. In *Proceedings of ISIE'00*, Vol. 2, 2000, pp. 453–458.
[11] Malinowski, M., Kazmierkowski, M. P., Hansen, S., Blaabjerg, F. and Marques, G. D., 'Virtual-Flux-Based Direct Power Control of Three-Phase PWM Rectifiers'. *IEEE Transactions on Industry Applications*, **37**(4), July/August 2001, 1019–1027.
[12] Noguchi, T., Tomiki, H., Kondo, S. and Takahashi, I., 'Direct Power Control of PWM Converter Without Power-Source Voltage Sensors'. *IEEE Transactions on Industry Applications*, **34**(3), May/June 1998, 473.
[13] Malinowski, M., Jasinski, M. and Kazmierkowski, M. P., 'Simple Direct Power Control of Three-Phase PWM Rectifier Using Space-Vector Modulation (DPC-SVM)'. *IEEE Transactions on Industrial Electronics*, **51**(2), April 2004, 447–454.
[14] Kaldellis, K. and Kavadias, K. A., 'Cost–Benefit Analysis of Remote Hybrid Wind–Diesel Power Stations: Case Study of Aegean Sea Islands'. *Energy Policy*, **35**(3), March 2007, 1525–1538, ISSN 0301-4215.
[15] Teodorescu, R. and Blaabjerg, F., 'Flexible Control of Small Wind Turbines with Grid Failure Detection Operating in Stand-Alone and Grid-Connected Mode", *IEEE Trans. on Power Electronics*, Vol. 19, No. 5, 2004, pp. 1323–1332.
[16] Cárdenas, R., Peña, R., Pérez, Clare, J., Asher, G. and Vargas, F., 'Vector Control of Front-End Converters for Variable-Speed Wind–Diesel Systems'. *IEEE Transactions on Industrial Electronics*, **53**(4), August 2006, 1127.
[17] Guerrero, J. M., de Vicuña, L. G., Matas, J., Castilla, M. and Miret, J., 'A Wireless Controller to Enhance Dynamic Performance of Parallel Inverters in Distributed Generation Systems'. *IEEE Transactions on Power Electronics*, **19**(5), September 2004, 1205–1213.
[18] Karlsson, P., Bjornstedt, J. and Strom, M., 'Stability of Voltage and Frequency Control in Distributed Generation Based on Parallel-Connected Converters Feeding Constant Power Loads'. In *Proceedings of EPE 2005*, Dresden, 2005.

[19] Guerrero, J. M., Matas, J., de Vicuña, L. G., Castilla, M. and Miret, J., 'Decentralized Control for Parallel Operation of Distributed Generation Inverters Using Resistive Output Impedance'. *IEEE Transactions on Industrial Electronics*, **54**(2), April 2007, 994–1004.
[20] De Brabandere, K., Bolsens, B., Van den Keybus, J., Woyte, A., Driesen, J. and Belmans, R., 'A Voltage and Frequency Droop Control Method for Parallel Inverters'. *IEEE Transactions on Power Electronics*, **22**(4), July 2007, 1107–1115.
[21] Kundur, P., *Power system stability and control*, EPRI, McGraw-Hill, 1993.

10

Control of Grid Converters under Grid Faults

10.1 Introduction

The electrical network is a dynamical system, whose behaviour depends upon many factors, as for instance constraints set by power generation systems, the occurrence of grid faults and other contingencies, the excitation of resonances or the existence of nonlinear loads. As a consequence, grid-connected power converters should be designed bearing in mind that they should guarantee a proper operation under generic grid voltage conditions, being especially important to design control algorithms that ensure a robust and safe performance under abnormal grid conditions.

The ever-increasing integration of distributed generation systems, which should fullfil the tight requirements imposed by the grid operator, mainly regarding low-voltage ride-through and grid support during transient grid faults, has encouraged engineers and researchers to improve the conventional control solutions for grid-connected power converters. Despite the fact that the control of power converters under abnormal grid conditions is not a new issue, most of the studies within this field were mainly focused on the control of active rectifiers. The main concern in such applications was to guarantee a proper performance on the DC side of the converter under grid faults. In the grid connection of distributed generators, the interaction between the power converter and the networks under balanced and unbalanced faults is a crucial matter, since it is not only necessary to guarantee that any protection of the converter would not trip but also to support the grid voltage under such faulty operating conditions.

Occurrences of grid faults usually give rise to the appearance of unbalanced grid voltages at the point of connection of the power converter. Under unbalanced conditions, the currents injected into the grid lose their sinusoidal and balanced appearance. The interaction between such currents and the unbalanced grid voltages may give rise to uncontrolled oscillations in the active and reactive power delivered to the network. The proper operation of the power converter under such conditions is a challenging control issue. However, as will be shown in this chapter, the injection of such unbalanced currents may also give rise to other useful effects. For instance, the injection of a proper set of unbalanced currents under unbalanced grid voltage conditions allows attenuating power oscillations, maximizing the instantaneous

---

Grid Converters for Photovoltaic and Wind Power Systems   Remus Teodorescu, Marco Liserre, and Pedro Rodríguez
© 2011 John Wiley & Sons, Ltd

power delivery, or balancing the grid voltage at the point of connection. However, the injection of unbalanced currents into the grid cannot be accurately achieved by using most of the conventional current controllers currently implemented in the industry. For this reason, some improved control structures specifically designed to inject unbalanced currents into the grid will be presented in this chapter.

Depending on the objective of the control strategy used to generate the reference currents during grid faults, the overall performance of the power converter and its interaction with the electrical grid will vary considerably. Moreover, the grid codes regulating the connection of PV and WT systems state specific requirements regarding the injection of active and reactive power during grid faults. Therefore, reference current generation under grid faults is another crucial issue in the control of power converters. Since different current set-points can be found to deliver a given amount of active and reactive power, a detailed study about implementation of different reference current generation strategies, together with an analysis of their performance under generic conditions, will be presented in this chapter. The design of such strategies will be performed considering not only the shape of the currents but also the behaviour of the instantaneous power delivered by the converter.

The currents injected by the power converter into the phases of the grid should always be under control, even though the grid voltage experiences strong variations. Therefore, the control algorithms setting the reference currents should estimate the instantaneous performance of these phase currents at any time, even during transient faults, in order to avoid any overcurrent tripping. For this reason, the last but not the least important issue tackled in this chapter is related to the calculation of the maximum power that can be delivered to the grid, without overpassing the current limits of the power converter.

The control of grid-connected power converters under grid faults studied in this chapter represents an essential complement to the methods treated previously regarding the control of power converters under balanced grid voltage conditions.

10.2 Overview of Control Techniques for Grid-Connected Converters under Unbalanced Grid Voltage Conditions

The fast penetration of renewable energy sources and distributed generation systems has boosted the connection of power converters to the electrical network. Most of the conventional power electronics applications, as for instance motor drives, were mainly focused on processing active power absorbed from the grid to achieve an optimal performance of the electrical motor under control. In modern applications, the connection of the motor drive to the grid is made by using an active rectifier, which provides enhanced features, such as regenerative breaking, power factor correction or active filtering. Likewise, active front-end converters play a decisive role in interfacing renewable energy sources to the grid. In fact, a proper control of the grid-side converter under generic grid conditions is a crucial issue in achieving an effective integration of renewable energy sources into the electrical networks.

Far away from being perfectly constant, balanced and stable, the electrical network behaves as an 'alive' system, with its own dynamical performance, that is influenced by faults, resonances, overloads, etc. Therefore, the control of grid-connected power converters should be carefully tackled in order to guarantee a proper performance under such operating conditions.

In the last years, the operation of power converters under abnormal grid conditions, mainly under voltage sags, has become a challenge for the distributed generation industry due to the

Control of Grid Converters under Grid Faults

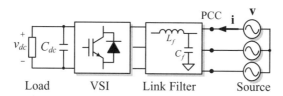

Figure 10.1 Simplified diagram of a grid-connected active rectifier

increasing demands of the grid connection codes regarding low-voltage ride-through (LVRT) and reactive power injection during grid faults. Despite the fact that the first developments were mainly oriented to provide solutions for balanced grid faults, the attention of engineers and researchers has lately moved towards controlling the current injection of grid-connected power converters under unbalanced grid voltage conditions.

Nevertheless, the first analysis oriented to control grid-connected power converters under unbalanced voltage conditions are previous to the grid integration of renewable energy technologies. Actually, the first studies were focused on improving the performance of active rectifiers used in power supplies and motor drives. However, the results of this research have paved the way for further developments that can now be applied to renewable energy systems.

Some relevant studies on the control of active rectifiers under abnormal grid conditions date from the early 1990s. These studies were mainly focused on regulating the AC currents drawn by an active rectifier, like the one shown in Figure 10.1, to reduce its sensitivity when affected by typical grid disturbances.

Under generic grid conditions, the voltage at the point of connection of this active rectifier can be written as

$$\mathbf{v} = \sum_{n=1}^{\infty} \left(\mathbf{v}^{+n} + \mathbf{v}^{-n} + \mathbf{v}^{0n} \right)$$

$$= \sum_{n=1}^{\infty} \left\{ V^{+n} \begin{bmatrix} \cos(n\omega t + \phi^{+n}) \\ \cos(n\omega t - \frac{2\pi}{3} + \phi^{+n}) \\ \cos(n\omega t + \frac{2\pi}{3} + \phi^{+n}) \end{bmatrix} + V^{-n} \begin{bmatrix} \cos(n\omega t + \phi^{-n}) \\ \cos(n\omega t + \frac{2\pi}{3} + \phi^{-n}) \\ \cos(n\omega t - \frac{2\pi}{3} + \phi^{-n}) \end{bmatrix} \right.$$

$$\left. + V^{0n} \begin{bmatrix} \cos(n\omega t + \phi^{0n}) \\ \cos(n\omega t + \phi^{0n}) \\ \cos(n\omega t + \phi^{0n}) \end{bmatrix} \right\} \quad (10.1)$$

where superscripts $+n$, $-n$ and $0n$ represent respectively the positive-, negative- and zero-sequence components of the nth harmonic of the voltage vector $\mathbf{v}$.

Likewise, the current drawn by the three-phase active rectifier of Figure 10.1 can be generically expressed as

$$\mathbf{i} = \sum_{n=1}^{\infty} \left\{ I^{+n} \begin{bmatrix} \sin(n\omega t + \delta^{+n}) \\ \sin(n\omega t + \delta^{+n} - \frac{2\pi}{3}) \\ \sin(n\omega t + \delta^{+n} + \frac{2\pi}{3}) \end{bmatrix} + I^{-n} \begin{bmatrix} \sin(n\omega t + \delta^{-n}) \\ \sin(n\omega t + \delta^{-n} + \frac{2\pi}{3}) \\ \sin(n\omega t + \delta^{-n} - \frac{2\pi}{3}) \end{bmatrix} \right\} \quad (10.2)$$

According to the instantaneous power theory [1], the instantaneous active and reactive powers resulting from the interaction of these generic voltages and currents can be obtained by respectively calculating their inner and cross product, as follows:

$$p = v \cdot i; \quad q = |v \times i| \tag{10.3}$$

As explained in reference [1], the instantaneous powers resulting from (10.3) when the voltage and current is expressed in terms of sequence components consist of both constant and oscillatory terms, given by

$$\bar{p} = \frac{3}{2} \sum_{n=1}^{\infty} \left[V^{+n} I^{+n} \cos(\phi^{+n} - \delta^{+n}) + V^{-n} I^{-n} \cos(\phi^{-n} - \delta^{-n}) \right] \tag{10.4}$$

$$\begin{aligned}
\tilde{p} = \frac{3}{2} \Bigg\{ & \sum_{\substack{m=1 \\ m \neq n}}^{\infty} \left[\sum_{n=1}^{\infty} V^{+m} I^{+n} \cos((\omega_m - \omega_n)t + \phi^{+m} - \delta^{+n}) \right] \\
+ & \sum_{\substack{m=1 \\ m \neq n}}^{\infty} \left[\sum_{n=1}^{\infty} V^{-m} I^{-n} \cos((\omega_m - \omega_n)t + \phi^{-m} - \delta^{-n}) \right] \\
+ & \sum_{m=1}^{\infty} \left[\sum_{n=1}^{\infty} - V^{+m} I^{-n} \cos((\omega_m + \omega_n)t + \phi^{+m} + \delta^{-n}) \right] \\
+ & \sum_{m=1}^{\infty} \left[\sum_{n=1}^{\infty} - V^{-m} I^{+n} \cos((\omega_m + \omega_n)t + \phi^{-m} + \delta^{+n}) \right] \Bigg\}
\end{aligned} \tag{10.5}$$

$$\bar{q} = \frac{3}{2} \sum_{n=1}^{\infty} \left[V^{+n} I^{+n} \sin(\phi^{+n} - \delta^{+n}) - V^{-n} I^{-n} \sin(\phi^{-n} - \delta^{-n}) \right] \tag{10.6}$$

$$\begin{aligned}
\tilde{q} = \frac{3}{2} \Bigg\{ & \sum_{\substack{m=1 \\ m \neq n}}^{\infty} \left[\sum_{n=1}^{\infty} V^{+m} I^{+n} \sin((\omega_m - \omega_n)t + \phi^{+m} - \delta^{+n}) \right] \\
+ & \sum_{\substack{m=1 \\ m \neq n}}^{\infty} \left[\sum_{n=1}^{\infty} - V^{-m} I^{-n} \sin((\omega_m - \omega_n)t + \phi^{-m} - \delta^{-n}) \right] \\
+ & \sum_{m=1}^{\infty} \left[\sum_{n=1}^{\infty} - V^{+m} I^{-n} \sin((\omega_m + \omega_n)t + \phi^{+m} + \delta^{-n}) \right] \\
+ & \sum_{m=1}^{\infty} \left[\sum_{n=1}^{\infty} V^{-m} I^{+n} \sin((\omega_m + \omega_n)t + \phi^{-m} + \delta^{+n}) \right] \Bigg\}
\end{aligned} \tag{10.7}$$

These generic power expressions are evidence that constant terms in the instantaneous active and reactive powers supplied by the grid, $\bar{p}$ and $\bar{q}$, result from the interaction of voltage and current components with the same frequency and sequence, while oscillations in these instantaneous powers, $\tilde{p}$ and $\tilde{q}$, result from the interaction of voltage and current components with either different frequencies or sequences.

The operation of the active rectifier of Figure 10.1 under unbalanced grid conditions can be studied by considering only the fundamental frequency in the previous power expressions, i.e. making $m = n = 1$. Therefore, the instantaneous active and reactive powers associated with the active rectifier of Figure 10.1 under such unbalanced grid conditions can be also written as [2,3]

$$p = P_0 + P_{c2} \cos(2\omega t) + P_{s2} \sin(2\omega t) \tag{10.8}$$

$$q = Q_0 + Q_{c2} \cos(2\omega t) + Q_{s2} \sin(2\omega t) \tag{10.9}$$

where P_0 and Q_0 are the average values of the instantaneous active and reactive powers associated with the active rectifier respectively, whereas P_{c2}, P_{s2}, Q_{c2} and Q_{s2} represent the magnitude of the oscillating terms in these instantaneous powers. In most of the existing studies dealing with power flow in power converters under unbalanced grid conditions [2,4], the voltages and the currents to calculate these power magnitudes are expressed on synchronous reference frames. Hence, the amplitude of these power magnitudes can be calculated as

$$P_0 = \frac{3}{2}\left(v_d^+ i_d^+ + v_q^+ i_q^+ + v_d^- i_d^- + v_q^- i_q^-\right) \tag{10.10}$$

$$P_{c2} = \frac{3}{2}\left(v_d^- i_d^+ + v_q^- i_q^+ + v_d^+ i_d^- + v_q^+ i_q^-\right) \tag{10.11}$$

$$P_{s2} = \frac{3}{2}\left(v_q^- i_d^+ - v_d^- i_q^+ - v_q^+ i_d^- + v_d^+ i_q^-\right) \tag{10.12}$$

$$Q_0 = \frac{3}{2}\left(v_q^+ i_d^+ - v_d^+ i_q^+ + v_q^- i_d^- - v_d^- i_q^-\right) \tag{10.13}$$

$$Q_{c2} = \frac{3}{2}\left(v_q^- i_d^+ - v_d^- i_q^+ + v_q^+ i_d^- - v_d^+ i_q^-\right) \tag{10.14}$$

$$Q_{s2} = \frac{3}{2}\left(-v_d^- i_d^+ - v_d^- i_q^+ + v_d^+ i_d^- + v_q^+ i_q^-\right) \tag{10.15}$$

where v_d^+, v_q^+ and i_d^+, i_q^+ are calculated by means of the Park transform [5] and represent the dq components of the positive-sequence voltage and current vectors expressed on a synchronous reference frame rotating at the fundamental grid frequency ω, whereas v_d^-, v_q^- and i_d^-, i_q^- are the components of the negative-sequence voltage and current vectors lying on a synchronous reference frame rotating at $-\omega$ respectively.

One of the main objectives in the control of active rectifiers is to provide a constant DC output voltage. Oscillations in the DC output voltage are directly associated with the energy variation in the DC-bus capacitor, which is linked to the difference between the input and the output active powers associated with the active rectifier. Therefore, to guarantee a constant DC output voltage under constant load conditions, it is necessary to calculate the proper set of AC currents to be drawn by the active rectifier in order to guarantee that the active power absorbed by the active rectifier has a constant value with no power oscillations under generic grid conditions. As four degrees of freedom exist in the calculation of the currents to be injected by the power converter, namely $[i_d^+, i_q^+, i_d^-, i_q^-]$, four of the six power magnitudes defined by (10.10) to (10.15) can be controlled for given grid voltage conditions defined by $[v_d^+, v_q^+, v_d^-, v_q^-]$. Many of the studies on the control of active rectifiers under unbalanced grid voltage conditions [3,6,7] have collected such power terms in the following matrix expression:

$$\begin{bmatrix} P_0 \\ Q_0 \\ P_{c2} \\ P_{s2} \end{bmatrix} = \frac{3}{2} \underbrace{\begin{bmatrix} v_d^+ & v_q^+ & v_d^- & v_q^- \\ v_q^+ & -v_d^+ & v_q^- & -v_d^- \\ v_d^- & v_q^- & v_d^+ & v_q^+ \\ v_q^- & -v_d^- & -v_q^+ & v_d^+ \end{bmatrix}}_{M_{4\times 4}} \begin{bmatrix} i_d^+ \\ i_q^+ \\ i_d^- \\ i_q^- \end{bmatrix} \qquad (10.16)$$

By inverting the matrix $M_{4\times 4}$ in the system depicted in (10.16), it is possible to find the current set-point that gives rise to a certain value of the active and reactive power components for given grid voltage conditions, i.e.

$$\begin{bmatrix} i_d^{+*} \\ i_q^{+*} \\ i_d^{-*} \\ i_q^{-*} \end{bmatrix} = M_{4\times 4}^{-1} \times \frac{2}{3} \begin{bmatrix} P_0 \\ Q_0 \\ P_{c2} \\ P_{s2} \end{bmatrix}. \qquad (10.17)$$

Therefore, from (10.17) it is possible to calculate the currents to be drawn by the converter that would give rise to a certain value of P_0 and Q_0, while also cancelling out the active power oscillations terms, P_{c2} and P_{s2}, under unbalanced grid voltage conditions. These power requirements can be fulfilled by finding the current references as shown in (10.18), where Q_0 has been intentionally considered equal to zero:

$$\begin{bmatrix} i_d^{+*} \\ i_q^{+*} \\ i_d^{-*} \\ i_q^{-*} \end{bmatrix} = M_{4\times 4}^{-1} \times \frac{2}{3} \begin{bmatrix} P_0 \\ 0 \\ 0 \\ 0 \end{bmatrix}. \qquad (10.18)$$

The expression shown in (10.18) for setting the current references has been extensively used in the literature [7–9] and, although other alternative approaches have been presented more recently [10], the conclusions obtained give rise to analogous results. In other studies, the power losses associated with the link filter between the power converter and the grid are also considered in the calculation of the reference currents [6, 11]. Nevertheless, all these works mainly focus on attenuating active power oscillations but the analysis of the reactive power oscillations is normally beyond their scope.

In parallel with the discussion regarding the regulation of the currents to be injected by grid-connected converters in order to keep the active power constant during unbalanced voltage conditions, several studies started to stress the importance of proposing new operation modes in order to achieve additional features. Some of the first works in this area were focused on proposing control laws oriented to determine the currents to be drawn by active rectifiers for delivering active and reactive power according to a certain power factor, while cancelling out the instantaneous active power ripple [6]. Actually, there exist infinite combinations for the currents to be injected into the grid by a three-phase grid-connected power converter in order to develop certain net values of active and reactive powers, P_0 and Q_0, under given grid voltage conditions. Therefore, depending on the control objective, e.g. perfect control of the instantaneous active and reactive powers, cancellation of injection of active and reactive power oscillations, injection of sinusoidal and balanced currents into the grid, etc., different expressions can be proposed to calculate the currents to be drawn by the grid-connected power converter to exchange given averaged values of active and reactive powers [12–16]. The calculation of these currents can be conducted from a generic point of view, considering any power flow direction, which covers power converters used in both active rectifiers and inverters, expressing voltage and current vectors on any generic reference frame (dq, $\alpha\beta$, abc, $m\angle\theta$, ...) and paying equivalent attention to the evolution of the instantaneous active and reactive powers, which allows controlling not only the quality of the voltage on the DC side of the power converter but also the interaction with the grid on its AC side. This last feature is a key issue in the control of distributed generators during unbalanced grid faults [14], as pointed out at the beginning of this chapter.

Once it is assumed that the reference currents to exchange a given power with the grid under generic voltage conditions are properly calculated, it is necessary to have a suitable current controller that is able to inject such currents into the grid. The current control loop structures proposed for tracking the current references obtained in (10.18) have been significantly improved throughout the years. The first works presented in the 1990s proposed the implementation of a single PI current controller on the positive-sequence synchronous reference frame for tracking both positive- and negative-sequence reference currents, which was not an optimal solution [6, 11].

Later, by the end of the 1990s, other works introduced two control loops, one for the positive-sequence and another one for the negative-sequence currents [4]. As a difference with the single-loop controller, this structure permitted an increase in the performance of the control and a reduction in the steady state error, without hindering the stability of the system. In this kind of double reference frame regulator, the feedback currents are projected on both the positive and the negative reference frames, by means of the Park transformation, in order to control the positive- and negative-sequence current vectors independently. As will be explained in Section 10.3, the interaction between current vectors and reference frames with different sequences gives rise to oscillations at twice the fundamental frequency in the dq signals obtained from the Park transformation. Since this oscillatory signal cannot be suitably controlled by the PI regulators of the conventional synchronous controllers, some authors proposed using a notch filter tuned at twice the fundamental frequency to attenuate such oscillations [4]. Another solution for reducing the effect of these oscillations is based on modifying the conventional loop control by adding some additional resonant regulators specifically focused on reducing the steady-state error generated by the PI controller at twice the fundamental frequency [6, 11]. Some other decoupled controllers suitable to work under unbalanced conditions are discussed in Section 10.3.

As an alternative to the current controllers based on the double synchronous reference frame, resonant controllers operating on the stationary reference frame have been shown to be an effective solution to regulate the currents injected by the grid-connected power converters under unbalanced and distorted grid operating conditions [17, 18]. These controllers are based on frequency adaptive filters, which exhibit the same performance for both positive and negative frequencies.

Other solutions not based on vector-oriented control (VOC), for instance based on hysteresis current controllers [10, 19], direct power control methods [20, 21] and model-based predictive control [22], can also be used to control grid-connected power converters operating under generic grid conditions. Among these solutions, those based on hysteresis are robust and offer a fast dynamic response [23].

In the framework of this short overview about control of grid-connected converters under unbalanced grid voltage conditions, it is mandatory to point out two relevant issues that all, classical and modern, techniques have in common. On the one hand, it is necessary to stress that the successful performance of any current controller of grid-connected converters greatly depends on the accuracy of the grid synchronization system used [24]. In single-phase systems, the amplitude and phase angle of the grid voltage should be accurately estimated to achieve the expected instantaneous exchange of power with the grid. In the case of three-phase systems, the synchronization requirements are even more severe, since the symmetrical components of the grid voltage during unbalanced grid faults should be perfectly estimated to achieve a proper control of the positive- and negative-sequence currents injected into the grid. Several advanced grid synchronization techniques, able to provide good results under generic grid voltage conditions, were presented in Chapters 4 and 8 for single- and three-phase systems respectively.

On the other hand, depending on the desired performance for the instantaneous active and reactive powers exchanged with the grid, the reference current generated by the current controller during unbalanced grid faults may be balanced, unbalanced or even distorted. As a consequence, the current injected by the grid-connected power converter is prone to be different from phase to phase. For this reason, one additional crucial issue to be controlled in grid-connected converters is the instantaneous evolution of the currents injected by the power converter in each phase of the grid in order to avoid any undesired overcurrent tripping.

10.3 Control Structures for Unbalanced Current Injection

The current injected into the grid by a power converter should keep a certain relationship with the voltage at the point of connection in order to deliver a given amount of active and reactive power. For this reason, the structure of the current controller is a key issue in the design of grid-connected power converters. Under grid faults, the injected currents are very different from the ones injected during regular operating conditions. In single-phase systems, grid faults give rise to variations in the amplitude and phase angle of the single-phase voltage at the point of connection of the power converter. As a consequence, the power converter should also change the amplitude and phase angle of the injected current in order to ride through the fault. Usually, conventional current controllers used in single-phase applications are able to withstand properly this kind of faults. As commented in Chapter 8, faults in three-phase systems create unbalanced voltages. Depending on the control objectives, the currents to be injected into the grid might include negative-sequence components to thwart the effects of the fault. Most conventional current controllers used in three-phase power converters are not suitable for injecting unbalanced currents, especially when the grid voltage is unbalanced.

Current controllers based on synchronous reference frames are the most extended solution for controlling current injection in distributed generation systems. As shown in Chapter 9, most of these controllers use PI regulators, operating on the synchronous dq axes, to control the injected currents. When balanced positive-sequence currents are injected into the grid, these control structures achieve a good performance. However, if unbalanced currents are injected the behaviour of these controllers is quite deficient, since there is not any specific control loop for the negative-sequence current components.

In the following, different control structures, specially designed to work with positive- and negative-sequence currents, will be presented.

10.3.1 Decoupled Double Synchronous Reference Frame Current Controllers for Unbalanced Current Injection

The most intuitive way to control a current vector, consisting of positive- and negative-sequence components, is to use a current controller based on two synchronous reference frames, rotating at the fundamental grid frequency in the positive and the negative directions respectively. Figure 10.2 shows the structure of this controller, which is based on a double synchronous reference frame (DSRF).

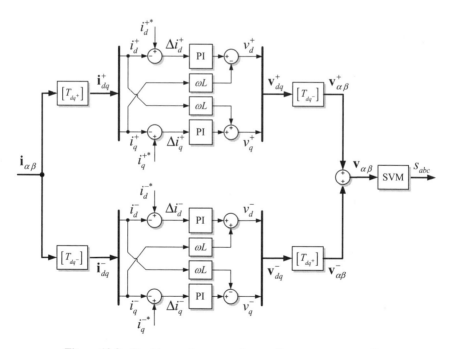

Figure 10.2 Double synchronous reference frame current controller

As shown in the figure, the measured currents are transformed into the positive and negative reference frames by using the Park transformation, $[T_{dq}{}^+]$ and $[T_{dq}{}^-]$, being

$$[T_{dq^+}] = [T_{dq^-}]^T = \begin{bmatrix} \cos(\theta') & \sin(\theta') \\ -\sin(\theta') & \cos(\theta') \end{bmatrix} \quad (10.19)$$

where θ' is the phase angle detected by the synchronization systems for the positive- and negative-sequence voltage vectors. It is worth pointing out that these phase angles should be estimated by an accurate grid synchronization system like the ones presented in Chapter 8. As depicted in the figure, the controller used in each synchronous reference frame is similar to those presented in Chapter 9, in which PI controllers were usually used to regulate the injected currents. It should be noted in this figure that the terms for decoupling the dq signals (ωL) on the positive and the negative sequences have different signs, due to their opposite rotation directions.

Considering that the current vector to be injected into the grid is given by the expression

$$\mathbf{i} = I^+ \begin{bmatrix} \sin(\omega t + \delta^+) \\ \sin(\omega t + \delta^+ - \frac{2\pi}{3}) \\ \sin(\omega t + \delta^+ + \frac{2\pi}{3}) \end{bmatrix} + I^- \begin{bmatrix} \sin(\omega t + \delta^-) \\ \sin(\omega t + \delta^- + \frac{2\pi}{3}) \\ \sin(\omega t + \delta^- - \frac{2\pi}{3}) \end{bmatrix} \quad (10.20)$$

its projection on the positive and negative synchronous reference frames, rotating at $+\omega$ and $-\omega$ respectively, can be written as

$$\mathbf{i}_{dq}^+ = \begin{bmatrix} i_d^+ \\ i_q^+ \end{bmatrix} = \begin{bmatrix} \bar{i}_d^+ \\ \bar{i}_q^+ \end{bmatrix} + \begin{bmatrix} \tilde{i}_d^+ \\ \tilde{i}_q^+ \end{bmatrix} = \underbrace{I^+ \begin{bmatrix} \cos(\delta^+) \\ \sin(\delta^+) \end{bmatrix}}_{\text{DC terms}}$$

$$+ \underbrace{I^- \cos(\delta^-) \begin{bmatrix} \cos(2\omega t) \\ -\sin(2\omega t) \end{bmatrix} + I^- \sin(\delta^-) \begin{bmatrix} \sin(2\omega t) \\ \cos(2\omega t) \end{bmatrix}}_{\text{AC terms}} \quad (10.21)$$

$$\mathbf{i}_{dq}^- = \begin{bmatrix} i_d^- \\ i_q^- \end{bmatrix} = \begin{bmatrix} \bar{i}_d^- \\ \bar{i}_q^- \end{bmatrix} + \begin{bmatrix} \tilde{i}_d^- \\ \tilde{i}_q^- \end{bmatrix} = \underbrace{I^- \begin{bmatrix} \cos(\delta^-) \\ \sin(\delta^-) \end{bmatrix}}_{\text{DC terms}}$$

$$+ \underbrace{I^+ \cos(\delta^+) \begin{bmatrix} \cos(2\omega t) \\ \sin(2\omega t) \end{bmatrix} + I^+ \sin(\delta^+) \begin{bmatrix} -\sin(2\omega t) \\ \cos(2\omega t) \end{bmatrix}}_{\text{AC terms}} \quad (10.22)$$

The expressions shown in (10.21) and (10.22) give evidence of the cross-coupling between the dq axis signals of both synchronous reference frames. This coupling effect is manifested by a 2ω oscillation overlapping the DC signals on the dq axes, being ω the fundamental grid frequency, as discussed in Chapter 8. The amplitude of the 2ω oscillations on the dq

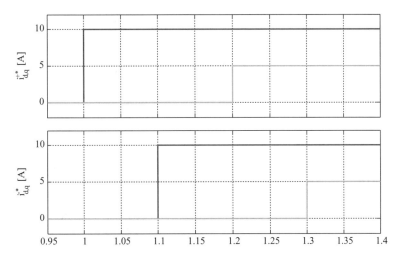

Figure 10.3 Reference currents for i_{dq+} and i_{dq-} (d current axis signal in black and q current axis signal in gray)

axes of the positive-sequence reference frame matches the mean value on the dq axes of the negative-sequence one, and vice versa.

To illustrate the effect of the cross-coupling between the positive and negative reference frames, a power converter injecting positive- and negative-sequence currents into the grid has been considered. The reference currents for this power converter, expressed on the positive and negative reference frames, dq^+ and dq^-, are shown in Figure 10.3.

When the system of Figure 10.2 is used for controlling the injection of the reference currents shown in Figure 10.3, the measured currents, expressed on the dq^+ and dq^- axes, are the ones shown in Figure 10.4. In this figure, it can be appreciated how the injection of any positive-sequence current gives rise to oscillations at 2ω on the negative reference frame, and vice

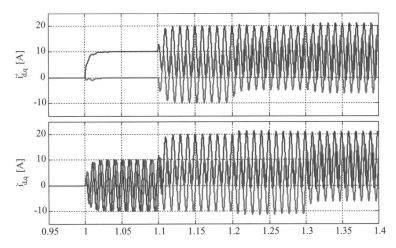

Figure 10.4 Measured currents on the dq^+ and dq^- reference frames (d current axis signal in black and q current axis signal in gray)

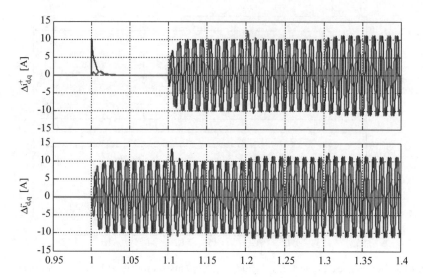

Figure 10.5 Error signals on the double synchronous reference frame controller (d error current signal in black and q error current signal in gray)

versa. It can be also proved from this figure that the amplitude of the oscillations in one of the synchronous reference frames matches the DC amplitude of the other one.

Oscillations at 2ω in the measured currents cannot be totally cancelled out by the PI controller, which gives rise to steady state errors when tracking the reference currents. This drawback is illustrated in Figure 10.5, where the error signals at the input of the PI controllers have been plotted, $\Delta i_{d,q+}$ and $\Delta i_{d,q-}$. It is important to stress that the mean value of the error signals is equal to zero for all the dq components, since the PI controller presents infinite gain for DC inputs.

Obviously, these 2ω oscillations on the synchronous reference frames should be cancelled out in order to achieve full control of the injected currents under unbalanced conditions. In the following, some solutions oriented to overcome this drawback are presented.

Maybe the most straightforward solution to attenuate the effects of the 2ω oscillations consist of filtering the measured currents by using a notch filter (NF), tuned at 2ω, as shown in Figure 10.6 [4].

The effect of including the aforementioned notch filter on the measured currents is illustrated in Figure 10.7, where the current errors on the positive and negative reference frames are plotted. It can be noticed in this figure that the performance of the controller is clearly improved, being the 2ω oscillations in the current error significantly reduced after a nonnegligible transient period.

It is worth pointing out that the notch filter affects the direct chain of the control loop, reducing the phase margin of the system, and ends by hindering the system stability if the damping factor of the notch filter is too high. In the example of Figure 10.7, the damping factor was set to 0.5. Lower values for the damping factor would reduce the settling time of the system. However, it would make the notch filter more selective, which may become a problem in the case of grid frequency deviations. For this reason, this notch filter should be frequency adaptive in order effectively to cancel out the 2ω oscillations at any grid frequency value.

Control of Grid Converters under Grid Faults

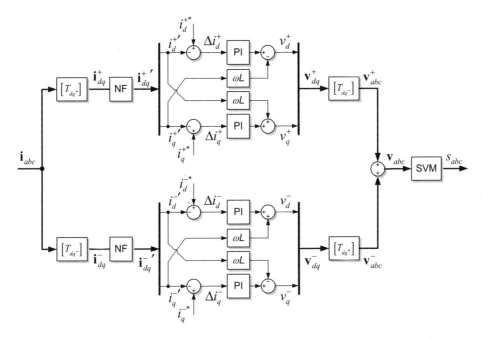

Figure 10.6 Double synchronous reference frame current controller using a notch filter

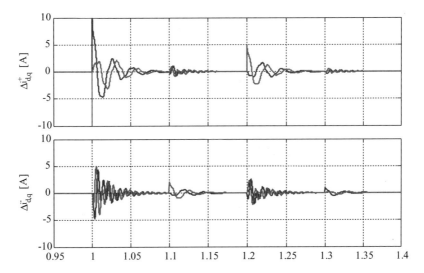

Figure 10.7 Error signals on the double synchronous reference frame controller using notch filters on the measured currents (d error current signal in black and q error current signal in gray)

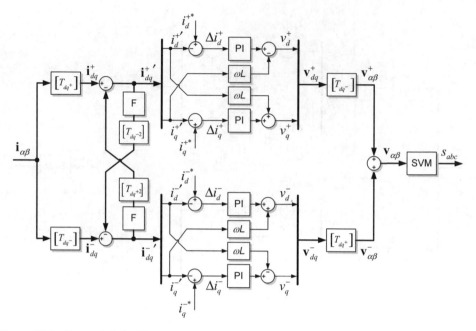

Figure 10.8 Decoupled double synchronous reference frame current controller using a decoupling network based on the measured signals

The same effect provided by the notch filter, regarding the attenuation of 2ω oscillations, can be achieved by adding the 2ω oscillations of the measured currents to the reference currents. In this way, the input signal to the PI controller would be free of oscillations. However, in order to avoid any mismatching in cancelling these oscillations, a resonant controller, also tuned at 2ω, is connected in parallel with the original PI controller [6].

In the previous techniques, the effect of the 2ω oscillation resulting from the cross-coupling between reference frames and current vectors with different sequences has been overcome by using filters or by modifying the controller. Nevertheless, as previously evidenced in the equations (10.21) and (10.22), a relationship exists between the amplitude of the AC oscillations in the positive reference frame and the DC values in the negative reference frame, and vice versa. Therefore, it is feasible to use a cross-decoupling network, like the one presented in Chapter 8, to make both reference frames independent from each other. Implementation of this decoupled double synchronous reference frame (DDSRF) current controller is shown in Figure 10.8, where F represents a first-order low-pass filter.

As a difference from the previous case, where the 2ω oscillations were just attenuated by using a notch filter, the cross-feedback decoupling network allows a perfect estimation of the amplitude of the oscillations in the measured currents, achieving their complete cancellation at the input of the PI controllers. Moreover, as evidenced in the current error signals shown in Figure 10.9, where the cut-off frequency of the low-pass filter was set to $\omega_f = \omega/\sqrt{2}$, the DDSRF current controller exhibits a better dynamics than the one based on notch filters.

Since the amplitude of the AC oscillations in the measured currents on the positive reference frame matches the DC values in the measured currents on the negative reference frame, and

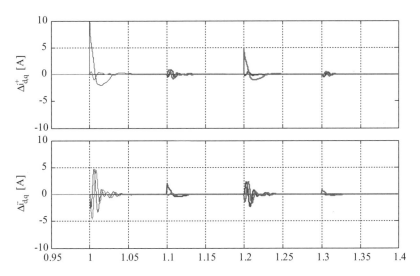

Figure 10.9 Error signals on the double synchronous reference frame current controller using a decoupling network based on the measured signals (d error current signal in black and q error current signal in gray)

taking into account that ideal PI controllers are able to track DC references accurately, it seems logical to use the reference for the negative-sequence currents as feedforward signals for cancelling out the 2ω oscillations in the measured currents on the positive reference frame. A similar reasoning can be followed for cancelling out the oscillations in the negative reference frame. This decoupling network would work properly provided that the PI controllers are able to track the DC references perfectly. However, if any tracking error happens, there will be a mismatch between the DC reference currents and the DC measured currents on the synchronous reference frames. Therefore, under such conditions, it would not be possible to achieve a full cancellation of the oscillations in the measured currents. However, any DC error in tracking the references currents can be detected at the input of the PI controllers by using simple low-pass filters. Thus, the output signal of these filters can be used to compensate the error made by the feedforward loops in the cancellation of the 2ω oscillations. Implementation of this decoupled double synchronous reference frame current controller, based on feedforward plus additional feedback loops, is shown in Figure 10.10.

The error signals on the decoupled double synchronous reference frame, shown in Figure 10.11, shows the good performance of this last controller. In this case, the cut-off frequency of the low-pass filter was set to $\omega_f = \omega/\sqrt{2}$. It can be seen from the figure that this control structure shows very fast dynamics while achieving a zero steady-state error when injecting unbalanced currents.

10.3.2 Resonant Controllers for Unbalanced Current Injection

When a synchronous controller based on PI is transformed into the stationary reference frame, a proportional resonant (PR) controller is obtained [17, 25]. Therefore, both are equivalent to each other, but they work on different reference systems. The synchronous controller is

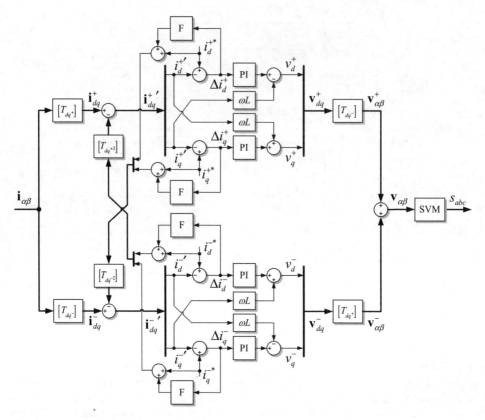

Figure 10.10 Decoupled double synchronous reference frame current controller using a decoupling network based on the reference and the error signals

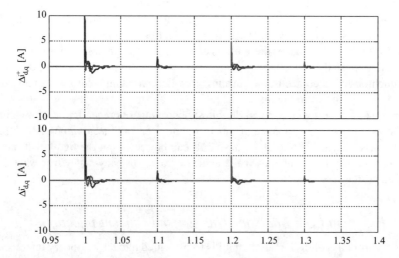

Figure 10.11 Error signals on the double synchronous reference frame current controller using a decoupling network based on the reference and the error signals (*d* error current signal in black and *q* error current signal in gray)

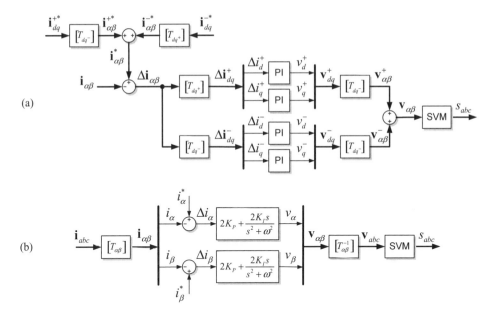

Figure 10.12 Equivalent controllers on: (a) the synchronous and (b) the stationary reference frames

based on an infinite gain DC controller rotating at ω frequency on a synchronous reference frame; meanwhile, the PR controller on the stationary frame presents an infinite gain at $\pm\omega$ frequencies.

The synchronous controller uses the phase angle of the grid voltage as a basis magnitude, whereas the resonant controller uses the grid frequency. As a consequence, the synchronous controller needs to build two synchronous control loops to control unbalanced currents, one for the positive-sequence component and another for the negative-sequence component, as the phase angles of both sequence components are not correlated. However, since the value of the grid frequency is the same for both sequence components, just one resonant controller is necessary to control both positive- and negative-sequence currents simultaneously. Therefore, no decoupling networks are needed to deal with both sequences simultaneously.

The equivalence between the PI controller on the double synchronous reference frame and the PR controller on the stationary reference frame is demonstrated in the following. When only the positive reference frame controller is considered in the system shown in Figure 10.12(a), the next expression can be written as

$$\mathbf{v}_{dq^+} = \begin{bmatrix} v_{d^+} \\ v_{q^+} \end{bmatrix} = [PI(t)] * \Delta \mathbf{i}_{dq^+} = \begin{bmatrix} k_p + k_i \int & 0 \\ 0 & k_p + k_i \int \end{bmatrix} * \begin{bmatrix} \Delta i_{d^+} \\ \Delta i_{q^+} \end{bmatrix} \quad (10.23)$$

where * represents the convolution product in the time domain. Likewise, the output of the negative reference frame controller of Figure 10.12(a) is given by

$$\mathbf{v}_{dq^-} = \begin{bmatrix} v_{d^-} \\ v_{q^-} \end{bmatrix} = [PI(t)] * \Delta \mathbf{i}_{dq^-} = \begin{bmatrix} k_p + k_i \int & 0 \\ 0 & k_p + k_i \int \end{bmatrix} * \begin{bmatrix} \Delta i_{d^-} \\ \Delta i_{q^-} \end{bmatrix} \quad (10.24)$$

In the Laplace domain, the transfer functions of these two controllers on the stationary reference frame are given by

$$\mathbf{v}(s)_{\alpha\beta+} = [PI(s)_{\alpha\beta+}] \Delta\mathbf{i}(s)_{\alpha\beta+} \qquad (10.25)$$

$$\mathbf{v}(s)_{\alpha\beta-} = [PI(s)_{\alpha\beta-}] \Delta\mathbf{i}(s)_{\alpha\beta-} \qquad (10.26)$$

where the stationary transfer functions for the PI controller on the positive and negative reference frames are defined as

$$\begin{aligned}
\left[PI(s)_{\alpha\beta+}\right] &= [T_{dq^-}][PI(s)][T_{dq^+}] \\
&= \frac{1}{2} \begin{bmatrix} (PI(s+j\omega)+PI(s-j\omega)) & j(PI(s+j\omega)-PI(s-j\omega)) \\ j(-PI(s+j\omega)+PI(s-j\omega)) & (PI(s+j\omega)+PI(s-j\omega)) \end{bmatrix} \\
&= \begin{bmatrix} k_p + \dfrac{k_i s}{s^2+\omega^2} & \dfrac{k_i \omega}{s^2+\omega^2} \\ -\dfrac{k_i \omega}{s^2+\omega^2} & k_p + \dfrac{k_i s}{s^2+\omega^2} \end{bmatrix},
\end{aligned} \qquad (10.27)$$

$$\begin{aligned}
\left[PI_{\alpha\beta-}(s)\right] &= [T_{dq^+}][PI(s)][T_{dq^-}] \\
&= \frac{1}{2} \begin{bmatrix} (PI(s+j\omega)+PI(s-j\omega)) & j(-PI(s+j\omega)+PI(s-j\omega)) \\ j(PI(s+j\omega)-PI(s-j\omega)) & (PI(s+j\omega)+PI(s-j\omega)) \end{bmatrix} \\
&= \begin{bmatrix} k_p + \dfrac{k_i s}{s^2+\omega^2} & -\dfrac{k_i \omega}{s^2+\omega^2} \\ \dfrac{k_i \omega}{s^2+\omega^2} & k_p + \dfrac{k_i s}{s^2+\omega^2} \end{bmatrix}.
\end{aligned} \qquad (10.28)$$

The transformation matrixes $[T_{dq+}]$ and $[T_{dq-}]$ in (10.27) and (10.28) correspond to the Park transform, as indicated in (10.19).

Adding the two partial transfer functions defined in (10.25) and (10.26) for the positive and negative reference frame controllers, the following expression is obtained:

$$\mathbf{v}(s)_{\alpha\beta} = \mathbf{v}(s)_{\alpha\beta+} + \mathbf{v}(s)_{\alpha\beta-} \qquad (10.29)$$

$$\begin{bmatrix} v(s)_\alpha \\ v(s)_\beta \end{bmatrix} = 2 \begin{bmatrix} k_p + \dfrac{k_i s}{s^2+\omega^2} & 0 \\ 0 & k_p + \dfrac{k_i s}{s^2+\omega^2} \end{bmatrix} \begin{bmatrix} \Delta i_\alpha \\ \Delta i_\beta \end{bmatrix} \qquad (10.30)$$

As shown in (10.30), the joint action of a positive and a negative synchronous reference frame controller, both using PI controllers, is equivalent to implementing two PR controllers to cancel out the current error, one on the α axis and another on the β axis. The diagonal terms in the transfer function matrix of (10.30) are equal to zero, which indicates that there is no cross-coupling between the signals on the α and β stationary axes. The analysis carried out in this section demonstrates that PR controllers on the stationary reference frame are able to cancel out the error on the positive- and negative-sequence components simultaneously when unbalanced currents are injected by the power converter. Moreover, the simplicity of the

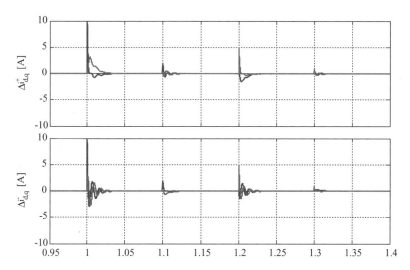

Figure 10.13 Error signals on the double synchronous reference frame using the current controller of Figure 10.12(a) (*d* error current signal in black and *q* error current signal in gray)

resultant PR controller makes it very attractive to be implemented in digital control platforms because of its low computational burden. Figure 10.12(b) shows the diagram for implementing a current controller based on PR on the $\alpha\beta$ stationary reference frame.

The *dq* error signals resulting from the synchronous controller when injecting the same reference currents as in previous study cases are shown in Figure 10.13, whereas the $\alpha\beta$ error signals for the PR controller are shown in Figure 10.14. As shown in this figure, the PR controller and its equivalent synchronous controller achieve a zero steady-state error and

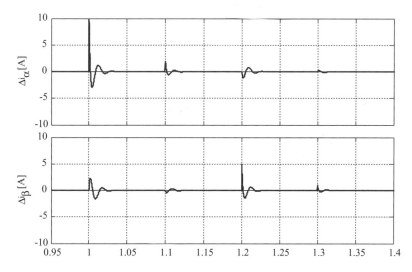

Figure 10.14 Error signals on the stationary reference frame using the PR current controller of Figure 10.12(b)

exhibit a fast dynamic response, comparable to some of the best responses obtained with the decoupled synchronous current controllers previously studied.

10.4 Power Control under Unbalanced Grid Conditions

In the previous sections of this chapter, some important issues regarding the operation of grid-connected power converters under unbalanced grid conditions have been pointed out.

As shown in the overview section, the control of the instantaneous active and reactive power exchanged with the grid, and mainly the power oscillating terms that appear due to the interaction between voltages and currents with different sequences, requires the design of specific strategies for calculating the current that should be injected into the grid by the power converter. However, implementation of these strategies gives rise to the injection of unbalanced currents into the network. Therefore, specific current control structures, able to properly regulate the positive- and negative-sequence components, are necessary in order to obtain satisfactory results. Several control solutions regarding this issue have been discussed in Section 10.3.

Once it is assumed that it is possible to design a suitable current controller to properly inject unbalanced currents during grid faults, it is necessary to tackle calculation of the proper reference currents to be applied to the input of such controllers. In the framework of this section, different methods for determining the reference currents to be tracked by the grid-connected power converter to achieve a specific performance will be presented. The analysis that will be carried out in the following will consider generic three-phase grid-connected power converters, working as either rectifiers or inverters, to regulate the instantaneous evolution of the active and the reactive power exchanged with the grid.

In the forthcoming study, the different techniques for calculating reference currents will be developed using a generic vector approach. This kind of analysis permits a generalized study to be carried out that is valid in either a stationary or a synchronous reference frame.

Before conducting further developments, some assumptions, which will be considered from this point on, should be introduced:

- The energy source supplying power through the inverter exhibits slow dynamics and hence the energy yield can be assumed as a constant throughout a grid period.
- The reference for the instantaneous active and reactive powers to be supplied by the grid-connected converter can also be considered constant throughout each grid cycle, i.e. $p^* = P$ and $q^* = Q$.
- The distributed power generator delivers power into the electrical network through a three-phase three-wire connection; hence there is no active power contribution from zero-sequence current components. Thus, the zero-sequence voltage component of the grid voltage will be neglected.
- The positive- and negative-sequence components of the grid voltage have been accurately estimated using a precise grid synchronization system.

Considering these assumptions, and according to the instantaneous power theory [1], the instantaneous active power, p, supplied or drained by a grid-connected three-phase power converter can be calculated as

$$p = v \cdot i \tag{10.31}$$

where $v = (v_a, v_b, v_c)$ is the voltage vector in the point of common coupling (PCC), $i = (i_a, i_b, i_c)$ is the injected current vector in such a point and '·' represents the dot product of both vectors. Considering the symmetrical components of the voltage and the current, the previous expression can be rewritten as

$$p = (v^+ + v^-) \cdot (i^+ + i^-) \tag{10.32}$$

$$p = \underbrace{v^+ \cdot i^+ + v^- \cdot i^-}_{P} + \underbrace{v^+ \cdot i^- + v^- \cdot i^+}_{\tilde{p}} = P + \tilde{p} \tag{10.33}$$

where v^+, v^-, i^+ and i^- are the positive- and negative-sequence vector components of the voltage and the current vectors while P and $\tilde{p}$ are the average value and the oscillatory term of the active power respectively.

On the other hand, the instantaneous reactive power q generated by the power converter, due to the interaction between the current vector i and the generic voltage vector v, can be written as

$$q = |v \times i| \tag{10.34}$$

Hence, the instantaneous reactive power can be defined as the module of the cross-product between v and i. However, the instantaneous reactive power can also be calculated by means of the following dot product:

$$q = v_\perp \cdot i \tag{10.35}$$

where $v_\perp$ is an orthogonal version (90° leaded) of the original grid voltage vector v. The orthogonal voltage $v_\perp$ expressed on different reference frames can be found by means of applying the following transformations:

$$v_{\perp \atop abc} = \frac{1}{\sqrt{3}} \begin{bmatrix} 0 & 1 & -1 \\ -1 & 0 & 1 \\ 1 & -1 & 0 \end{bmatrix} v_{abc}; \quad v_{\perp \atop \alpha\beta} = \begin{bmatrix} 0 & -1 \\ 1 & 0 \end{bmatrix} v_{\alpha\beta}; \quad v_{\perp \atop dq} = \begin{bmatrix} 0 & -1 \\ 1 & 0 \end{bmatrix} v_{dq}. \tag{10.36}$$

The reactive power shown in (10.35) can be written as well as a function of the voltage and the current symmetrical components, giving rise to

$$q = (v_\perp^+ + v_\perp^-) \cdot (i^+ + i^-) \tag{10.37}$$

$$q = \underbrace{v_\perp^+ \cdot i^+ + v_\perp^- \cdot i^-}_{Q} + \underbrace{v_\perp^+ \cdot i^- + v_\perp^- \cdot i^+}_{\tilde{q}} = Q + \tilde{q} \tag{10.38}$$

In (10.38) it is shown that the reactive power can also be split into a constant component, Q, and an oscillatory term $\tilde{q}$.

It is worth highlighting the point that the expressions from (10.31) to (10.38) are valid in any stationary (a, b, c or α, β), rotational (d, q) or polar ($m\angle\theta$) reference frames.

In the following, five different strategies for determining the reference current vector, i^*, to deliver given active and reactive power set-points, P and Q, under unbalanced grid voltage conditions, will be discussed. It will be shown that these strategies intend to provide different features, such as the cancellation of active power oscillations, the injection of balanced sinusoidal currents or the flexible regulation of the injected positive and negative sequence currents. In the next sections, the performance of these strategies for calculating reference currents will be compared and discussed, stressing the advantages and drawbacks of each one from the grid integration and the current injection control points of view.

10.4.1 Instantaneous Active–Reactive Control (IARC)

Considering the equations (10.31) and (10.34), and according to that stated in instantaneous power theories [1, 26, 27], any current vector aligned with the voltage vector v will give rise to active power, while any current vector aligned with $v_\perp$ will generate reactive power. This concept can be represented by the following expressions, which constitute the basis of the instantaneous active–reactive control (IARC) strategy to determine the reference currents:

$$i_p^* = g\, v \qquad (10.39)$$

$$i_q^* = b\, v_\perp \qquad (10.40)$$

where i_p^* and i_q^* can be considered as the active and the reactive currents vectors respectively, representing g an instantaneous conductance and b an instantaneous susceptance, both real terms, which sets the proportion between the voltage and the current vectors. The value of these terms that give rise to the exchange of a certain amount of power with the grid, P and Q, under given voltage conditions, defined by v, can be calculated as

$$i_p^* \cdot v = g\, v \cdot v = P \Rightarrow g\, |v|^2 = P \Rightarrow g = \frac{P}{|v|^2} \qquad (10.41)$$

$$i_q^* \cdot v_\perp = b\, v_\perp \cdot v_\perp = Q \Rightarrow b\, |v|^2 = Q \Rightarrow b = \frac{Q}{|v|^2} \qquad (10.42)$$

Considering these values of instantaneous conductance and susceptance seen from the output of the inverter, and according to (10.39) and (10.40), the reference current vectors to deliver the P and Q powers to the grid are given by

$$i_p^* = \frac{P}{|v|^2} v \qquad (10.43)$$

$$i_q^* = \frac{Q}{|v|^2} v_\perp \qquad (10.44)$$

where i_p^* is the active component of the reference current vector and i_q^* is the reactive one. Therefore, the final reference current can be calculated by just adding (10.43) and (10.44), i.e.

$$i^* = i_p^* + i_q^* \qquad (10.45)$$

These reference currents are the most efficient set of currents that exactly deliver the instantaneous active and reactive powers, P and Q, to the grid under generic voltage conditions. In comparison to other control strategies, which will be explained in the following, the IARC strategy exhibits the highest degree of control on the instantaneous powers exchanged with the grid.

In (10.43) and (10.44), the norm of the grid voltage vector, $|v|^2$, can be calculated as follows for different reference frames:

$$|v|^2 = v_a^2 + v_b^2 + v_c^2 = v_\alpha^2 + v_\beta^2 = v_d^2 + v_q^2 = m^2 \qquad (10.46)$$

Under balanced sinusoidal conditions, the resulting current references from (10.45) are perfectly sinusoidal, since the module of the voltage, $|v|$, and g and b are constants. However, as explained in Chapter 8, v and $v_\perp$ are formed by positive- and negative-sequence voltage components when unbalanced grid faults occur. Under such operating conditions, the module $|v|^2$ has oscillations at twice the fundamental grid frequency, i.e.

$$|v|^2 = |v^+|^2 + |v^-|^2 + 2|v^+||v^-|\cos(2\omega t + \phi^+ - \phi^-) \qquad (10.47)$$

When the expression of (10.47) is processed in the denominator of (10.43) and (10.44), the resulting reference currents i_p^* and i_q^* are not sinusoidal, but consist of high-order harmonics, giving rise to distorted reference signals for the currents to be injected in the phases of the grid. This issue is a serious drawback of the IARC strategy due to the fact that injecting distorted currents requires the implementation of more complex control systems than the ones presented in Section 10.3 of this chapter. In the case of a high-power grid-connected converter, the capability for injecting harmonic currents into the grid is significantly limited by the maximum value of the switching frequency of the power converter's semiconductors and the reduced bandwidth set by the link filter. Moreover, injection of harmonics into the grid can give rise to additional problems such as excitation of resonances or extra deterioration of the grid voltage at the point of connection of the power converter.

Therefore, in spite of its obvious advantages in terms of power controllability, the application of IARC seems not to be the most suitable strategy for generating reference currents in many applications of grid-connected power converters, especially those in the range of several megawatts. Nevertheless, by means of introducing particular constraints in the calculation of the reference currents, it is possible to deduct other alternative strategies that would give rise to reference currents more suitable to be injected into the grid by using positive- and negative-sequence currents controllers like the ones shown in Section 10.3, which are closer to the synchronous controller conventionally used by the industry.

10.4.2 Positive- and Negative-Sequence Control (PNSC)

The name of this strategy introduces itself as one of its main features. The positive- and negative-sequence control (PNSC) strategy deals with the calculation of a reference current vector, containing a proper set of positive- and negative-sequence components, that is able to cancel out oscillations in the instantaneous powers injected into the grid. This strategy is reminiscent of some of the first techniques proposed to control active rectifiers under unbalanced grid conditions. However, the formulation used in the PNSC strategy is more general, allowing the use of both synchronous and stationary reference frames while considering the attenuation of power oscillations in both the active and the reactive instantaneous powers. The current reference provided with this technique takes as a requirement that the resulting currents to be injected into the grid solely consist of positive- and negative-sequence components at the fundamental frequency, i.e.

$$i^* = i^{*+} + i^{*-} \tag{10.48}$$

where i^{*+} and i^{*-} represent such positive- and negative-sequence components respectively.

To determine the expressions for the reference currents generated by the PNSC strategy it is initially assumed that only active power is delivered to the grid. Moreover, it is imposed as a condition that the delivered active power is free of oscillations. Considering both constraints, the following expressions can be written:

$$v^+ \cdot i_p^{*+} + v^- \cdot i_p^{*-} = P \tag{10.49}$$

$$v^+ \cdot i_p^{*-} + v^- \cdot i_p^{*+} = 0 \tag{10.50}$$

Operating (10.50), the following expression for the negative-sequence reference current vector can be obtained:

$$v^+ \cdot i_p^{*-} = -i_p^{*+} \cdot v^- \Rightarrow |v^+|^2 \cdot i_p^{*-} = -v^+ \cdot i_p^{*+} \cdot v^- \Rightarrow i_p^{*-} = -\frac{v^+ \cdot i_p^{*+}}{|v^+|^2} v^- \tag{10.51}$$

Therefore, the following equation can be found by substituting (10.51) in (10.49) and regrouping terms:

$$P = v^+ \cdot i_p^{*+} \left(1 - \frac{|v^-|^2}{|v^+|^2} \right) \tag{10.52}$$

Operating (10.52), the reference for the positive-sequence reference current vector is given by

$$v^+ \cdot P = |v^+|^2 \cdot i_p^{*+} \left(\frac{|v^+|^2 - |v^-|^2}{|v^+|^2} \right) \Rightarrow i_p^{*+} = \frac{P}{|v^+|^2 - |v^-|^2} v^+ \tag{10.53}$$

Finally, adding (10.51) and (10.53), the reference for the active current vector is given by

$$i_p^{\,*} = g^{\pm}\left(v^+ - v^-\right); \quad g^{\pm} = \frac{P}{|v^+|^2 - |v^-|^2} \tag{10.54}$$

Similar constraints to the ones set to arrive at (10.54) can be set to determine the reference for the reactive current vector as follows:

$$v_{\perp}^+ \cdot i_q^{*+} + v_{\perp}^- \cdot i_q^{*-} = Q \tag{10.55}$$

$$v_{\perp}^+ \cdot i_q^{*-} + v_{\perp}^- \cdot i_q^{*+} = 0 \tag{10.56}$$

From these expressions, and making the same steps as in the case of the active current vector, the following expression can be found to calculate the reference for the reactive current vector:

$$i_q^{\,*} = b^{\pm}\left(v_{\perp}^+ - v_{\perp}^-\right); \quad b^{\pm} = \frac{Q}{|v^+|^2 - |v^-|^2} \tag{10.57}$$

Therefore, the final reference current can be calculated by adding (10.54) and (10.57) as

$$i^{\,*} = i_p^{\,*} + i_q^{\,*} = g^{\pm}\left(v^+ - v^-\right) + b^{\pm}\left(v_{\perp}^+ - v_{\perp}^-\right) \tag{10.58}$$

In order to study the performance of the instantaneous powers delivered using the PNSC strategy, the positive- and negative-sequence currents injected into the grid are written as the addition of their active and reactive component as follows:

$$i^+ = i_p^+ + i_q^+ \tag{10.59}$$

$$i^- = i_p^- + i_q^- \tag{10.60}$$

Considering that the current controller used in the power converter guarantees the injection into the grid of the reference currents set by (10.54) and (10.57) with no error, the following expressions can be written for the instantaneous active and reactive powers delivered by the power converter:

$$p = \underbrace{v^+ \cdot i_p^+ + v^- \cdot i_p^-}_{P} + \underbrace{v^+ \cdot i_q^+ + v^- \cdot i_q^-}_{0} + \underbrace{v^+ \cdot i_p^- + v^- \cdot i_p^+}_{0} + \underbrace{v^+ \cdot i_q^- + v^- \cdot i_q^+}_{\tilde{p}} \tag{10.61}$$

$$q = \underbrace{v_{\perp}^+ \cdot i_q^+ + v_{\perp}^- \cdot i_q^-}_{Q} + \underbrace{v_{\perp}^+ \cdot i_p^+ + v_{\perp}^- \cdot i_p^-}_{0} + \underbrace{v_{\perp}^+ \cdot i_q^- + v_{\perp}^- \cdot i_q^+}_{0} + \underbrace{v_{\perp}^+ \cdot i_p^- + v_{\perp}^- \cdot i_p^+}_{\tilde{q}} \tag{10.62}$$

In both expressions the average value of the active and reactive power, P and Q, and the oscillatory components of both, $\tilde{p}$ and $\tilde{q}$, have been indicated. The third and fourth terms have been cancelled out in both (10.61) and (10.62), as the dot product between two components with the same sequence and 90° shifted is equal to zero. On the other hand, the fifth and the sixth terms in (10.61) and (10.62) are zero due to the conditions stated in (10.50) and (10.56), where both oscillating terms where forced to be zero when calculating the reference for the active and reactive currents.

As evidenced in (10.61) and (10.62), the instantaneous active and reactive power delivered to the grid by applying the PNSC strategy, p and q, differ from the ones provided as a reference, P and Q, by the oscillatory power terms $\tilde{p}$ and $\tilde{q}$. This is due to the interaction between in-quadrature voltage and currents with different sequences.

Nevertheless, when one of the power references is null, either P or Q, the performance of the instantaneous power when using the PNSC is slightly different. For instance, if just the injection of active power into the network is considered under unbalanced conditions, while the reactive power set-point is equal to zero, the active power oscillations are cancelled. This is due to the fact that the remaining oscillating component in the active power, $\tilde{p}$, depends upon the reactive current components i_q^{*+} and i_q^{*-}, as stated in (10.61). Hence, if the reactive current i_q^* is cancelled, the oscillating term, $\tilde{p}$, does not appear. A similar phenomenon occurs with the reactive power oscillations when the active power set-point is set to zero. In this case, the reactive power oscillations are completely cancelled, as can be deduced from (10.62).

10.4.3 Average Active–Reactive Control (AARC)

As commented previously, during unbalanced grid faults, the reference currents obtained by using the IARC strategy present harmonics in their waveform because g and b do not remain constant throughout the grid period, T. Since P and Q have been assumed to be constant, such harmonics come from the effect of the second-order component of $|v|^2$ on the calculation of the instantaneous conductance and susceptance, g and b, as shown in (10.41) and (10.42). Therefore, if the effect of the $|v|^2$ oscillations is cancelled out in the calculation of such conductance and susceptance, the harmonics in the reference currents will be cancelled as well. To achieve this goal, the average active–reactive control (AARC) strategy calculates the average value of the instantaneous conductance and susceptance, throughout one grid period, and then determines the reference for the active and reactive current vectors i_p^* and i_q^*, operating as

$$i_p^* = G\, v\,; \quad G = \frac{P}{V_\Sigma^2} \quad (10.63)$$

$$i_q^* = B\, v_\perp\,; \quad B = \frac{Q}{V_\Sigma^2} \quad (10.64)$$

where V_Σ is the collective rms value of the grid voltage, which is defined as

$$V_\Sigma = \sqrt{\frac{1}{T}\int_0^T |v|^2\, dt} = \sqrt{|v^+|^2 + |v^-|^2} \quad (10.65)$$

Since G and B are constants in the AARC strategy, the voltage and the current waveforms are monotonously proportional. According to the studies conducted by Buchholz, one of the main precursors of the study on the time domain of active and nonactive currents in polyphase systems, for a given grid voltage v, the current references calculated by (10.63) and (10.64) lead to the smallest possible collective rms value of such currents, I_Σ, delivering a constant active power P over one grid period, T [28]. The lower value of I_Σ, the lower are the conduction losses in the system and the higher the efficiency. An analogous conclusion can be reached for the reactive current case.

Considering the active power injection, the current vector of (10.63) has the same direction as the grid voltage vector; hence it will not give rise to any reactive power component. However, the instantaneous active power delivered to the unbalanced grid will not be equal to P but will be given by

$$p = i_p^* \cdot v = \frac{|v|^2}{V_\Sigma^2} P = P + \tilde{p} \qquad (10.66)$$

where $\tilde{p}$ is an oscillating term added to the average value P.

Substituting (10.63) and (10.65) into (10.66), the instantaneous active power delivered to the unbalanced grid can be written as

$$p = P \left[1 + \frac{2|v^+||v^-|}{|v^+|^2 + |v^-|^2} \cos(2\omega t + \phi^+ - \phi^-) \right] \qquad (10.67)$$

where ϕ^+ and ϕ^- are the phase angles of the positive- and negative-sequence voltage vector components, v^+ and v^-.

In an analogous way, the reactive current found in (10.64) has the same direction as the in-quadrature voltage vector $v_\perp$. As in the active power case, the reactive power consists of an average value equal to Q plus an oscillating term at twice the grid frequency, i.e.

$$q = v_\perp \cdot i_q^* = \frac{|v|^2}{V_\Sigma^2} Q = Q + \tilde{q} \qquad (10.68)$$

$$q = Q \left[1 + \frac{2|v^+||v^-|}{|v^+|^2 + |v^-|^2} \cos(2\omega t + \phi^+ - \phi^-) \right] \qquad (10.69)$$

After obtaining the equations for p and q, given in (10.67) and (10.69), the same particular cases analysed for the PNSC can be discussed here. If just reactive power is injected into the grid by using the AARC, the residual instantaneous active power delivered to the network will be equal to zero. Likewise, if only active power is delivered, the residual instantaneous reactive power will be equal to zero.

10.4.4 Balanced Positive-Sequence Control (BPSC)

Considering the same principle used in the AARC strategy, it is possible to find other ways for modifying the value of the conductance and susceptance in the expressions used to calculate the reference currents in order to achieve other objectives. In the case of the balanced positive-sequence control (BPSC) strategy, the goal is to inject into the grid a set of balanced sinusoidal currents with only positive-sequence components. This method can be useful if the quality of the currents injected becomes a preferential issue. Moreover, the balanced currents generated by the BPSC strategy can be injected by using simple synchronous controllers, provided that the synchronization system is able to estimate accurately the phase angle of the positive-sequence component of the grid voltage.

The BPSC strategy calculates the active and reactive reference currents as

$$i_p^* = G^+ v^+; \quad G^+ = \frac{P}{|v^+|^2} \qquad (10.70)$$

$$i_q^* = B^+ v_\perp^+; \quad B^+ = \frac{Q}{|v^+|^2} \qquad (10.71)$$

The current vectors of (10.70) and (10.71) consist of a set of perfectly balanced positive-sequence sinusoidal waveforms. Under unbalanced operating conditions, the instantaneous active and reactive power delivered to the grid will differ from P and Q because of the interaction between the positive-sequence injected current and the negative-sequence grid voltage, i.e.

$$p = v \cdot i_p^* = \underbrace{v^+ \cdot i_p^*}_{P} + \underbrace{v^- \cdot i_p^* + v^- \cdot i_q^*}_{\tilde{p}} \qquad (10.72)$$

$$q = v \cdot i_q^* = \underbrace{v_\perp^+ \cdot i_q^*}_{Q} + \underbrace{v_\perp^- \cdot i_q^* + v_\perp^- \cdot i_p^*}_{\tilde{q}} \qquad (10.73)$$

where $\tilde{p}$ and $\tilde{q}$ are power oscillations at twice the fundamental utility frequency.

In the BPSC, both the instantaneous active and reactive powers will be affected by oscillations under unbalanced grid conditions, as demonstrated in (10.72) and (10.73). In the BPSC, the null value of either the P or Q set-point does not give rise to the cancellation of any power oscillation, which is a difference from previous strategies. On the other hand, this method is the only one that permits not only sinusoidal but also balanced currents to be obtained.

10.4.5 Performance of the IARC, PNSC, AARC and BPSC Strategies

The four strategies presented earlier to generate the reference currents are compared in this section by considering their implementation in a STATCOM; hence just reactive power will be injected into the grid by the power converter. A layout of the study case considered is depicted in Figure 10.15.

In the system of Figure 10.15, an unbalanced fault occurs at the electrical line in which the STATCOM is connected. The reference currents for the STATCOM set by the IARC, PNSC, AARC and BPSC strategies under such grid fault conditions will be presented in the following. Since only the performance of the reference currents is of interest in this simple example, it is assumed that the short-circuit power at the point of connection is much higher than the reactive power injected by the STATCOM. Therefore, it is assumed that the voltage at the point of connection is practically not affected by the current injected by the STATCOM.

For comparison purposes, the responses obtained with the four control strategies are shown in a single plot in Figure 10.16. The unbalanced grid voltages are shown in Figure 10.16(a). In the next two plots, Figures 10.16(b) and (c), the instantaneous active and reactive powers

Control of Grid Converters under Grid Faults

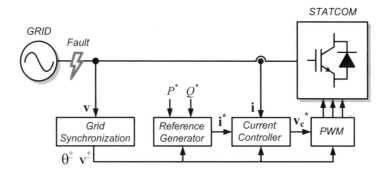

Figure 10.15 Layout of the study case considering a STATCOM application

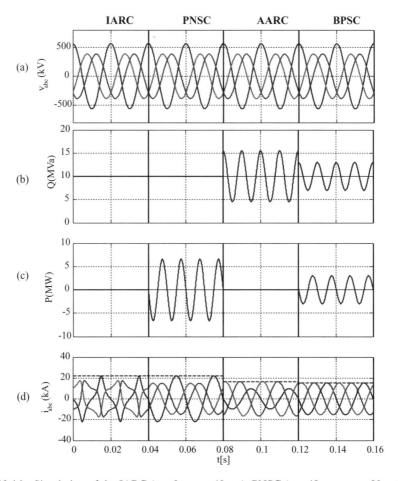

Figure 10.16 Simulation of the IARC ($t = 0$ to $t = 40$ ms), PNSC ($t = 40$ ms to $t = 80$ ms), AARC ($t = 80$ ms to $t = 120$ ms) and BPSC ($t = 120$ ms to $t = 160$ ms): (a) unbalanced voltage measurement, (b) injected reactive power, (c) delivered active power and (d) injected currents

delivered to the grid are shown. Finally, the current waveforms obtained with each control strategy are depicted in Figure 10.16(d).

Between $t = 0$ and $t = 40$ ms the IARC strategy is applied, giving rise to flat instantaneous active and reactive powers while the currents obtained are highly distorted. After $t = 40$ ms the PNSC strategy is enabled. In this case, the reference currents consist on a set of unbalanced three-phase signals, but without harmonic distortion, as expected. Regarding the power behaviour, it is proven how the reactive power remains constant, while the active power presents oscillations around the zero mean value. The particular performance of the PNSC strategy in these specific conditions, where one of the power references is zero, has already been discussed in Section 10.4.2.

From $t = 80$ ms to $t = 120$ ms the current reference generator implements the AARC strategy. As can be seen in the figure, the obtained currents are again sinusoidal. Moreover, the maximum amplitude of these currents is lower than those obtained from the IARC and the PNSC strategies, keeping the same value of Q. This feature points out that by means of using the AARC a higher value in the reactive power injected into the grid under unbalanced conditions can be reached, which is a remarkable advantage. Additionally, the instantaneous active power drawn by the AARC strategy is free of oscillations, which atenuates voltage fluctuations at the DC-bus of the STATCOM.

Finally, the behaviour of the BPSC is depicted at the end of the simulation plots, from $t = 120$ ms to $t = 160$ ms. In this case, since no negative-sequence components are injected into the grid, oscillations in both active and reactive power occur. However, this technique improves even more the advantages of the AARC regarding the reduction of the maximum current. It can be stated from Figure 10.16 that the BPSC is the one that gives rise to the lowest peak value in the injected current among all the four compared strategies, while injecting the same average reactive power set-point.

After evaluating the different strategies for determining the current reference under unbalanced grid conditions, it can be stated that the main concerns when calculating the current reference are focused on:

- the magnitude of the active power oscillations,
- the magnitude of the reactive power oscillations,
- the harmonic content of the signal to be injected.

The influence of these items in the proposed strategies has been summarized in Table 10.1.

In this table, the different active and reactive power injection combinations have been considered for all the strategies previously presented to determine the reference currents. Most

Table 10.1 Features of the current reference calculation strategies

Strategy	$P \neq 0; \tilde{p} = 0$		$Q \neq 0; \tilde{q} = 0$		$P \neq 0; \tilde{q} = 0$		$Q \neq 0; \tilde{p} = 0$		I_{harm}
	$Q^* = 0$	$Q^* \neq 0$	$P^* = 0$	$P^* \neq 0$	$Q^* = 0$	$Q^* \neq 0$	$P^* = 0$	$P^* \neq 0$	
IARC	Y	Y	Y	Y	Y	Y	Y	Y	Y
PNSC	Y	N	Y	N	Y	N	N	N	N
AARC	N	N	N	N	N	Y	Y	N	N
BPSC	N	N	N	N	N	N	N	N	N

of the columns in this table are focused on determining whether active or reactive power oscillations, $\tilde{p}$ and $\tilde{q}$, are produced, where Y and N stand for 'yes' and 'no' respectively. The I_{harm} column indicates whether the injected currents are polluted by harmonics or not.

It can be concluded from the presented analysis that the strategies that achieve better results regarding the power injection capability are mainly those that inject sinusoidal currents. However, Table 10.1 shows that these kinds of strategy give rise to power oscillations, and only for the PNSC and the AARC is it possible to find particular cases where these oscillations are cancelled. On the other hand, the IARC is the most advantageous strategy in terms of power controllability, but unfortunately permits the lowest amount of power injection under unbalanced conditions, while injecting the maximum amount of harmonics among the presented methods.

10.4.6 Flexible Positive- and Negative-Sequence Control (FPNSC)

In the strategies already presented in this chapter for determining the reference currents for the power converter under grid faults, the relationship between the positive- and negative-sequence components of such reference currents is set by the respective forming equation, which is mainly defined by the voltage vectors v^+ and v^-. However, and taking advantage of the previous analysis, in the following it will be shown that it is possible to implement a method for adjusting, in a more flexible way, the relationship between the symmetrical components of the reference currents, giving rise to a more flexible strategy to calculate such references, namely the flexible positive- and negative-sequence control (FPNSC) strategy.

Expanding the principle introduced in Section 10.4.4, the instantaneous conductance, g, can be divided into a positive-sequence and a negative-sequence conductance value, G^+ and G^-. Through this reasoning, the active reference current vector can be written as

$$i_p^* = G^+ v^+ + G^- v^- \tag{10.74}$$

As a difference from the BPSC strategy previously presented, this active power current also contains a negative-sequence component. If just either positive- or negative-sequence currents had to be injected, the value of G^+ and G^- would be found as

$$G^+ = \frac{P}{|v^+|^2}; \quad G^- = \frac{P}{|v^-|^2} \tag{10.75}$$

respectively. However, if the current injected into the grid to deliver a given active power P had to be composed by both sequence components simultaneously, it would be necessary to regulate the relationship between them in order to keep constant the total amount of active power delivered to the grid. In order to do that, a scalar parameter, k_1, regulates the contribution of each sequence component on the active reference currents in the form

$$i_p^* = k_1 \frac{P}{|v^+|^2} v^+ + (1 - k_1) \frac{P}{|v^-|^2} v^- \tag{10.76}$$

By means of regulating k_1 in (10.76) within the range from 0 to 1, it is possible to change the proportion in which the positive- and negative-sequence components of the active currents injected into the grid participate in delivering a given amount of active power P to the grid. For instance, by making $k_1 = 1$, the same behaviour as that of the BPSC is obtained, while by making $k_1 = 0$, perfectly balanced negative-sequence currents will be injected into the grid to deliver the active power P. In some special cases, k_1 might be out of the [0, 1] range. In such cases, one of the sequence components of the injected currents would absorb active power from the grid, whereas the other sequence component would deliver as much active power as necessary to balance the system and make the total active power delivered to the grid equal to P.

An analogous reasoning can be followed for finding the reference for the reactive currents, which can be calculated as

$$i_q^* = k_2 \frac{Q}{|v^+|^2} v_\perp^+ + (1 - k_2) \frac{Q}{|v^-|^2} v_\perp^- \qquad (10.77)$$

In this case, another scalar parameter, namely k_2, has been used to control the proportion between the positive- and the negative-sequence components in the reference currents and to inject a given reactive power Q into the grid.

Finally, after rearranging some terms, the reference currents provided to the current controller of the power converter can be found through the following expression:

$$\mathbf{i}^* = P \left(\frac{k_1}{|v^+|^2} \cdot v^+ + \frac{(1 - k_1)}{|v^-|^2} \cdot v^- \right) + Q \left(\frac{k_2}{|v^+|^2} \cdot v_\perp^+ + \frac{(1 - k_2)}{|v^-|^2} \cdot v_\perp^- \right) \qquad (10.78)$$

By means of changing the value of k_1 and k_2 in (10.78), the relationship between positive- and negative-sequence current components in both the active and the reactive currents can be easily modified. This feature is very interesting when the interaction between the power converter and the grid during faults is studied. For instance, the positive-sequence voltage component at the point of connection (PCC) of an inductive line will be boosted if some amount of reactive current is injected into the grid making $k_2 = 1$, as just positive-sequence reactive currents are injected. On the other hand, the negative-sequence voltage component at such PCC will be reduced if $k_2 = 0$, since just negative-sequence reactive currents are injected into the grid.

The performance of the instantaneous active power delivered to the grid when the FPNSC strategy is used to set the reference currents can be written as

$$p = P + \tilde{p} \qquad (10.79)$$

where both power components are given by

$$P = \frac{Pk_1}{|v^+|^2} \cdot v^+ \cdot v^+ + \frac{P(1 - k_1)}{|v^-|^2} \cdot v^- \cdot v^- \qquad (10.80)$$

$$\tilde{p} = \left(\frac{Pk_1}{|v^+|^2} + \frac{P(1 - k_1)}{|v^-|^2} \right) v^+ \cdot v^- + \left(\frac{Qk_2}{|v^+|^2} - \frac{Q(1 - k_2)}{|v^-|^2} \right) v_\perp^+ \cdot v^- \qquad (10.81)$$

Although not presented in this chapter for simplicity reasons, similar conclusions can be found for the instantaneous reactive power components.

If the active power reference, P, is not null, the only way to cancel out the first term of the power oscillation $\tilde{p}$ in (10.81) is to make k_1 higher than 1, as shown in the following expression:

$$\frac{-Pk_1}{|\boldsymbol{v}^+|^2} = \frac{P(1-k_1)}{|\boldsymbol{v}^-|^2} \Rightarrow k_1 = \frac{|\boldsymbol{v}^+|^2}{|\boldsymbol{v}^+|^2 - |\boldsymbol{v}^-|^2} \geq 1 \qquad (10.82)$$

However, for any reactive power reference, Q, the cancellation of the second power oscillation $\tilde{p}$ term in (10.81) is feasible by setting a value for k_2 within the range [0, 1], i.e.

$$\frac{Qk_2}{|\boldsymbol{v}^+|^2} = \frac{Q(1-k_2)}{|\boldsymbol{v}^-|^2} \Rightarrow k_2 = \frac{|\boldsymbol{v}^+|^2}{|\boldsymbol{v}^+|^2 + |\boldsymbol{v}^-|^2} \leq 1 \qquad (10.83)$$

This feature is interesting, even though not all the power oscillations might be eliminated because of some specific restriction set on the range of values for k_1 and k_2. It is worth pointing out that, by means of using (10.82) and (10.83), the FPNSC strategy would behave as the PNSC one. In the next sections other important aspects regarding the FPNSC strategy are discussed.

10.5 Flexible Power Control with Current Limitation

Despite the fact that the analysis of the reference current generation strategies discussed in this chapter has been mainly centred on the performance of each proposal regarding the injection of active and reactive powers into the network, a detailed study focused on the performance of the currents associated with each strategy is also necessary.

As stated in the previous section, several of the proposed strategies refer to the injection of unbalanced currents through the power converter. As a consequence, the instantaneous value of these currents may be different from phase to phase in some cases. Under such conditions, accurate control of the power converter is necessary in order to avoid an undesired trip, since an overcurrent in any of the phases of the power converter usually results in the instantaneous disconnection of the system from the network.

Therefore, controlling the performance of the currents injected into the grid is a mandatory issue that should be considered when designing control strategies for grid-connected power converters operating under unbalanced grid voltage conditions. Controlling the power converter in such a way that any phase current never exceeds a given instantaneous admissible limit permits:

- LVRT capabilities under unbalanced grid faults to be achieved,
- the integrity of the power converter to be protected,
- the P and Q injection according to the converter rating to be maximized,
- participation in the mitigation of the grid voltage unbalance.

Nevertheless, the relationship between the power delivered by the converter and the associated currents depends strongly on the selected power strategy. Hence, there is not a standard expression for calculating the maximum value of the currents for all the power control strategies discussed previously. Depending on the intended objective, i.e. the cancellation of active power oscillations, the cancellation of both active and reactive power oscillations or the reduction of harmonic currents injection, the expression for finding the maximum peak value in the phase currents will be different. As a consequence, each technique requires a specific analysis, something that cannot be afforded in the scope of this chapter for all the previously presented strategies for reference current generation.

In this section, however, a method for determining the maximum peak value of the currents injected by the power converter into the phases of the three-phase system, as well as the maximum active and reactive power set-points that give rise to such currents, will be presented for the FPNSC strategy.

10.5.1 Locus of the Current Vector under Unbalanced Grid Conditions

Considering the definition of the FPNSC given in the previous section, the reference current obtained with this strategy can be split into an active current term, i_p^*, and a reactive current term, i_q^*, as follows:

$$i^* = \underbrace{P\left(\frac{k_1}{|v^+|^2} \cdot v^+ + \frac{(1-k_1)}{|v^-|^2} \cdot v^-\right)}_{i_p^*} + \underbrace{Q\left(\frac{k_2}{|v^+|^2} \cdot v_\perp^+ + \frac{(1-k_2)}{|v^-|^2} \cdot v_\perp^-\right)}_{i_q^*} \quad (10.84)$$

Considering the injection of a certain value of P and Q under steady-state grid voltage unbalanced conditions, with a fixed set of parameters k_1 and k_2, the instantaneous values of v^{+1}, v^{-1}, $v_\perp^{+1}$ and $v_\perp^{-1}$ in (10.84) are multiplied by constant factors, namely C_1, C_2, C_3 and C_4, to simplify formulation, i.e.

$$i^* = \underbrace{C_1 \cdot v^+ + C_2 \cdot v^-}_{i_p^*} + \underbrace{C_3 \cdot v_\perp^+ + C_4 \cdot v_\perp^-}_{i_q^*} \quad (10.85)$$

where

$$C_1 = \frac{Pk_1}{|v^+|^2}; C_2 = \frac{P(1-k_1)}{|v^-|^2}; C_3 = \frac{Qk_2}{|v^+|^2}; C_4 = \frac{Q(1-k_2)}{|v^-|^2} \quad (10.86)$$

As the equations that permit calculating i_p^* and i_q^* can be developed considering a stationary $\alpha\beta$ reference frame, their evolution can be graphically described in these coordinates, in the same way as the voltage vector, v, was presented in Figure 8.2(a) in Chapter 8. In that chapter, it was demonstrated that the addition of a positive-sequence voltage vector, v^+, and a negative-sequence voltage vector, v^-, results in a ellipse in the $\alpha\beta$ domain. Therefore, the addition of $C_1 \cdot v^+$ and $C_2 \cdot v^-$ will give rise as well to an ellipse for i_p^*. Likewise, the same can be concluded for $C_3 \cdot v_\perp^+$ and $C_4 \cdot v_\perp^-$, which will generate the i_q^* ellipse. An example of the graphical loci for both i_p^* and i_q^* have been depicted in Figure 10.17 for a generic case.

The active current ellipse of Figure 10.17(a) will be aligned with the locus of v. The values of the constant terms C_1 and C_2 will just scale it. Likewise, the same can be concluded for the reactive current ellipse in Figure 10.17(b), which is 90° shifted and aligned with the in-quadratue components of the voltage $v_\perp$.

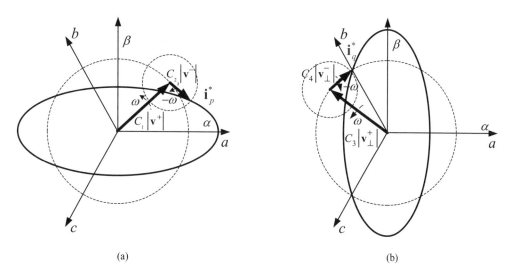

Figure 10.17 Loci of (a) active currents, i_p^*, and (b) reactive currents, i_q^*

Once the loci of i_p^* and i_q^* have been depicted, their addition will permit the locus of i^* to be obtained, as shown in Figure 10.18.

For the sake of clarity, the influence of the positive and negative sequence phase of the voltage, ϕ^+ and ϕ^-, was not considered in the locus description of Figure 10.18. However, both angles have a significant influence on the evolution of i^*. If $\phi^+ = \phi^-$, the $\alpha\beta$ representation of i_p^* is an ellipse whose focus is aligned with the α axis, while i_q^* results in an orthogonal ellipse centred on the β axis. On the other hand, if $\phi^+ \neq \phi^-$, the main ellipses i_p^* and i_q^* are

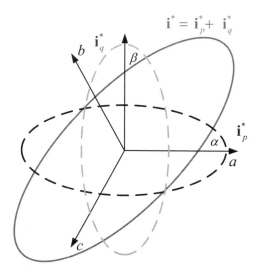

Figure 10.18 Locus of the resulting current $i^* = i_p^* + i_q^*$

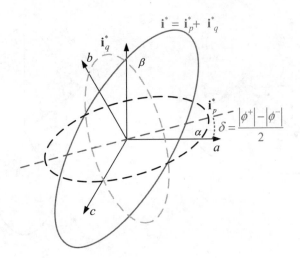

Figure 10.19 Loci of i^*, i_p^* and i_q^* when $\phi^+ = \pi/6$ and $\phi^- = 0$

not aligned with the $\alpha\beta$ axis, but shifted a certain angle, δ. This angle can be calculated as the difference between the absolute value of the positive and negative phase angles divided by two, i.e.

$$\delta = \frac{|\phi^+| - |\phi^-|}{2} \tag{10.87}$$

In Figure 10.19 the shape of the evolution of i^*, i_p^* and i_q^* are displayed considering $\phi^+ = \pi/6$ and $\phi^- = 0$.

10.5.2 Instantaneous Value of the Three-Phase Currents

The analysis performed above, regarding the evolution of the current vector on the $\alpha\beta$ reference frame, is useful to find an expression that permits the instantaneous value of its $\alpha\beta$ components to be determined. However, it is first necessary to introduce some changes in their current formulation.

The active current ellipse, shown in Figure 10.17(a), can be mathematically defined on the $\alpha\beta$ coordinates as

$$i_p^* = \begin{bmatrix} i_{p\alpha}^* \\ i_{p\beta}^* \end{bmatrix} = \begin{bmatrix} I_{pL} \cos \omega t \\ I_{pS} \sin \omega t \end{bmatrix} \tag{10.88}$$

where I_{pL} and I_{pS} are the modulus of two rotating vectors, with I_{pL} equal to the value of the large axis of the i_p^* ellipse while I_{pS} is the magnitude of its short axis. As written in (10.88), the $\alpha\beta$ components of the ellipse can be found as the horizontal and vertical projections of the large and short vectors respectively, for each value of the angle ωt. The graphical representation of this concept is depicted in Figure 10.20(a).

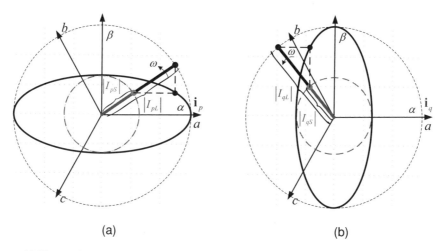

Figure 10.20 (a) Graphical representation of $i^*_{p\alpha}$ and $i^*_{p\beta}$ (b) Graphical representation of $i^*_{q\alpha}$ and $i^*_{q\beta}$.

The values of both I_{pL} and I_{pS}, can be found through the following equations:

$$I_{pL} = P\left(\frac{k_1}{|v^+|} + \frac{(1-k_1)}{|v^-|}\right) \tag{10.89}$$

$$I_{pS} = P\left(\frac{k_1}{|v^+|} - \frac{(1-k_1)}{|v^-|}\right) \tag{10.90}$$

Following the same reasoning, the reactive current ellipse can also be expressed in this alternative way. However, in this case it must be taken into account that the origin of the angle ωt is aligned with the β axis. Hence, the components in the $\alpha\beta$ axis for i^*_q can be written as

$$i^*_q = \begin{bmatrix} i^*_{q\alpha} \\ i^*_{q\beta} \end{bmatrix} = \begin{bmatrix} -I_{qS}\sin\omega t \\ I_{qL}\cos\omega t \end{bmatrix} \tag{10.91}$$

The resulting reactive current ellipse when applying (10.91) is depicted in Figure 10.20(b). In this case, the value of the long and short axes of the ellipse can be found as

$$I_{qL} = Q\left(\frac{k_2}{|v^+|} + \frac{(1-k_2)}{|v^-|}\right) \tag{10.92}$$

$$I_{qS} = Q\left(\frac{k_2}{|v^+|} - \frac{(1-k_2)}{|v^-|}\right) \tag{10.93}$$

Finally, the instantaneous $\alpha\beta$ components of the reference current vector i^*, which considers the effect of both the active and the reactive currents, can be found by means of adding (10.88) to (10.91), which results in the expression

$$i^* = i_p^* + i_q^* = \begin{bmatrix} i_\alpha^* \\ i_\beta^* \end{bmatrix} = \begin{bmatrix} I_{pL} \cos \omega t - I_{qS} \sin \omega t \\ I_{pS} \sin \omega t + I_{qL} \cos \omega t \end{bmatrix} \quad (10.94)$$

The $\alpha\beta$ components in (10.94) can be rewritten and simplified as shown in the following:

$$i_\alpha^* = k_\alpha \cos(\omega t + \theta_\alpha); \quad k_\alpha = \sqrt{I_{pL}^2 + I_{qS}^2}; \quad \theta_\alpha = \tan^{-1}\left(\frac{I_{qS}}{I_{pL}}\right) \quad (10.95)$$

$$i_\beta^* = k_\beta \sin(\omega t + \theta_\beta); \quad k_\beta = \sqrt{I_{qL}^2 + I_{pS}^2}; \quad \theta_\beta = \tan^{-1}\left(\frac{I_{qL}}{I_{pS}}\right) \quad (10.96)$$

By means of these last expressions, the instantaneous values of the three-phase currents to be injected into the grid by the power converter, expressed on the $\alpha\beta$ reference frame, can be found.

10.5.3 Estimation of the Maximum Current in Each Phase

Once the expressions to determine the instantaneous evolution of the currents on the $\alpha\beta$ reference frame have been found, the next step is to determine the value of the peak current at each phase, in order to find out which phase will be limiting the injection of power into the network as a function of the specific unbalanced grid conditions. Moreover, the main purpose of the technique presented in this section is not just to estimate the value of the peak currents injected in the three phases of the grid, but also to deduct an expression that allows setting the active and reactive powers set-points while considering the ratings of the power converter, the grid conditions and the control parameters.

As shown in Figure 10.21, the reference for the current vector resulting from the FP-NSC strategy, i^*, describes an ellipse in the $\alpha\beta$ reference frame. Taking advantage of this

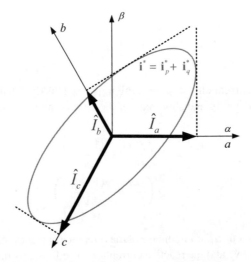

Figure 10.21 Maximum current at each phase

representation, the maximum current at each phase of the three-phase system, $\hat{I}_a$, $\hat{I}_b$ and $\hat{I}_c$ can be calculated by finding the maximum projection of the current ellipse on the *abc* axis, as graphically shown in Figure 10.21.

Considering that phase *a* of the system is aligned with the α axis, the peak value of the current in this phase, $\hat{I}_a$, is equal to the maximum value of the i_α^* component. Considering the expression written in (10.95), which permits the instantaneous value of i_α^* to be found, it can be concluded that the maximum value of this component arises when the trigonometric term is equal to one. Therefore, the maximum value of i_α^* can be found as

$$\hat{I}_\alpha = \sqrt{I_{pL}^2 + I_{qS}^2} \underbrace{\cos\left(\omega t + \tan^{-1}\left(\frac{I_{qS}}{I_{pL}}\right)\right)}_{1} \Rightarrow \hat{I}_\alpha = \sqrt{I_{pL}^2 + I_{qS}^2}. \qquad (10.97)$$

and hence, as stated previously, $\hat{I}_\alpha = \hat{I}_a$.

The same method can be used as well for finding $\hat{I}_b$ and $\hat{I}_c$. Nevertheless, both maximum currents on the *abc* reference frame depend upon α and β components, hence determining their value is not as straightforward. However, if the resulting ellipse of Figure 10.21 is rotated in such a way that the expressions giving $\hat{I}_b$ and $\hat{I}_c$ have only a single component on the α axis, the expression for finding $\hat{I}_a$ could be applied to determine the values of $\hat{I}_b$ and $\hat{I}_c$ as well. Thus, the maximum for $\hat{I}_b$ can be found by leading the original ellipse $\pi/3$ rad. Under these conditions, the maximum value in the α axis is equal to $\hat{I}_b$. Likewise, $\hat{I}_c$ can be found in an analogous way, but lagging the ellipse $-\pi/3$ rad.

The mechanism for rotating the ellipse in the $\alpha\beta$ reference frame can be made by using the following rotation matrix:

$$\mathbf{i}^{*\prime} = \begin{bmatrix} i_\alpha^{*\prime} \\ i_\beta^{*\prime} \end{bmatrix} = \begin{bmatrix} \cos\gamma & -\sin\gamma \\ \sin\gamma & \cos\gamma \end{bmatrix} \begin{bmatrix} k_\alpha \cos(\omega t + \theta_\alpha) \\ k_\beta \sin(\omega t + \theta_\beta) \end{bmatrix} \qquad (10.98)$$

In this equation, γ is the angle to be rotated, while the new components of the rotated ellipse are $i_\alpha^{*\prime}$ and $i_\beta^{*\prime}$. The value of γ is different depending on phase *a*, *b* or *c*. Considering the general case, where $\phi^+ \neq \phi^-$, as in Figure 10.19, γ is equal to δ for phase *a*, while for phases *b* and *c* an additional $\pi/3$ and $-\pi/3$ angle should be added respectively. In a nutshell, the corresponding γ angles for each phase are

$$\hat{I}_a = \hat{I}_\alpha'(\gamma = \delta)$$
$$\hat{I}_b = \hat{I}_\alpha'(\gamma = \delta + \pi/3) \qquad (10.99)$$
$$\hat{I}_c = \hat{I}_\alpha'(\gamma = \delta - \pi/3)$$

where δ is the angle defined by (10.87). Once the ellipse is properly rotated, only the maximum value of the current on the α axis is of interest, as the maximum on this axis is equal to the peak value of the phase current. For this reason only this component will be analyzed in the following. By means of expanding (10.98), $i_\alpha^{*\prime}$ can be rewritten as

$$i_\alpha^{*\prime} = k_\alpha \cos\gamma \left(\cos\omega t \cos\theta_\alpha - \sin\omega t \sin\theta_\alpha\right) \\ - k_\beta \sin\gamma \left(\sin\omega t \cos\theta_\beta + \cos\omega t \sin\theta_\beta\right) \quad (10.100)$$

Regrouping (10.100) as sine and cosine terms, the following expression can be found:

$$i_\alpha^{*\prime} = A_1 \cos\omega t + B_1 \sin\omega t \quad (10.101)$$

where

$$A_1 = \left(k_\alpha \cos\gamma \cos\theta_\alpha - k_\beta \sin\gamma \sin\theta_\beta\right) \\ B_1 = \left(-k_\alpha \cos\gamma \sin\theta_\alpha - k_\beta \sin\gamma \cos\theta_\beta\right) \quad (10.102)$$

Moreover, the values of A_1 and B_1 can be further simplified, obtaining

$$A_1 = I_{pL} \cos\gamma - I_{qL} \sin\gamma \\ B_1 = -I_{qS} \cos\gamma - I_{pS} \sin\gamma \quad (10.103)$$

where the values of I_{pL}, I_{pS}, I_{qL} and I_{qS} are detailed in (10.89), (10.90), (10.92) and (10.93). Finally, the maximum value on the α axis of (10.101) can be written as

$$\hat{I}_\alpha^\prime = \sqrt{A_1^2 + B_1^2} \quad (10.104)$$

As a conclusion of the study carried out until this point, the value of the peak currents injected in each phase, for given grid voltage and power conditions, can be determined by using the expressions shown in the Table 10.2.

According to (10.104) the square value of the maximum current, $\hat{I}_\alpha^\prime$, of each phase is equal to

$$\left(\hat{I}_\alpha^\prime\right)^2 = A_1^2 + B_1^2 \quad (10.105)$$

Table 10.2 Peak values for the three-phase injected currents

Phase	γ	Coefficients
$\hat{I}_a$	$\gamma = \dfrac{\lvert\phi^+\rvert - \lvert\phi^-\rvert}{2} + 0$	$\hat{I}_\alpha^\prime = \sqrt{A_1^2 + B_1^2}$
$\hat{I}_b$	$\gamma = \dfrac{\lvert\phi^+\rvert - \lvert\phi^-\rvert}{2} + \pi/3$	$A_1 = I_{pL} \cos\gamma - I_{qL} \sin\gamma$ $B_1 = -I_{qS} \cos\gamma - I_{pS} \sin\gamma$
$\hat{I}_c$	$\gamma = \dfrac{\lvert\phi^+\rvert - \lvert\phi^-\rvert}{2} - \pi/3$	

By writing $\hat{I}'_\alpha$ as simply $\hat{I}$, denoting the maximum current admissible by the power converter, and expanding (10.105), the following expression can be written:

$$\hat{I}^2 = P^2 \left[\frac{k_1^2 \cdot |\boldsymbol{v}^-|^2 + (1-k_1)^2 \cdot |\boldsymbol{v}^+|^2 + 2k_1(1-k_1)\cos 2\gamma \cdot |\boldsymbol{v}^+| \cdot |\boldsymbol{v}^-|}{|\boldsymbol{v}^+|^2 \cdot |\boldsymbol{v}^-|^2} \right]$$
$$+ Q^2 \left[\frac{k_2^2 \cdot |\boldsymbol{v}^-|^2 + (1-k_2)^2 \cdot |\boldsymbol{v}^+|^2 - 2k_2(1-k_2)\cos 2\gamma \cdot |\boldsymbol{v}^+| \cdot |\boldsymbol{v}^-|}{|\boldsymbol{v}^+|^2 \cdot |\boldsymbol{v}^-|^2} \right]$$
$$- PQ \left[\frac{(2k_1 + 2k_2 - 4k_1 k_2) \cdot |\boldsymbol{v}^+| \cdot |\boldsymbol{v}^-| \cdot \sin 2\gamma}{|\boldsymbol{v}^+|^2 \cdot |\boldsymbol{v}^-|^2} \right] \quad (10.106)$$

The peak current for each phase, given by $\hat{I}$, can take three values, one for each phase, matching the three possible values of the angle γ according to (10.99).

The equation shown in (10.106) is a key expression, as it permits an estimation to be made of the maximum value of the current that will arise at each phase under given grid conditions, active and reactive powers references and control parameters selection, when implementing the FPNSC strategy. It will be shown in the next section that this expression also allows determination of the maximum active and reactive powers that can be delivered to the grid by the power converter, under generic gird voltage conditions, without reaching the maximum admissible current in any of its phases.

10.5.4 Estimation of the Maximum Active and Reactive Power Set-Point

In the previous sections, the equations of the FPNSC current strategy have been presented. Moreover, the analytical relationship between the maximum current, the grid voltage components and the control parameters have been determined. Hence, now it is possible to estimate easily the maximum currents that will be obtained in the three phases of the power converter under different operating conditions.

In PV and WT generators, a high-level control layer is normally responsible for setting the reference for the active and reactive powers to be delivered by the power converter. Therefore, it is very important to find an expression that allows setting the maximum active and reactive power set-points that can be provided by the power converter, without exceeding its nominal current ratings, when it works under unbalanced grid voltage conditions.

In the following, two practical cases will be presented:

1. Simultaneous active and reactive power delivery. The power converter should deliver a given amount of active power and maximize the injection of reactive power, and vice versa.
2. Injection of maximum reactive power. The power converter should inject the maximum amount of reactive power to boost the voltage at the PCC under unbalanced grid voltage conditions.

10.5.4.1 Simultaneous Active and Reactive Power Delivery

Equation (10.107) can be found by operating the expression shown in (10.106). This expression allows the power developed by a power converter to be maximized, while both active and reactive powers should be delivered, being one of those two power magnitudes given as a reference and having to calculate the maximum magnitude for the other power term, without overpassing in any of the phases the maximum instantaneous current that can be drawn by the power converter:

$$\begin{aligned}
0 =\ & Q^2 \left[k_2^2 \cdot |\boldsymbol{v}^-|^2 + (1-k_2)^2 \cdot |\boldsymbol{v}^+|^2 - 2k_2(1-k_2)\cos 2\gamma \cdot |\boldsymbol{v}^+| \cdot |\boldsymbol{v}^-| \right] \\
& - PQ\left[(2k_1 + 2k_2 - 4k_1 k_2) \cdot |\boldsymbol{v}^+| \cdot |\boldsymbol{v}^-| \sin 2\gamma \right] \\
& + P^2 \left[k_1^2 \cdot |\boldsymbol{v}^-|^2 + (1-k_1)^2 \cdot |\boldsymbol{v}^+|^2 + 2k_1(1-k_1)\cos 2\gamma \cdot |\boldsymbol{v}^+| \cdot |\boldsymbol{v}^-| \right] \\
& - \hat{I}^2 \cdot |\boldsymbol{v}^+|^2 \cdot |\boldsymbol{v}^-|^2
\end{aligned} \qquad (10.107)$$

As an example, when the active power to be delivered is set to P^*, the reactive power can be found by solving the following equation:

$$0 = Q^2 \underbrace{\left[k_2^2 \cdot |\boldsymbol{v}^-|^2 + (1-k_2)^2 \cdot |\boldsymbol{v}^+|^2 - 2k_2(1-k_2)\cos 2\gamma \cdot |\boldsymbol{v}^+| \cdot |\boldsymbol{v}^-| \right]}_{a}$$
$$- QP^* \underbrace{\left[(2k_1 + 2k_2 - 4k_1 k_2) \cdot |\boldsymbol{v}^+| \cdot |\boldsymbol{v}^-| \sin 2\gamma \right]}_{b}$$
$$+ \underbrace{P^{*2} \left[k_1^2 \cdot |\boldsymbol{v}^-|^2 + (1-k_1)^2 \cdot |\boldsymbol{v}^+|^2 + 2k_1(1-k_1)\cos 2\gamma \cdot |\boldsymbol{v}^+| \cdot |\boldsymbol{v}^-| \right] - \hat{I}^2 \cdot |\boldsymbol{v}^+|^2 \cdot |\boldsymbol{v}^-|^2}_{c}$$
$$(10.108)$$

and

$$0 = aQ^2 + bQ + c \qquad (10.109)$$

The resolution of this system gives rise to three possible solutions, since the angle γ can take three different values, as indicated in the Table 10.2. Between these three values the minimum Q solution should be selected in order to give the final set-point for the power converter, i.e. $P = P^*$ and $Q = Q_{min}\,(\hat{I}, P^*, k_1, k_2)$.

In an analogous way, the same reasoning can be followed when the reactive power set-point is given as a reference and the maximum active power to be delivered should be calculated, without triggering the overcurrent protection in any of the phases.

10.5.4.2 Injection of Maximum Reactive Power

In this case, the objective is to inject into the grid the highest amount of Q, while setting the active power reference equal to zero, without exceeding the maximum current admitted by the power converter in any of the phases. Considering these constraints and operating (10.106), the following expression can be found, which allows calculation of the maximum value for the reactive power to be injected, Q, as a function of the sequence components of the grid voltage, the control parameters and the limit current in any of the phases of the power converter:

$$Q = \sqrt{\frac{\hat{I}^2 \cdot |v^+|^2 \cdot |v^-|^2}{k_2^2 \cdot |v^-|^2 + (1-k_2)^2 \cdot |v^+|^2 - 2k_2(1-k_2)\cos 2\gamma \cdot |v^+| \cdot |v^-|}} \quad (10.110)$$

The value of $\hat{I}$ in (10.110) can be changed according to the capability of the converter to withstand transient overcurrents.

Considering that γ can take three possible values, as shown in the Table 10.2, the expressions (10.110) will give rise to three different values for Q. Therefore, the final set-point for reactive power should be the minimum of the three possible values obtained by operating (10.110). This minimum value will limit the maximum reactive power injected by the power converter, for over this value of Q single-phase currents will trigger the overcurrent protection.

10.5.5 Performance of the FPNSC

This section presents the performance of the FPNSC strategy to generate the reference currents for the power converters in a small power plant. In this application example, the power plant has been modelled by using an aggregated model of its generators. Therefore, the whole power plant has been modelled as a 10 MVA power converter connected to the grid. The electrical network considered in this study case is shown in Figure 10.22.

In this network the unbalanced voltage at the PCC of the power converter arises due to the occurrence of a phase-to-ground fault at the overhead line L_2. After passing through two transformers, this kind of single-phase fault becomes a type D voltage sag at BUS 2.

The 10 MVA power converter will inject only reactive power into the network during the fault. The FPNSC strategy allows adjustment of the values of the k_1 and k_2 parameters to set the ratio between the positive- and negative-sequence currents for given values of active and reactive powers delivered to the grid. However, the parameter k_1 can be discarded in this study case, since it only affects the active currents, which are considered equal to zero in this application, since only reactive currents should be injected into the grid.

The value of k_2 plays an important role in this study case, as it permits setting the performance, in term of sequence components, of the reactive currents injected into the grid.

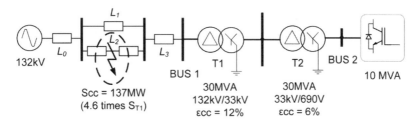

Figure 10.22 Layout of the study case for evaluating the FPNSC strategy.

Although the parameter k_2 can take different values, just three cases in the range of [0,1] will be considered in this example:

- Case A. Injection of positive-sequence reactive power ($k_2 = 1$). Grid voltage support.
- Case B. Injection of negative-sequence reactive power ($k_2 = 0$). Grid voltage unbalance compensation.
- Case C. Injection of simultaneous positive- and negative-sequence reactive power ($0 < k_2 < 1$). Cancellation of active power oscillations.

In addition to implementation of these three strategies for generating the reference currents in the system of Figure 10.22, the control algorithm also contains the equations for limiting the maximum reactive power to be injected by the power converter in order not to trip the overcurrent protection in any of the phases.

However, in order to make visible the influence of the FPNSC strategy, a particular sequence will be applied to control the power converter in all the cases. Figure 10.23 shows a diagram with the time sequence of the events considered in this study case, as well as operation of the power converter in each period.

According to the time sequence shown in Figure 10.23, the power converter injects a certain amount of active and reactive powers into the network at $t = 0$. After 400 ms, the unbalanced fault occurs. However, the control mode of the converter does not change until $t = 500$ ms, where the FPNSC is applied.

It should be pointed out that the following simulation results have been obtained working with a realistic weak grid model, with a short-circuit ratio in the range of 4.5. Therefore, the appearance of the fault will give rise not only to unbalance voltage components but also to

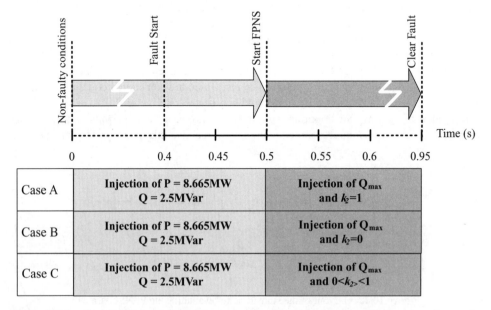

Figure 10.23 Control sequence of the power converter during the fault for different control strategies

transient oscillations in the voltage, which are produced mainly by the dynamical response of the weak electrical network. Regarding the operation of the power converter, it should be mentioned that a current control structure based on resonant controllers have been implemented in this study case.

10.5.5.1 Case A. Injection of Positive-Sequence Reactive Power

In the first simulation test, case A, the FPNSC is controlled in order to inject the highest positive-sequence reactive current that the power converter is able to deliver. The plot of Figure 10.24(a) shows the value of the grid voltage at BUS 1. On the other hand, the plot of Figure 10.24(b) shows the performance of the grid voltage at the output of the power converter, BUS 2, which is divided into its positive- and negative-sequence at the plot of Figure 10.24(c). The instantaneous active and reactive powers delivered to the grid are shown in the plot of

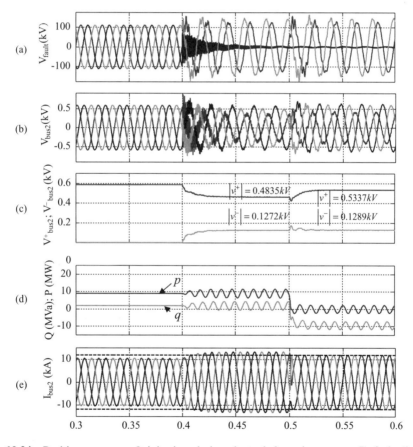

Figure 10.24 Positive-sequence Q injection during the unbalanced sag type D fault ($k_2 = 1$): (a) grid voltage at the fault point, (b) grid voltage at the power converter, (c) positive and negative voltage sequences at the power converter, (d) instantaneous active–reactive power delivered by the converter and (e) currents injected by the converter

Figure 10.24(d). Finally, the currents injected by the converter are shown in the plot of Figure 10.24(e). In this last plot, the two black dashed lines represent the maximum current that can be injected by the converter according to its nominal ratings.

Figure 10.24 shows how the full injection of positive-sequence reactive current during the fault permits boosting the voltage at the point of connection of the power converter, and hence supporting the grid during the fault. As shown in the plot of Figure 10.24(c), the positive-sequence voltage at BUS 2 increases from 0.4835 kV to 0.5337 kV. On the other hand, the negative sequence is not affected, as no negative-sequence reactive current is injected in this case.

The plot of Figure 10.24(d) shows the oscillations in the active and reactive powers, which is produced by the interaction between the positive-sequence currents and the negative-sequence voltages that exist in the grid during the unbalanced voltage sag.

One of the most relevant conclusions that can be obtained from this study case is reflected in the plot of Figure 10.24(e). At $t = 400$ ms, the voltage sag occurs but the FPNSC, together with the embedded current limitation algorithm, is not still enabled. At this time, the reference currents are generated by a control algorithm that is focused on generating a set of positive-sequence reference currents to maintain the prefault power delivery. Since the amplitude of the grid voltage decreased during the grid fault, the reference currents provided to the current controller by such an algorithm overpasses the limit value. At $t = 500$ ms, the FPNSC, together with the embedded current limitation algorithm, is activated, which makes the currents remain below the maximum admissible value, while the maximum reactive power is injected into the grid.

10.5.5.2 Case B. Injection of Negative-Sequence Reactive Power

In the second experiment, case B, the injection of only negative-sequence reactive current has been considered ($k_2 = 0$). The performance of the different electrical variables in this case is shown in Figure 10.25. As in the previous case, between $t = 0$ s and $t = 500$ ms, the control strategy that operates the power converter is focused on delivering a certain active and reactive power set-point by injecting positive-sequence currents. After that, the FPNSC is enabled at $t = 500$ ms.

In this study case, due to the injection of balanced negative-sequence currents, the voltage at BUS 2 becomes almost balanced. This effect can be clearly noticed in the voltage waveforms shown in the plot of Figure 10.25(b). This improvement in the voltage balance is due to the reduction in the negative-sequence voltage at BUS 2. However, this feature is even more noticeable in the plot of Figure 10.25(c), where it can be appreciated how the value of $|v^-|$ goes down from 0.1272 kV to 0.02964 kV.

As a difference from the previous case, the value of $|v^+|$ drops a bit when the FPNSC is enabled. However, this is not due to the injection of negative-sequence currents, but to the cancellation of the injection of positive-sequence active current into the weak grid at $t = 500$ ms, which was boosting the positive-sequence voltage at the point of connection of the power converter.

Oscillations in the instantaneous active and reactive powers when using the FPNSC strategy with $k_2 = 0$ are specially relevant, as shown in Figure 10.25(d). In this operation mode, as the magnitude of both v^+ and i^- are quite high, the oscillations are also important. Nevertheless, this is the price to be paid when using a power strategy focused on balancing the grid voltage at the PCC.

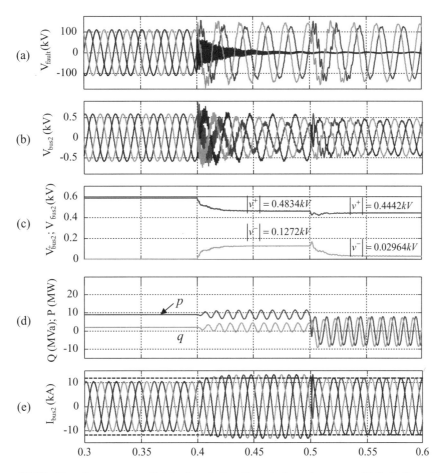

Figure 10.25 Negative-sequence Q injection during the unbalanced sag type D fault ($k_2 = 0$): (a) grid voltage at the fault point, (b) grid voltage at the power converter, (c) positive and negative voltage sequences at the power converter, (d) instantaneous active–reactive power delivered by the converter and (e) currents injected by the converter

In the plot of Figure 10.25(e), the current waveforms are displayed, showing again the capability of the FPNSC for limiting the currents injected by the converter below a given limit.

10.5.5.3 Case C. Injection of Simultaneous Positive- and Negative-Sequence Reactive Power

In order to show the performance of the FPNSC strategy when both positive- and negative-sequence reactive currents are injected into the grid, the plots of Figure 10.26 show the effect of injecting the maximum amount of reactive power into the grid when the parameter k_2 linearly varies from 1 to 0 under unbalanced grid voltage conditions. When the FPNSC strategy is started at $t = 500$ ms, the change in k_2 gives rise to a linear variation of the symmetrical components of the voltage at the point of connection of the power converter. This can be

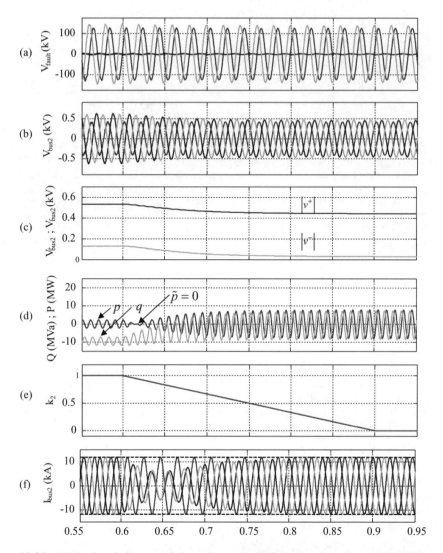

Figure 10.26 Q injection during an unbalanced sag type D fault ($0 < k_2 < 1$): (a) grid voltage at the fault point, (b) grid voltage at the power converter, (c) positive and negative voltage sequences at the power converter, (d) instantaneous active–reactive power delivered by the converter, (e) k_2 evolution and (f) currents injected by the converter

noticed in the plots of Figures 10.26(b) and (c). It can be concluded from this figure that the higher the value of k_2 the better for boosting the $|v^+|$ value, while reducing it contributes to attenuation of $|v^-|$.

One interesting feature in this study case to be stressed is the cancellation of the active power oscillation for a certain value of k_2. This point, which is highlighted in the plot of Figure 10.26(d), can be found as a function of $|v^+|$ and $|v^-|$, as shown in (10.83).

10.6 Summary

The occurrence of unbalanced grid faults gives rise to the appearance of negative-sequence voltages in the network that may affect the operation of grid-connected power converters. As presented in this chapter, implementation of specific control structures, able to deal with the injection of both symmetrical components of the current under unbalanced grid voltage conditions, is a key issue.

In the framework of this chapter, different control structures oriented to achieve independent control in the injection of positive- and negative-sequence currents have been presented. Among the different controllers addressed in this chapter, those based on the implementation of decoupling concepts have been studied in more detail.

On the other hand, the interaction of the symmetrical components produces active and reactive power oscillations at the output of the converter, which should be carefully controlled. In order to overcome this drawback, several strategies for creating the current reference to be tracked by the controller, in order to manage the P and Q oscillations in different ways, have been discussed.

As shown in this chapter, the behaviour of the currents depends on the selected strategy, but the major part of them gives rise to unbalanced current references and even distorted signals in some cases. This feature introduces an additional issue in the operation of grid-connected power converters, which is the control of the maximum current drawn at the output, which should be limited in order not to exceed the converter's nominal value. Moreover, the control of these currents should be carried out through the calculation of power references that consider the maximum current injection as a constraint. A detailed study has been performed for the most generalized structure presented in the chapter, the FPNSC.

Despite the fact that the presented methods can be extended, considering other control constraints, and also new ones can be proposed, the presented strategies have pointed out the main issues in the operation of grid-connected converters under grid unbalanced conditions, which are mainly related to the structure of the current controllers and the calculation of current reference, as a function of the desired power delivery performance and the ratings of the converter.

References

[1] Akagi, H., Watanabe, E. and Aredes, M., *Instantaneous Power Theory and Applications to Power Conditioning*, Wiley–IEEE Press, April 2007. ISBN 978-0-470-10761-4.

[2] Rioual, P., Pouliquen, H. and Louis, J. P., 'Regulation of a PWM Rectifier in the Unbalanced Network State'. In *Power Electronics Specialists Conference, 1993, PESC '93 Record. 24th Annual IEEE*, 20–24 June 1993, pp. 641–647.

[3] Rioual, P., Pouliquen, H. and Louis, J.-P., 'Regulation of a PWM Rectifier in the Unbalanced Network State Using a Generalized Model'. *IEEE Transactions on Power Electronics*, **11**(3), May 1996, 495–502.

[4] Hong-Seok Song and Kwanghee Nam, 'Dual Current Control Scheme for PWM Converter under Unbalanced Input Voltage Conditions'. *IEEE Transactions on Industrial Electronics*, **46**(5), October 1999, 953–959.

[5] Park, R.H., 'Tow Reaction Theory of Synchronous Machines. Generalized Method of Analysis – Part I'. In *Proceedings of Winter Convention of AIEE*, 1929, pp. 716–730.

[6] Yongsug Suh, Tijeras, V. and Lipo, T. A., 'Control scheme in hybrid synchronous–stationary frame for PWM AC/DC converter under generalized unbalanced operating conditions'. *IEEE Transactions on. Industry Applications*, **42**(3), June 2006, 825–835.

[7] Etxeberria-Otadui, I., Viscarret, U., Caballero, M., Rufer, A. and Bacha, S., 'New Optimized PWM VSC Control Structures and Strategies under Unbalanced Voltage Transients'. *IEEE Transactions on. Industrial Electronics,.* **54**(5), October 2007, 2902–2914.

[8] Bo Yin, Oruganti, R., Panda, S. K. and Bhat, A. K. S., 'An Output-Power-Control Strategy for a Three-Phase PWM Rectifier under Unbalanced Supply Conditions'., *IEEE Transactions on Industrial Electronics*, **55**(5), May 2008, 2140–2151.

[9] Lie Xu, Andersen, B. R. and Cartwright, P., 'VSC Transmission Operating under Unbalanced AC Conditions – Analysis and Control Design'. *IEEE Transactions on Power Delivery*, **20**(1), January 2005, 427–434.

[10] Stankovic, A. V. and Lipo, T. A., 'A Novel Control Method for Input Output Harmonic Elimination of the PWM Boost Type Rectifier under Unbalanced Operating Conditions'. *IEEE Transactions on Power Electronics*, **16**(5), September 2001, 603–611.

[11] Yongsug Suh, Tijeras, V. and Lipo, T. A., 'A Control Method in dq Synchronous Frame for PWM Boost Rectifier under Generalized Unbalanced Operating Conditions'. In *Power Electronics Specialists Conference, 2002, PESC 02. 2002. IEEE 33rd Annual*, Vol. 3, 2002, pp. 1425–1430.

[12] Rodriguez, P., Timbus, A. V., Teodeorescu, R., Liserre, M. and Blaabjerg, F., 'Independent PQ Control for Distributed Power Generation Systems under Grid Faults.' In *IEEE 32nd Annual Conference on Industrial Electronics, IECON 2006*, 6–10 November 2006, pp. 5185–5190.

[13] Rodriguez, P., Timbus, A. V., Teodorescu, R., Liserre, M. and Blaabjerg, F., 'Flexible Active Power Control of Distributed Power Generation Systems During Grid Faults.' *IEEE Transactions on Industrial Electronics*, **54**(5), October 2007, 2583–2592.

[14] Rodriguez, P., Timbus, A., Teodorescu, R., Liserre, M. and Blaabjerg, F., 'Reactive Power Control for Improving Wind Turbine System Behavior Under Grid Faults'. *IEEE Transactions on Power Electronics*, **24**(7), July 2009, 1798–1801.

[15] Fei Wang, Duarte, J. L. and Hendrix, M. A. M., 'Active Power Control Strategies for Inverter-Based Distributed Power Generation Adapted to Grid-Fault Ride-Through Requirements'. In *13th European Conference on Power Electronics and Applications, 2009, EPE '09.*, 8–10 September 2009, pp. 1–10.

[16] Fei Wang, Duarte, J. L. and Hendrix, M., 'Active and Reactive Power Control Schemes for Distributed Generation Systems under Voltage Dips'. In *IEEE Energy Conversion Congress and Exposition, 2009, ECCE 2009*, 20–24 September 2009, pp. 3564–3571.

[17] Teodorescu, R., Blaabjerg, F., Liserre, M. and Loh, P. C., 'Proportional-Resonant Controllers and Filters for Grid-Connected Voltage-Source Converters'. In *Electric Power Applications, IEE Proceedings*, September 2006, Vol. 153, No. 5, pp. 750–762.

[18] Lascu, C., Asiminoaei, L., Boldea, I. and Blaabjerg, F., 'High Performance Current Controller for Selective Harmonic Compensation in Active Power Filters'. *IEEE Transactions on Power Electronics*, **22**(5), September 2007, 1826–1835.

[19] Serpa, L. A., Round, S. D. and Kolar, J. W., 'A Virtual-Flux Decoupling Hysteresis Current Controller for Mains Connected Inverter Systems'. *IEEE Transactions on Power Electronics*, **22**(5), September 2007, 1766–1777.

[20] Serpa, L.A., Ponnaluri, S., Barbosa, P. M. and Kolar, J. W., 'A Modified Direct Power Control Strategy Allowing the Connection of Three-Phase Inverters to the Grid Through LCL Filters'. *IEEE Transactions on Industry Applications*, **43**(5), September/October 2007, 1388–1400.

[21] Cortés, P., Rodríguez, J., Antoniewicz, P. and Kazmierkowski, M., Direct Power Control of an AFE Using Predictive Control'. *IEEE Transactions on Power Electronics*, **23**(5), September 2008, 2516–2553.

[22] Cortés, P., Ortiz, G., Yuz, J. I., Rodríguez, J., Vazquez, S. and Franquelo, L. G., 'Model Predictive Control of an Inverter with Output LC Filter for UPS Applications'. *IEEE Transactions on Industrial Electronics*, **56**(6), June 2009, 1875–1883.

[23] Buso, S., Malesani, L. and Mattavelli, P., 'Comparison of Current Control Techniques for Active Filter Applications'. *IEEE Transactions on. Industrial Electronics*, **45**(5), October 1998, 722–729.

[24] Blaabjerg, F., Teodorescu, R., Liserre, M. and Timbus, A. V., 'Overview of Control and Grid Synchronization for Distributed Power Generation Systems'. *IEEE Transactions on Industrial Electronics*, **53**, October 2006, 1398–1409.

[25] Zmood, D. N. and Holmes, D. G., 'Stationary Frame Current Regulation of PWM Inverters with Zero Steady-State Error'. *IEEE Transactions on Power Electronics*, **18**(3), May 2003, 814–822. DOI: 10.1109/TPEL.2003.810852.

[26] Aredes, M., Watanabe, E. H. and Akagi, H., *Instantaneous Power Theory and Applications to Power Conditioning*. IEEE Press Series on Power Engineering, March 2007. ISBN 0470107618.
[27] Willems, J. L., 'A New Interpretation on the Akagi–Nabae Power Components for Nonsinuoidal Three-Phase Situations'. *IEEE Transactions on Instrumentation and Measurement*, **41**, August 1992, 523–527.
[28] Buchholz, F., *Das Begriffsystem Rechtleistung. Wirkleistung, totale Blindleistung*. Munich, Germany: Selbstverlag, 1950.

11

Grid Filter Design

11.1 Introduction

The role of the grid filters in VSC-based grid converter operation is twofold. On one side the grid filter should have a dominant inductive behaviour to guarantee the proper operation of the voltage source converter if connected to a voltage source type system such as the utility grid. In this sense grid converters replicate the well-known behaviour of synchronous generators and of transmission lines where the control of active and reactive power exchange is related to the control of phase and magnitude of the electromagnetic force. On the other side, VSC-based grid converters generate PWM carrier and side-band voltage harmonics. These voltages may lead to current flowing into the grid, which can disturb other sensitive loads/equipment and increase losses if proper grid filters are not adopted to prevent them flowing. A grid filter made by a simple inductor is the simplest solution to comply with the two aforementioned requirements.

For applications around and above several hundreds of kW, like the wind turbine systems, the switching frequency is low, to limit losses. Hence to attenuate the harmonics in the current enough to meet the demands of standards and grid codes the use of a high value of input inductance could be not enough alone (Figure 11.1) and it becomes quite expensive to realize higher value filter reactors and sometimes also the encumbrance of the inductor could be an issue [1]. Moreover, the system dynamic response may become poorer.

For applications around a few kW, like the most widespread photovoltaic systems, the switching frequency is higher. Hence even smaller inductors could help in meeting the requirements, but the encumbrance of the inductors is certainly an issue since the converter and its passive elements are integrated.

At a system level, like in the case of wind or photovoltaic parks, the main concern is related to the disturbances produced by some specific harmonics. Hence a possibility is to use a bank of tuned LC trap filters, which have the advantage to stop specific harmonics that could deteriorate the voltage quality.

However, typically standards and grid codes recommend compliance with limitations that are very stringent for frequencies above a certain threshold. Hence a low-pass filter attenuation is needed and the preferred solution becomes the use of high-order filters like LCL, which provide 60 dB per decade attenuation for the PWM carrier and side-band voltage harmonics.

Grid Converters for Photovoltaic and Wind Power Systems   Remus Teodorescu, Marco Liserre, and Pedro Rodríguez
© 2011 John Wiley & Sons, Ltd

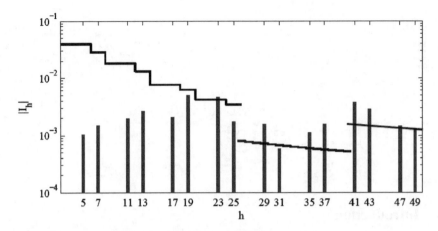

Figure 11.1 Worst-case weighted harmonic voltage over the modulation index range (0.8–1.15) compared to the German VDEW per-unit harmonic current injection limits (base voltage 3.3 kV, base power 6 MVA, SCR = 20, 1.05 kHz, 1 p.u. inductor)

With this solution, optimum results can be obtained using quite small values of inductors and capacitors [2, 3].

A further issue for a VSC is high-frequency EMI (differential mode and common mode) [4], which needs specific filters in frequency ranges above 150 kHz and rated at lower power levels. Of course an LCL filter that is effective in the reduction of switching frequency harmonics may also be effective for differential mode EMI if the filter inductors are built using chokes that can mitigate high frequency (using ferrite cores, for example). Similarly for common mode EMI, a common mode inductor could be included in the differential mode filter as suggested in reference [5]. However, conducted EMI is a very complex problem: depending on the frequency range it needs different solutions and specific designed filters. Hence, even if filter integration is feasible in some cases, the use of one filter over a wide frequency range is often too expensive since the same reactive element must be designed to work over different frequency ranges and at different power levels.

A good criterion to choose filter parameters is to limit the size of the installed reactive elements (these can result in a poor power factor) and the power losses (due to the passive damping required to avoid resonance).

Hence in this chapter some design considerations on the filter are carried out, after having first reviewed the main filter topologies. Then the problems arising from the interaction with the grid and damping solutions are discussed. Finally, the possible nonlinear behaviour of the filter (mainly due to saturation) is discussed.

11.2 Filter Topologies

As already pointed out in the introduction, the two most adopted approaches to reduce PWM carrier and side-band harmonics are the use of a tuned LC filter (typically at a system level to meet requirements related to the voltage quality) or a low-pass LCL filter. In the first case a group of trap filters acts on selective harmonics that need to be reduced. This solution

Grid Filter Design

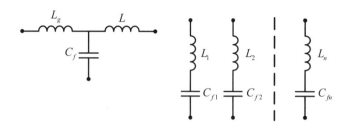

Figure 11.2 LCL filter (low-pass filter) and bank of LC filters (trap filters)

is particularly feasible for power converters switching at hundreds of Hz, producing PWM harmonics that are at a frequency so low that it is difficult to tune a low-pass filter like the LCL filter. This one acts on the whole harmonic spectrum and provides a 60 dB/dec attenuation after the resonance frequency. Hence the resonance frequency should be far enough from the switching frequency but not so much that it does not challenge the current control loop.

Figure 11.2 shows the two kinds of filter previously introduced. It is possible to have different kinds of trap filters (obtained with an inductor and a capacitor in parallel) and several other combinations of these two families of filters.

The design of the trap filter or of the part of the filter tuned to reduce a certain frequency can be developed on the basis of robustness considerations as described in reference [6]. On the other hand, the design of the low-pass LCL filter has some degree of freedom, as described in the following.

11.3 Design Considerations

The voltage source converter needs both AC and DC passive elements, as shown in Figure 11.3. The passive elements, such as capacitors or inductors have both storage and filtering functions.

The energy stored in the AC passive stage is usually less than 5 % of all the energy stored. Thus the main storage element is the DC capacitor charged to a voltage that is able to ensure the basic function of the VSC: the VSC can control the AC current i through the switching and by acting as the current source. Then through the AC current control, the VSC can change and control the DC value v_{dc} of the capacitor. Thus the filtering action, which is necessary because of the fast-switching PWM, is done both on the DC side and on the AC side. The passive elements are charged/discharged during the switching period, ensuring the smoothing of the AC currents and of the DC voltage. This filtering action is also the basis of the control performed.

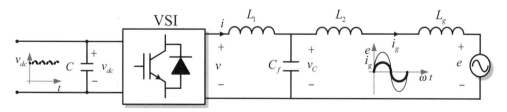

Figure 11.3 Voltage source converter with a sketch of the desired grid current i_g in phase with the grid voltage e and of the desired DC voltage v_{dc}

In fact, the dynamics of the AC current/DC voltage control depend on the time constants of the two filtering stages. Generally, the overall design, which should include filtering and control issues, is a trade-off between a high filtering and a fast dynamic performance.

Once clarified, the main function of the passive elements, some design rules regarding capacitors and inductors can be introduced.

Thus the reactance of L_1, $X_1 = \omega L_1$, of L_2, $X_2 = \omega L_2$, of L_g, $X_g = \omega L_g$, and of C_f, $X_C = -1/\omega C_f$, when reported in small letters, are considered in p.u. of the base resistance R_b that represents the active power injected by the power converter into the grid. The power converter is current controlled such as there is no displacement between the measured voltage and the measured current.

In order to chose the LCL-filter parameters it is worth noting that one of the main concerns is usually related to the possible power factor decrement due to the displacement between the voltage and current caused by the reactive elements installed in the filter. However, the equivalent impedance at the fundamental frequency at the PCC or at the converter terminals is strictly dependent on the position of the voltage and current sensors, assuming that the current control will guarantee that the current is in phase with the voltage.

In Figure 11.4 four cases are reviewed considering the equivalent vector diagram. On the basis of these vector diagrams, it is possible to calculate the grid side z_{Tgrid} and the converter side z_{Tconv} p.u. impedance at the grid frequency. The impedances are calculated neglecting the phase displacement introduced by the capacitor between the voltage drop on the grid side and on the converter side.

If the capacitor voltage is sensed and the converter current is controlled to be in phase with the voltage then

$$z_{Tgrid} = 1 + j(x_2 + x_g - x_C)$$
$$z_{Tconv} = 1 + jx_1 \tag{11.1}$$

If the capacitor voltage is sensed and the grid current is controlled to be in phase then

$$z_{Tgrid} = 1 + j(x_2 + x_g)$$
$$z_{Tconv} = 1 + j(x_1 - x_C) \tag{11.2}$$

If the grid voltage is sensed and the grid current is controlled to be in phase then

$$z_{Tgrid} = 1 + jx_g$$
$$z_{Tconv} = 1 + j(x_1 - x_2 - x_g - x_c) \tag{11.3}$$

If the grid voltage is sensed and the converter current is controlled to be in phase with the voltage then

$$z_{Tgrid} = 1 + j(x_g - x_C)$$
$$z_{Tconv} = 1 - j(x_1 + x_2) \tag{11.4}$$

The main design rules for inductors are on the choice of the core dimension, material and gap, and of the winding. The most used magnetic cores are ferrite, laminated steel and powdered metal [7, 8] (Table 11.1). The required energy must be stored in a nonmagnetic gap

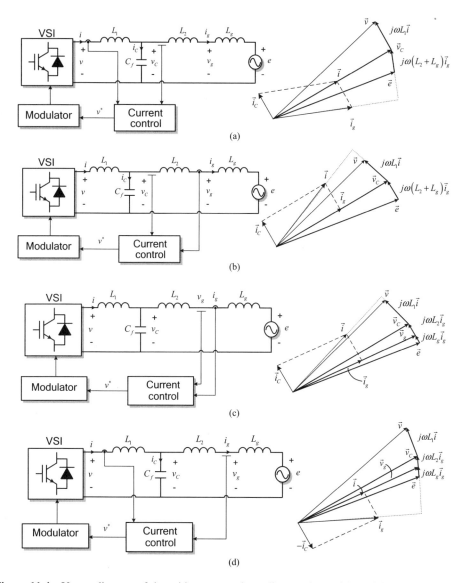

Figure 11.4 Vector diagram of the grid converter depending on the position of the sensors: (a) with voltage sensed on the capacitor and current sensed on the converter side, (b) with voltage sensed on the capacitor and current sensed on the grid side, (c) with voltage and current sensed on the grid side and (d) with voltage sensed on the grid side and current sensed on the converter side

distributed in the case of a powdered metal core or in a discrete gap in series in the case of a ferrite core.

The maximum flux density B_{max} has to be in all operating conditions lower than the saturation flux density B_{sat} of the core material for ensuring a linear behavior and a safe operation. The power losses have to be limited and an efficient power dissipation has to be ensured to avoid overheating of the inductor. In fact the thermal resistance between the core and the ambient as

Table 11.1 Magnetic properties (content, permeability, saturation, resistivity, curie temperature) of the most commonly used core materials

Material	Iaminated FeSi	Powdered iron	Ferrites
Contents	3–6% Si	95% Fe resin	MnZn, NiZn resin
μr	1000–10000	1–500	100–20000
B_{Sat} [T]	1.9	1–1.3	0.3–0.45
ρ [$\mu\Omega$m]	0.4–0.7		10^2–10^4 MnZN
			10^7–10^9 NiZN
Curie temp.	720	700	125–450

well as the inner windings and the ambient is high, due to the fact that the electrical isolation of the winding is also a not negligible thermal isolation. For differential mode filters materials with a low permeability are beneficial. However for economical reasons also laminated steel with a rather high permeability is used. Since this material can be processed in any shapes one or more air-gaps can be realized with low efforts. Laminated steel or electrical steel is an alloy of iron and silicon. To reduce the eddy current losses the core is made from stacks of many thin laminations, which are electrically insulated from each other by a thin insulating coating. For the core design it is important that the direction of magnetic flux is along the lamination to utilize the full benefits. The advantage of adding silicon to the iron is to reduce its conductivity and thus limiting the eddy current losses. Another point is the noise reduction caused by the magnetostriction. To achieve even lower losses the orientation of the grain is forced in a defined direction to achieve an anisotropic material there the losses are reduced when the magnetic flux is in the direction of the lamination.

In order to tune correctly the LCL filter the starting point is the current ripple on the converter side of the filter and the harmonic limit imposed on the grid current by the standards, recommendations and utility codes. The converter-side inductance is designed in order to limit the ripple of the converter-side current. Moreover, the inductor should be properly designed so as not to saturate and hence the correct inductor choice is a trade-off between ripple reduction and inductor cost. Accepting high values of the current ripple may lead to saturation problems in the core and as a consequence to the use of a core that could be used also for realizing higher value inductors. On the other side of the filter the grid pollution is evaluated in terms of harmonics rather than in terms of ripple amplitude. Hence the LCL-filter effectiveness in reducing them should be evaluated using a frequency domain approach. In brief, a time-domain analysis (ripple evaluation) drives the choice of the converter-side inductor while a frequency-domain analysis (harmonic evaluation) drives the choice of the LCL-filter resonance frequency and, as a consequence, of some couples of values of grid-side inductor and capacitor that could meet that requirement. The last degree of freedom in the choice among these couples of CL values that meet the resonance frequency requirement is used to minimize the installed reactive power of the LCL filter, evaluating cost, weight and encumbrances of the capacitor and inductive elements, the robustness of the LCL-filter attenuation capability with respect to the possible different grid impedance conditions and the effect of the damping resistor, if used, on the filter attenuation capability.

As previously described, the LCL-filter design can be organized in three steps, described in the following:

1. *Ripple analysis and converter-side inductor choice.* A rough approximation of the LCL-filter behaviour leads to a consideration of the capacitor short-circuit at the frequency of the ripple. However it should be noted that in case MW power converters, since the switching frequency is low and the resonance frequency should be even lower the approximation lead to underestimate the value of the converter side inductor [1]. The smaller this value the bigger will be the inductor cores of the filter and the lower will be the resonance frequency needed to guarantee the desired attenuation of the grid current harmonics. The ripple amplitude will depend only on the number of levels of the PWM voltage:

$$\Delta I_{MAX} = \frac{1}{n} \frac{V_{dc}}{L_1 f} \qquad (11.5)$$

where n is a coefficient that increases with the number of levels of the voltage waveform. The previous expression should be used to design the magnetic core in order to avoid saturation for the high-frequency ripple.

2. *Harmonic attenuation of the LCL filter and choice of the resonance frequency value.* Once the value of L_1 has been chosen according to the previous considerations the ripple attenuation should be considered:

$$\frac{i_g(\omega)}{i(\omega)} = \frac{z_{LC}^2}{|z_{LC}^2 - \omega^2|} \qquad (11.6)$$

where $\omega_{res}^2 = z_{LC}^2 (L_1 + L_2 + L_g)/L_1$, $z_{LC}^2 = [(L_g + L_2)C_f]^{-1}$ and ω is the frequency of interest, which are used to verify the compliance with standards, recommendations and grid codes. Equation (11.6), depicted in Figure 11.5, can be used to choose the resonance frequency and as a consequence the product $L_2 C_f$.

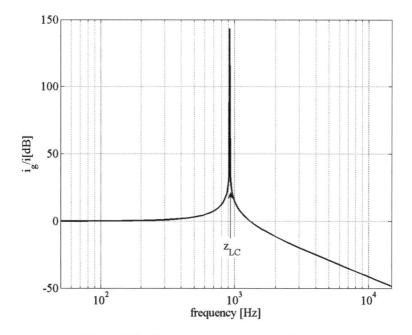

Figure 11.5 Ripple attenuation as reported in (11.6)

3. *LCL-filter optimization and choice of grid-side inductor, capacitor and damping method and value.* The final optimization of the filter consists in the choice of the values of L_2 and C_f and on the damping and its value. The following three criteria can be used:
   - Installed reactive power of the filter. This can be computed by considering the p.u. value of the impedance seen on the converter side. In other words, if the converter control maintains a unity power factor between the grid voltage and current which is the amount of reactive power that the converter has to manage? This p.u. value depends on the sensor position, in case the grid voltage is sensed, and the grid current is controlled, which is

$$z_{Tconv} = 1 + j(x_1 - x_2 - x_c) \quad (11.7)$$

   Hence L_2 and C_f contribute in the same way to the displacement between the converter voltage and the converter current.
   - Robustness of the resonance frequency, and as a consequence of the filter attenuation, to the grid impedance variation:

$$\Delta\omega_{res} = \frac{1}{2\omega_{res}C_f}\left(\frac{1}{L_2+L_{g1}} - \frac{1}{L_2+L_{g2}}\right) \quad (11.8)$$

   From (11.8) it is obvious that the higher is the capacitor of the LCL filter the less influent is the grid impedance on the system resonance.
   - The influence of damping on the LCL-filter attenuation should be calculated with and without damping.

11.4 Practical Examples of LCL Filters and Grid Interactions

The grid-filter design guidelines previously discussed are challenged by the large range of possible grid conditions in terms of background harmonic distortion and stiffness. The situation is particularly critical in rural areas where the distributed generation plant is connected due to the good wind resources (in case of WT) or more in general due to the need to supply more continuity to the local loads.

In Figure 11.6(a) a scheme of the grid connection to the PV home is depicted. The high-voltage/medium-voltage transformer as well as a three-phase cable to the MV transformer introduce only a small impedance. If the PV house is located in a remote area the medium-voltage line can be very extended (hundreds of km) and its cables can introduce a relevant reactive reactance [9]. The MV/LV transformer also introduces a reactance that could be considerably higher in locations where the transformer-rated power could be considerably lower. Then the low-voltage cable introduces a prevalently resistive impedance that varies with the distance of the PV inverter from the transformer. Moreover, the presence of capacitive loads (e.g. refrigerators) connected in the house could introduce a capacitive impedance that can create low-frequency resonances [9].

In Figure 11.6(b) a scheme of the connection of the grid to a WT back-to-back system is depicted. The system is similar to the previous one (hence also in this case a radial plant typical of a rural zone can introduce a very high impedance) up to the MV/LV transformer, to which the wind turbine converter is connected. Hence the impedance seen by the grid-connected

Grid Filter Design

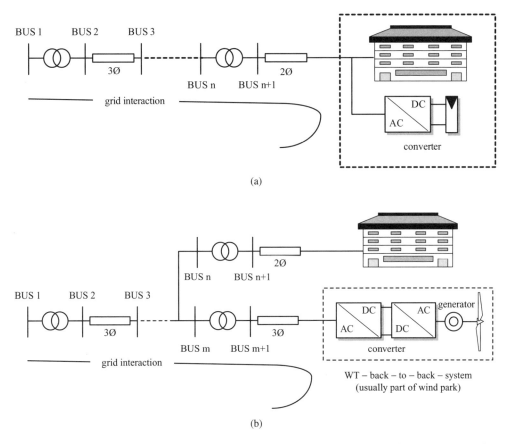

Figure 11.6 Scheme of the voltage network from the high-voltage (HV) distribution system: (a) to the home (which can be modelled as an RLC load) equipped with a PV inverter and (b) to a wind turbine back-to-back system that can also supply domestic loads connected to the same transformer

inverter can be mainly inductive and can be different, depending on the plant configuration. However, the system presents less interaction with domestic loads, avoiding possible low-frequency resonances.

A step-by-step procedure to design an LCL filter for grid-connected inverters has already been described in references [2] and [3]. This design procedure has been carried out in p.u. values in order to make it independent of the system power level.

However, two examples, a 500 kW WT system and a 3 kW PV system, demonstrate also how the power level and the type of converter play a role in the selection of the LCL-filter parameters (more inductive in the first case, more capacitive in the second case). The two sets of parameters found for a 500 kW WT plant and a 3 kW PV plant are respectively: 0.2 mH converter side, 83 µF and 0.03 mH grid side; 0.4 mH converter side, 5 µF and 0.2 mH grid side (Table 11.2). In the same table, typical grid impedance values have been added in the case of weak and stiff grid conditions.

Table 11.2 Parameters of the 500 kW WT and 3 kW PV LCL filters in absolute and p.u. values

		500 kW WT system		3 kW PV system	
LCL filter component	Boost inductance	0.2 mH	23 %	0.4 mH	1 %
	Grid-side inductance	0.03 mH		0.2 mH	
	Filter capacitor	83 µF	1 %	5 µF	3 %
Grid impedance	Maximum value (weak grid)	0.03 Ω (inductive)	10 %	2.7 Ω (41 % inductive)	15 %
	Minimum value (stiff grid)	0.003 Ω (inductive)	1 %	0.4 Ω (resistive)	2 %

In the first case the resonance frequency is around 3.4 kHz while in the second case it is 6.2 kHz. This can be explained by the fact that higher power converters should adopt a lower switching frequency and hence also the LCL-filter resonance should be lower in that case in order to allow a better filtering property. Particularly, the 500 kW system was designed to work with a 5 kHz switching frequency, while the 3 kW system was designed to work with a 17 kHz switching frequency.

However, this is not the only difference: while in the first case the LCL filter has been designed with 23 % inductance and 1 % capacitance, in the second case the LCL filter has been designed with 1 % inductance and 3 % capacitance. This is due to the fact that in the high-power converters the inductance saturation and the need to limit harmonic propagation in the plant are the most dangerous issues, while in the low-power mass-produced converters the packaging and cost issues are the most stringent. This leads to more p.u. inductance in the first case and, conversely, more p.u. capacitance in the second case (in order to compensate for the presence of too-low inductive elements and to limit the LCL-filter resonance frequency that otherwise would become too high).

Once these DG systems are connected to the grid, the resonance frequency of the LCL filter can vary depending on the grid reactance. The effect of an inductive grid reactance (long cable and low-power transformers) is a decrease of the resonance frequency, while the effect of a capacitive reactance (e.g. a refrigerator in home applications) leads to the creation of other resonant peaks in the LCL-filter frequency characteristic.

Hence the resonance frequency will be shown to vary as the grid inductance varies up to a maximum 0.1 p.u. (which corresponds to a short-circuit ratio of 10) in both cases. If 0.1 p.u. is applied to the 500 kW system it leads to 0.1 mH; if it is applied to a 3 kW system it leads to 5.6 mH.

Figure 11.7 shows the results of the analysis demonstrating that the resonance frequency can drop 40 % in both cases. The direct consequence of this reduction is that the LCL filter is more effective in harmonic reduction and also that the resonance frequency falls into the low-frequency domain. In this case the use of passive damping designed for the ideal condition (stiff grid case) can be ineffective. On the contrary, if the passive damping has been designed for the worst condition (weak grid case), there will be too high losses [3] when the system is working in the stiff grid case.

Hence the use of active damping is definitively more flexible since the change of the controller parameters, depending on the system inductance, is costless.

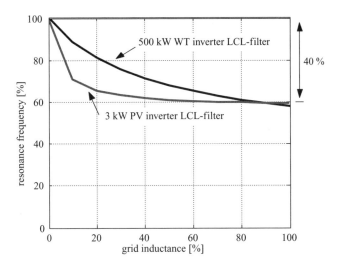

Figure 11.7 Resonance frequency variation (%) of the rated resonance frequency as a function of grid inductance (%) of 5.6 mH (0.1 p.u. for a 3 kW PV inverter with an LCL filter) and of 0.1 mH (0.1 p.u. for a 500 kW PV inverter with an LCL filter)

The next issue is to discuss the influence of capacitance impedance on the system and the effect on the filter frequency characteristic. Figure 11.8 shows the frequency characteristic of a 3 kW PV system with an LCL filter. From Figure 11.8 it is evident that the introduction of a 100 μF capacitance (cases C and D) generates an extra resonance peak while it attenuates the principal one. Also in this situation the use of passive damping can be ineffective and the use of a controller that actively damps the system is preferred due to its flexibility.

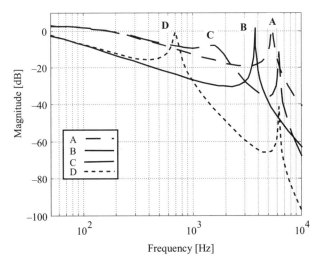

Figure 11.8 Frequency characteristic of the 3 kW PV system with an LCL filter in the following grid conditions: A: 0.1 mH, B: 3 mH, C: 0.1 mH, 100 μF, D: 3 mH, 100 μF

11.5 Resonance Problem and Damping Solutions

The possible instability of the current control loop is caused by the zero impedance that the LCL filter offers at its resonance frequency. A poor analysis made on qualitative considerations could lead to excessive damping (unnecessary increasing of the losses) or insufficient damping LCL filters have more state variables with respect to the simple L filter (as shown in Chapter 9, equation (9.9)). The proper damping of these dynamics can be achieved by modifying the filter structure with the addition of passive elements or by acting on the parameters or on the structure of the controller that manage the power converter. The first option is referred to as passive damping while the second is referred to as active damping.

Passive damping causes a decrease of the overall system efficiency because of the associated losses, which are partly caused by the low-frequency harmonics (fundamental and undesired pollution) present in the state variables and partly by the switching frequency harmonics, as will be shown in the following. Moreover, passive damping reduces the filter effectiveness since it is very difficult to insert the damping in a selective way at those frequencies where the system is resonating due to a lack of impedance. As a consequence, the passive damping is always present and the filter attenuation at switching frequency is compromised.

Active damping consists in modifying the controller parameters or the controller structure either by cutting the resonance peak and/or by proving phase lead around the resonance frequency range. Active damping methods are more selective in their action and they do not produce losses, but they are also more sensitive to parameter uncertainties. Moreover, the possibility of controlling the potential unstable dynamics is limited by the controller bandwidth. This is dependent on the controller sampling frequency. It has been demonstrated that the sampling frequency should be at least double with respect to the resonance frequency to perform active damping.

11.5.1 Instability of the Undamped Current Control Loop

In the following the analysis is developed in the synchronous reference frame. The cross-coupling between the d axis and the q axis is neglected and the current control loops are decoupled and linearized. Therefore, all the scheme are referred equivalently to the d axis or the q axis and the considered quantities like the converter voltage v, the filter capacitor voltage v_c, the grid voltage e, the converter current i and the grid current i_g could be d or q quantities. Moreover, the reference signals are addressed with the '*' as usual.

The configuration of the filter should be taken into account when the stability of the system is investigated. The design of the current controller has been done neglecting the zero and poles introduced by the capacitor presence.

Now the study is carried out in Laplace domain, being s the Laplace operator. If the whole LCL filter (Figure 11.9) is considered, neglecting all the filter losses and using for simplicity L as the converter-side inductance previously indicated with L_1 and L_g for the grid-side inductance previously indicated with $L_2 + L_g$, considering $z_{LC}^2 = [L_g C_f]^{-1}$ and $\omega_{res}^2 = L_T z_{LC}^2 / L$, the transfer function is

$$G_f(s) = \frac{i(s)}{v(s)} = \frac{1}{Ls} \frac{(s^2 + z_{LC}^2)}{(s^2 + \omega_{res}^2)} \quad (11.9)$$

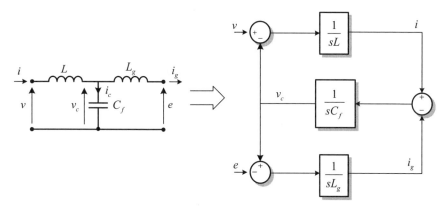

Figure 11.9 Input filter single-phase equivalent and the model for the active rectifier, neglecting filter losses

Thus the LCL filter has two zero and two poles more, in the open-loop transfer function, compared those of an L filter.

If the closed-loop root locus is considered, with the PI regulator tuned neglecting the capacitor C_f and considering only the total inductance L_T, the new zero and poles can make the system, shown in Figure 11.10, unstable if a proper damping is not adopted. The undamped closed-loop transfer function H_{ud} in the Z domain is

$$H_{ud}(z) = \frac{G_{PI}(z)G_f(z)}{z + G_{PI}(z)G_f(z)} \qquad (11.10)$$

where $G_f(z)$ is the zero-order hold (ZOH) equivalent of the $G_f(s)$ reported in (11.9) and $G_{PI}(z)$ is the controller designed neglecting C_f, and thus designed on a $G_f(s)$ in which $z_{LC} = 0$. Figure 11.10 shows that there are always two poles unstable on the left part of the plane and a resonant peak in the Bode diagram.

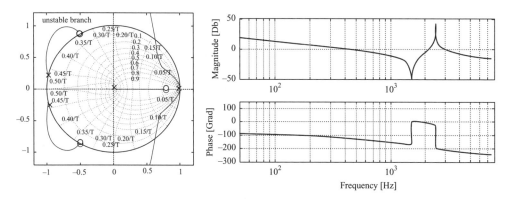

Figure 11.10 Pole-zero and Bode plots of the LCL-filter transfer function (11.10)

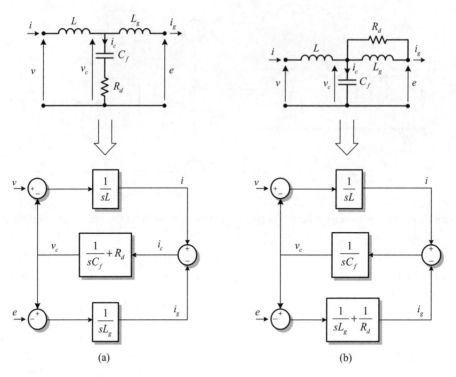

Figure 11.11 Input filter single-phase equivalent model: (a) for damping in series with the capacitor and (b) in parallel with the grid inductor

11.5.2 Passive Damping of the Current Control Loop

The introduction of a resistor in series with the filter capacitor or in parallel with the grid-side inductor can bring again the two unstable poles of the LCL-filter-based current-controlled system in the stability region (Figure 11.11).

If the damping R_d is connected in series with the filter capacitor, the plant transfer function becomes:

$$G(s) = \frac{i(s)}{v(s)} = \frac{1}{Ls} \frac{\left(s^2 + \frac{R_d}{L_g}s + z_{LC}^2\right)}{\left(s^2 + \frac{(L+L_g)}{L}\frac{R_d}{L_g}s + \omega_{res}^2\right)} \tag{11.11}$$

where the damping term $R_d/L_g \to 0$ for $R_d \to 0$. The effects of the damping on the LCL-filter transfer function are shown in Figure 11.12. It is possible to see clearly that the resonant peak is reduced, leading to the absence of zero-crossing where the phase margin could be critical.

Grid Filter Design

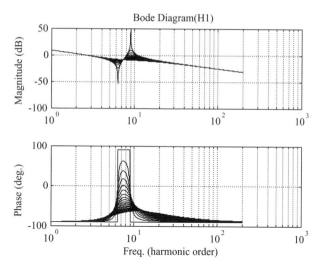

Figure 11.12 Effect of growing damping on the LCL filter $G(s)$ transfer function (11.11)

The consequent losses can be calculated as

$$P_d = 3R_d \sum_h \left[i(h) - i_g(h)\right]^2 \quad (11.12)$$

If the damping R_d is connected in parallel with the grid inductor, the plant transfer function becomes

$$G(s) = \frac{i(s)}{v(s)} = \frac{1}{Ls} \frac{\left(s^2 + \frac{1}{C_f R_d}s + z_{LC}^2\right)}{\left(s^2 + \frac{1}{C_f R_d}s + \omega_{res}^2\right)} \quad (11.13)$$

where the damping term $1/C_f R_d \to 0$ for $R_d \to \infty$. The consequent losses can be calculated as

$$P_d = \frac{3}{R_d} \sum_h \left[v_c(h) - e(h)\right]^2 \quad (11.14)$$

The main terms of the sums in (11.12) and (11.14) are for the index h near to the switching frequency and its multiples. In fact, the damping absorbs a part of the switching frequency ripple to avoid the resonance. The losses decrease as the damping resistor value increases but at the same time this reduces its effectiveness.

It is also possible to estimate the damping losses using only the converter voltage or the converter current and considering that the inverse of both (11.13) and (11.14) represent the LCL-filter impedance where the resistive part (and so the losses) is only due to damping.

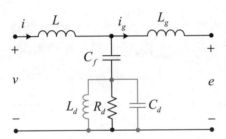

Figure 11.13 Selective passive damping

The selection of the best passive damping solution is a very challenging task when the resonance frequency is very low and the damping has not only influence on the stability and on the filter attenuation but also on the amplitude of the harmonics around the resonance frequency and hence on the overall harmonic content and on the losses that those harmonics can cause. The passive damping of a very low resonance frequency filter is a very nonlinear problem.

Both the losses and the harmonic content are strictly dependent on the voltage harmonic spectrum produced by the converter and this is changing as the modulation index changes. Hence a possible solution is to use selective high-pass damping and selective resonant damping [1]. They are an attempt to emulate with passive elements the selective effect of active damping. Two possibilities are considered: an inductor is inserted in parallel to the damping resistor (as shown in Figure 11.13 in grey) in order to cancel its effects at low frequencies where the inductor will be a short-circuit; an inductor and capacitor is inserted in parallel to the damping resistor (as shown in Figure 11.13 in grey) in order to cancel its effects at low and high frequencies where the inductor or the capacitor will be a short-circuit. In the first case it should be considered that the passive damping losses at low frequency can be even half of the overall losses, and hence the selective low-pass damping can cut the losses considerably. The second case is an attempt to minimize the losses due to PWM switching and preserve the filtering function that can be compromised by the impedance introduced by the passive damping.

From Figure 11.14 it is possible to note that the total damping has losses much higher with respect to the selective passive damping, mainly due to the first harmonic, which in the case of selective damping finds much higher impedance in the transverse row of the filter.

11.5.3 Active Damping of the Current Control Loop

Active damping methods can be classified in two main classes: multiloop-based and filter-based. In the first case the stability is guaranteed via the control of more system state variables that are measured or estimated [11]. Hence instead of a single current control loop, two or even three control loops are adopted. Among the controlled state-variables, the capacitor voltage can also be controlled in view of stand-alone or microgrid operation using droop control as discussed in Chapter 9. This approach can be formalized within the more general theoretical framework of the state-space control also using state-estimator [12]–[13]. A simple and straightforward example is the use of the so called "virtual resistor" where the capacitor current is feedback in the control loop multiplicated for a gain emulating the presence of a real

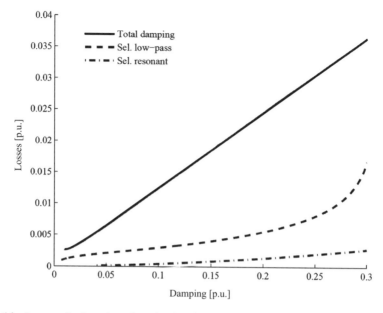

Figure 11.14 Losses of only resistor damping (total damping) compared with losses of selective passive damping

resistor connected to the LCL-filter [14]. The approach is depicted in Figure 11.15, where

$$H(s) = \frac{i_c(s)}{i(s)} = \frac{L_g C_f s^2}{1 + L_g C_f s^2}$$

The second class of active damping methods is based on the use of a higher order controller to regulate not only the low frequency dynamics but also to damp the high frequency ones (Figure 11.16):

$$G_{AD}(s) = \frac{s^2 + 2D_z \omega_z s + \omega_z^2}{s^2 + 2D_p \omega_p s + \omega_p^2}$$

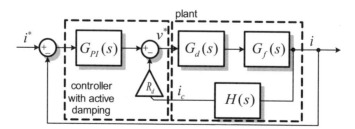

Figure 11.15 Virtual Resistor based active damping (neglecting the presence of grid voltage disturbance and its compensation).

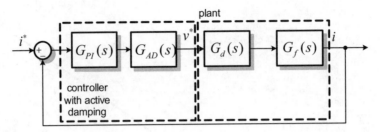

Figure 11.16 Filter-based active damping (neglecting the presence of grid voltage disturbance and its compensation).

Where D_z, ω_z and D_p, ω_p are respectively the damping and frequency of the zeros and poles of the filter. In case $\omega_z = \omega_p$ a notch filter is obtained (Figure 11.17).

The controller part responsible for active damping of the potentially unstable high frequency dynamics can be seen as a filter [15]–[17]. The filter can be designed using different approaches: two have been proposed in literature for active damping purposes. The first consists in designing an analog filter, typically a notch filter, then applying analog-to-discrete transformation based on a given set of specifications, called bilinear transformation [F]. In both cases this class of active damping methods has the advantage that it does not need more sensors. However, only in case of the first approach the straight forward physical meaning of "filtering" the resonance is preserved. The effect of filter-based active damping in digital domain, using the Z-transform, is shown in Figure 11.18, while the effect on the current in time-domain and frequency domain is shown in Figure 11.19.

11.6 Nonlinear Behaviour of the Filter

When the inductance has a nonlinear behaviour a distorted current waveform is produced (Figure 11.20); if the saturation is symmetric, it contains only odd harmonics. When an AC current drives an inductor into saturation for part of each cycle, the inductor presents two

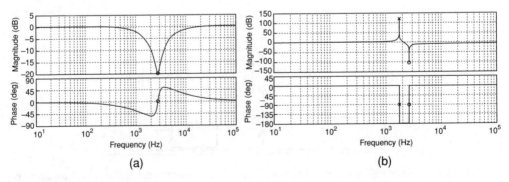

Figure 11.17 Filter-based active damping: a) notch filter and b) generic biquadratic filter.

Grid Filter Design

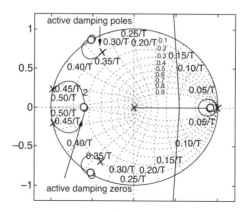

Figure 11.18 Effect of biquadratic-filter-based active damping (root locus in Z-domain)

values of inductance to the rest of the circuit, the rated value and the saturated value:

$$\phi(i) = sat(i) = \begin{cases} Li_{sat} & i > i_{sat} \\ Li & i_{sat} \geq i \geq -i_{sat} \\ -Li_{sat} & -i_{sat} > i \end{cases} \quad (11.15)$$

In many cases an appropriate value of the effective inductance is replaced in the circuit and it is sufficient to include the effects of saturation in the remainder of the examined system.

Distortion (Figure 11.20) caused by nonlinearities can be analysed in either time or frequency domains depending upon the characteristics of nonlinearities in the circuits. Few studies can

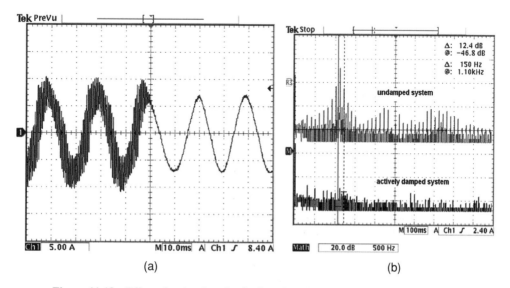

Figure 11.19 Effect of active damping in time-domain (a) and frequency-domain (b).

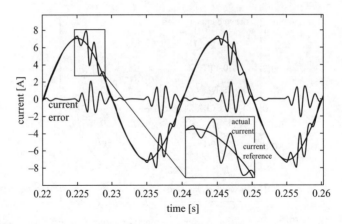

Figure 11.20 Grid current (reference, actual and error) as a consequence of the saturation of the inductor

be found in the literature that relate the output of a nonlinear circuit as a function of the input signals spectrum [10].

The frequency-domain methods are based on generalized power series and are better suited to study circuits containing a very limited number of nonlinear elements due to their high computational effort.

The mathematical model of the nonlinear inductance based on Volterra series expansion allows the frequency response of the nonlinear inductor to be derived and the harmonics generated as a consequence of saturation to be identified. The effect of multiple frequency input signals on the behaviour of the inductor can be determined.

The nonlinear inductance can be modelled as a nonlinear current-controlled source of flux:

$$\varphi(t) = \sum_{k=1}^{\infty} L_k i^k(t) \quad (11.16)$$

where $\varphi(t)$ and $i(t)$ are the flux and the current of the inductor and L_k are constants. The flux in (11.16) is defined through the Taylor series expansion; it can be truncated to the fifth-order term in order to simplify the analysis.

For example, Figure 11.21 reports the flux–current characteristic of the inductance for a rated value of $L = L_1 = 1.5$ mH and the values of the parameters of (11.16) are defined in the caption. They can be obtained using a trial-and-error procedure or an interpolation algorithm starting from the experimental data.

The frequency behaviour of the nonlinear inductance shown in Figure 11.21 can be studied by splitting the model into a linear part and a nonlinear part in accordance with the Volterra theory. The Volterra series expansion of $\varphi(t)$ defined in (11.16) is

$$\varphi(t) \approx \sum_{i=1}^{5} \varphi_i(t) \quad (11.17)$$

Grid Filter Design

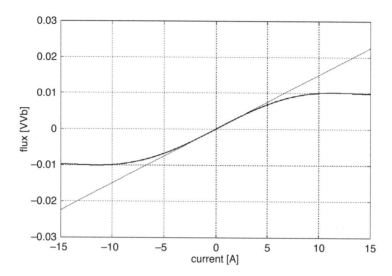

Figure 11.21 Flux–current characteristic of the inductance defined in (11.16): $L_1 = 1.5 \times 10^{-3}$ H, $L_2 = 1 \times 10^{-8}$ H, $L_3 = -8 \times 10^{-4}$ H, $L_4 = 1 \times 10^{-9}$ H, $L_5 = 5 \times 10^{-5}$ H

where the series expansion is truncated up to order five since the first few terms are sufficient to model the saturation of Figure 11.21; $\varphi_1(t)$ is the first-order response of the inductor, which describes its behaviour in the linear case; $\varphi_i(t)$ is the nonlinear response of the inductor obtained using an appropriate excitation:

$$\begin{cases} \varphi_1(t) = L_1 i_1(t) \\ \varphi_2(t) = L_2 i_1^2(t) \\ \varphi_3(t) = 2L_2 i_1(t) i_2(t) + L_3 i_1^3(t) \\ \varphi_4(t) = 2L_2 i_1(t) i_3(t) + L_2 i_2^2(t) + 3L_3 i_1^2(t) i_2(t) + L_4 i_1^4(t) \\ \varphi_5(t) = 2L_2 i_1(t) i_4(t) + 3L_3 i_1^2(t) i_3(t) + 3L_3 i_1(t) i_2^2(t) + 4L_4 i_1^3(t) i_2(t) + L_5 i_1^5(t) \end{cases} \quad (11.18)$$

Equations (11.18) represent a set of current-controlled sources of flux. They allow the equivalent Volterra circuit of the nonlinear inductor to be derived, as shown in Figure 11.22.

The proposed method allows the nonlinear element to be modelled as a linear element plus a set of nonlinear sources, as shown in Figure 11.23. For $i = 2, \ldots, 5$, the current $i_i(t)$ through the nonlinear inductor acts as an external source exciting the linear circuit and can be represented as an external source of current which is connected to the system between the converter and the grid.

The frequency domain response of the circuit is obtained by the use of the Fourier transform and the output spectrum of the circuit can be expressed as a function of all sources. Remembering that

$$i_1^5(t) = I_1^5 \sin^5 \omega_1 t = \frac{I_1^5}{16}(10 \sin \omega_1 t - 5 \sin 3\omega_1 t + \sin 5\omega_1 t) \quad (11.19)$$

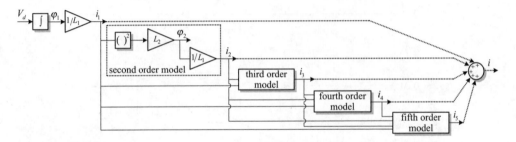

Figure 11.22 Implementation of the nonlinear inductance model defined by (11.17) and (11.18)

it can be easily explained that, when a 50 Hz sinusoidal current is applied to the inductance defined by equations (11.18), the first-, the third- and the fifth-order harmonics of flux are generated.

Similarly, applying a third harmonic current, the nonlinear inductance produces a third, a ninth and a fifteenth harmonic, which can be seen in the flux spectrum of Figure 11.24(a) and (b). Moreover, if the inductor is excited with a fifth harmonic current, the fifth, the fifteenth and the twenty-fifth harmonics are generated. Hence each current source $i_i(t)$ generates some harmonics, which act on the plant of the system as external disturbances.

In addition, when two sinusoids of different frequencies are applied simultaneously, intermodulation components are generated. This substantially increases both the frequency components in the response and the complexity of the analysis.

When considering applying simultaneously a first and third harmonic current of the same amplitude, the flux spectrum exhibits intermodulation components. Other than the first, the third, the fifth, the ninth and the fifteenth harmonics, the third-order intermodulation component at $\omega_1 + 2\omega_2$ and the fifth-order intermodulation components at $4\omega_2 - \omega_1$ and $4\omega_2 + \omega_1$ are not negligible, as shown in Figure 11.24(c).

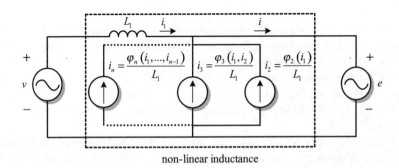

Figure 11.23 Volterra-based model of the nonlinear inductance connected between the inverter v and the grid e

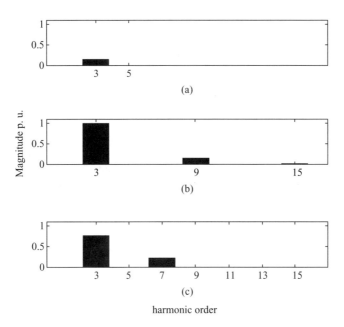

Figure 11.24 Flux spectrum of the nonlinear inductance defined by equations (11.18) in the case of: (a) input current at $\omega_1 = 314$ rad/s, (b) input current at $\omega_2 = 942$ rad/s and (c) input current at $(\omega_1 + \omega_2)$

11.7 Summary

The chapter has discussed the need for a higher-order filter to connect the grid converter to the utility. The design of the filter, the problems related to the variation of grid parameters, the stability problems and filter saturation have been discussed.

References

[1] Rockhill, A., Liserre, M., Teodorescu, R. and Rodriguez, P., 'Grid Filter Design for a Multi-Megawatt Medium-Voltage Voltage Source Inverter'. IEEE Transactions on Industrial Electronics 2011.

[2] Liserre, M., Blaabjerg, F. and Dell'Aquila, A., 'Step-by-Step Design Procedure for a Grid-Connected Three-Phase PWM Voltage Source Converter'. *International Journal of Electronics*, **91**(8), August 2004, 445–460.

[3] M. Liserre, M., Blaabjerg, F. and Hansen, S., 'Design and Control of an LCL-Filter Based Three-Phase Active Rectifier'. *IEEE Transactions on Industrial Applications*, **41**(5), September/October 2005, 1281–1291.

[4] Skibinski, G. L., Kerkman, R. J. and Schlegel, D., 'EMI Emissions of Modern PWM AC driver'. *IEEE Industry Applications Magazine*, **5**, November–December 1999, 47–80.

[5] Akagi, H., Hasegawa, H. and Doumoto, T., 'Design and Performance of a Passive EMI Filter for Use with a Voltage-Source PWM Inverter Having Sinusoidal Output Voltage and Zero Common-Mode Voltage'. *IEEE Transactions on Power Electronics*, **19**(4), July 2004, 1069–1076.

[6] Arrillaga, J. and Watson, N. R., Power System Harmonics, Hardcover, 3 October 2003.

[7] van den Bossche, A., and Valchev, V. C., *Inductors and Transformers for Power Electronics*. Taylor and Francis: CRC Press, 2009.

[8] Franke, W.-T., Dannehl, J., Fuchs, F. W. and Liserre, M., 'Characterization of Differential-Mode Filter for Grid-Side Converters'. IECON 2009, Oporto, Portugal, November 2009.

[9] Enslin J. H. R. and Heskes, P. J. M., 'Harmonic Interaction between a Large Number of Distributed Power Inverters and the Distributed Network'. *IEEE Transactions on Power Electronics*, **19**(6), November 2004, 1586–1593.

[10] Mastromauro, R. A., Liserre, M. and Dell'Aquila, A., 'Study of the Effects of Inductor Nonlinear Behavior on the Performance of Current Controllers for Single-Phase PV Grid Converters'. *IEEE Transactions on Industrial Electronics*, **55**(5), May 2008, 2043–2052.

[11] Twining, E. and Holmes, D., 'Grid current regulation of a three-phase voltage source inverter with an LCL input filter,' IEEE Transactions on Power Electronics, **18**(3), May 2003, 888–895.

[12] Draou, A., Sato, Y. and Kataoka, T., 'A new state feedback based transient control of PWM ac to dc voltage type converters,' IEEE Transactions on Power Electronics, **10**(6), Nov 1995, 716–724.

[13] Dannehl, J., Fuchs, F. and Thøgersen, P., 'PI state space current control of grid- connected PWM converters with lcl filters,' IEEE Transactions on Power Electronics, 2010.

[14] Dahono, P.A., 'A control method to damp oscillation in the input LC filter,' IEEE 33rd Annual Power Electronics Specialists Conference, 2002, **4**, 2002, 1630–1635.

[15] Liserre, M., Dell'Aquila, A. and Blaabjerg, F., 'Genetic algorithm-based design of the active damping for an lcl-filter three-phase active rectifier,' IEEE Transactions on Power Electronics, **19**(1), January 2004, 76–86.

[16] Dick, C., Richter, S., Rosekeit, M., Rolink, J. and De Doncker, R., 'Active damping of lcl resonance with minimum sensor effort by means of a digital infinite impulse response filter,' in Power Electronics and Applications, 2007 European Conference on, Sept. 2007, 1–8.

[17] Dannehl, J., Liserre, M. and Fuchs, F., "Filter-based Active Damping of Voltage Source Converters with LCL-filter", IEEE Transactions on Industrial Electronics, 2011.

12

Grid Current Control

12.1 Introduction

When wind and photovoltaic systems are grid-connected using a voltage source converter as the front-end, the grid current is usually controlled in order to control active and reactive power exchange, as discussed in previous chapters [1]. The position of voltage and current sensors can be different with respect to Figure 12.1, as already discussed in Chapter 11.

Moreover, in this chapter the influence of the C_f capacitor of the filter will be neglected since it only deals with the switching ripple frequencies. In fact, at frequencies lower than half of the resonance frequency the LCL-filter inverter model and the L-filter inverter models are practically the same. The LCL-filter-based inverter (Figure 12.2 shows the frequency characteristic i/v of the LCL filter) behaves as if the capacitor is not present, $i = i_g$ (hence in the following i will be used in all the equations and schemes of the reviewed current control techniques) and the frequency characteristic is equivalent to the frequency characteristic of a filter made by the sum of the inductances downstream of the converter (L_1, L_2 and L_g).

This chapter reviews basic current control techniques and modulation strategies. Only linear controllers employing a separated pulse width modulator are discussed. The current control will be discussed without making a distinction between single-phase and three-phase systems, since similar control schemes can be adopted even if the implementation of a specific current control strategy to single-phase and three-phase voltage source converters could present some differences, which are highlighted at the end of each section of the chapter.

12.2 Current Harmonic Requirements

One of the demands present in all standards regarding grid-tied systems is the quality of the distributed power. Demands are also made on grid-connected PV and WT systems to contribute to the preservation of the power quality [2]. Power quality requirements have been mainly developed in order to preserve the quality of the grid voltage waveform in amplitude, frequency and phase. The main perturbation to the voltage waveform are due to system transient operation (e.g. at startup) or to power fluctuations (due to the stochastic nature of the source). However, the quality of the current is also of concern and in this sense the grid converter is the sole agent

---

Grid Converters for Photovoltaic and Wind Power Systems   Remus Teodorescu, Marco Liserre, and Pedro Rodríguez
© 2011 John Wiley & Sons, Ltd

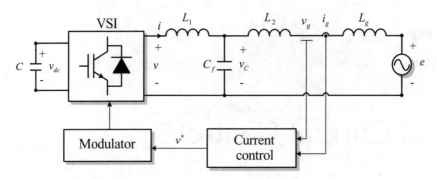

Figure 12.1 Block diagram of a typical three-phase distributed inverter (in PV system L_2 is usually not present)

responsible for compliance with power quality international recommendations and standards as well as the requirements imposed by the transmission system operators [3–5]. Particularly in WT systems asynchronous and synchronous generators directly connected to the grid have no limitations with respect to current harmonics [5].

The injected current in the grid should not have a total harmonic distortion (THD) larger than 5 %. A detailed image of the harmonic distortion regarding each harmonic is given in Table 12.1 as recommended by reference [3], which is valid for all distributed resource technologies with an aggregate capacity of 10 MVA or less at the point of common coupling interconnected with electrical power systems at typical primary and/or secondary distribution voltages. Similar limitations are recommended for PV systems in a European standard [4].

As regards WT systems, European standards [5] recommend application of the standards valid for polluting loads requiring the current THD to be smaller than 6–8 % depending on the type of network [6,7]. A detailed image of the harmonic distortion regarding each harmonic

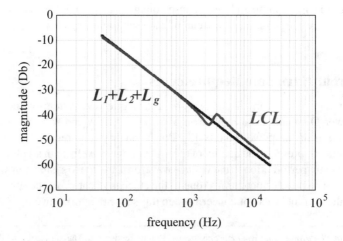

Figure 12.2 Frequency characteristic (i/v) of the LCL filter

Table 12.1 Distortion limits for distributed generation systems as a percentage of the fundamental [3, 4]

Odd harmonics	Distortion limit
3rd through 9th	Less than 4.0 %
11th through 15th	Less than 2.0 %
17th through 21st	Less than 1.5 %
23rd through 33rd	Less than 0.6 %

is given in Table 12.2. In case there are several wind turbines connected to the same PCC, the hth harmonic can be computed as

$$I_{h\Sigma} = \sqrt[\beta]{\sum_{i=1}^{N} \left(\frac{I_{hi}}{v_i}\right)^\beta} \tag{12.1}$$

where β is 1 for $h < 5$, 1.4 for $5 < \beta < 10$ and 2 for $h > 10$.

As mentioned previously, one of the responsibilities of the current controller is the power quality issue. Therefore, different methods to compensate for the grid harmonics in order to obtain an improved power quality are addressed in the following section.

12.3 Linear Current Control with Separated Modulation

The AC current control (CC) has become very popular because the current-controlled converter exhibits, in general, better safety, better stability and faster response [8]. This solution ensures several additional advantages but optimal techniques, which use precalculated switching patterns within the AC period, cannot be used, as they are not oriented to ensure current waveform control [8]. A classification of these methods is reported in Figure 12.3.

12.3.1 Use of Averaging

The AC current controllers that will be presented in the following are designed on the basis of an average model [9] of the converter based on a continuous switching vector whose

Table 12.2 Distortion limits for WTsystems set by IEC standard as a percentage of the fundamental [5]

Harmonic	Limit
5th	5–6 %
7th	3–4 %
11th	1.5–3 %
13th	1–2.5 %

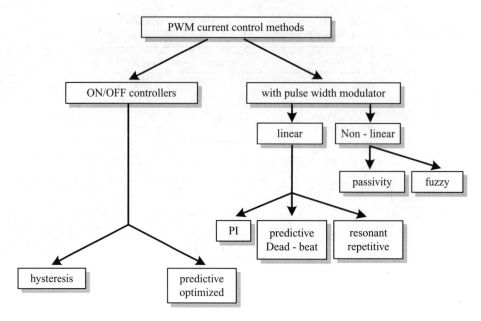

Figure 12.3 Classification of current control methods

components are the duty cycle of each converter leg $0 \leq d_j(t) \leq 1$ $(j = a, b, c)$:

$$\bar{d}(t) = \frac{2}{3}\left(d_a(t) + \alpha d_b(t) + \alpha^2 d_c(t)\right) \quad (12.2)$$

where $\alpha = e^{j2\pi/3}$.

The separated PWM block is responsible for the transformation of the continuous switching functions back into the discrete switching functions usable for driving the switches of the converter.

The AC voltage equations expressed in the dq frame, which is synchronous with the grid voltage vector, are

$$\begin{cases} \dfrac{di_d(t)}{dt} - \omega i_q(t) = \dfrac{1}{L}\left[-Ri_d(t) - e_d(t) + v_d(t)\right] \\ \dfrac{di_q(t)}{dt} + \omega i_d(t) = \dfrac{1}{L}\left[-Ri_q(t) - e_q(t) + v_q(t)\right] \end{cases} \quad (12.3)$$

The transformations from the natural frame abc to the synchronous dq frame are reported in Chapter 9. L is the overall inductance downstream with respect to the inverter. The current controllers calculate the desired d and q components of the switching vector ($d_d(t)$ and $d_q(t)$) and a modulator selects the converter switching states and their time of applications. The

Grid Current Control

average AC voltage produced by the converter (the voltage command for the PWM) can be expressed as

$$\begin{cases} v_d(t) = d_d(t)v_{dc}(t) \\ v_q(t) = d_q(t)v_{dc}(t) \end{cases} \quad (12.4)$$

Equation (12.3) can be linearized by considering $v_{dc}(t) = V_{dc}$ with V_{dc} constant:

$$\frac{d}{dt}\begin{bmatrix} i_d(t) \\ i_q(t) \end{bmatrix} = \begin{bmatrix} -\frac{R}{L} & \omega \\ -\omega & -\frac{R}{L} \end{bmatrix}\begin{bmatrix} i_d(t) \\ i_q(t) \end{bmatrix} + \begin{bmatrix} \frac{V_{dc}}{L} & 0 \\ 0 & \frac{V_{dc}}{L} \end{bmatrix}\begin{bmatrix} d_d(t) \\ d_q(t) \end{bmatrix} + \begin{bmatrix} -\frac{1}{L} & 0 \\ 0 & -\frac{1}{L} \end{bmatrix}\begin{bmatrix} e_d(t) \\ e_q(t) \end{bmatrix} \quad (12.5)$$

Then (12.5) is written in the form $\dot{i} = \mathbf{A} \cdot \mathbf{i} + \mathbf{B} \cdot \mathbf{d} + \mathbf{C} \cdot \mathbf{e}$, where the matrix $\mathbf{B}$ does not depend on the system state and so (12.5) is linear.

The average model is useful for designing the controller for slow processes with a bandwidth in the range of a few hundred Hz, like reactive power control and a stability analysis of WT.

12.3.2 PI-Based Control

Classical PI control with grid voltage feed-forward (v_g) [8], as depicted in Figure 12.4, is commonly used for current-controlled inverters, but this solution exhibits two well-known drawbacks: the inability of the PI controller to track a sinusoidal reference without steady-state error and a poor disturbance rejection capability (Figure 12.5). This is due to the poor performance of the integral action when the disturbance is a periodic signal.

The PI current controller $G_{PI}(s)$ is defined as

$$G_{PI}(s) = k_P + \frac{k_I}{s} \quad (12.6)$$

$G_d(s)$ is the $1.5T_s$ delay due to elaboration of the computation device (T_s) and to the PWM ($0.5T_s$), indicating with T_s the sampling period

$$G_d(s) = \frac{1}{1 + 1.5T_s s} \quad (12.7)$$

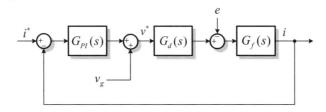

Figure 12.4 The current loop of a PI controller: $G_{PI}(s)$ is the controller, $G_d(s)$ is the delay due to elaboration of the computation device and to the PWM and $G_f(s)$ is the transfer function of the filter (i.e. the plant of the control loop)

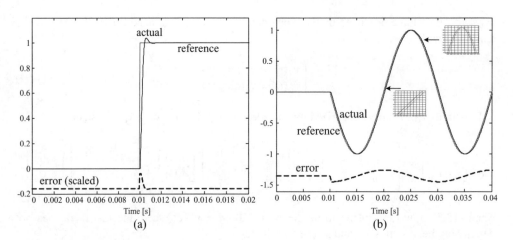

Figure 12.5 Typical behaviour of a PI controller (without feed-forward): (a) for a step reference and (b) for a sinusoidal reference

and $G_f(s)$ is the transfer function of the filter (i.e. the plant of the control loop)

$$G_f(s) = \frac{i(s)}{v(s)} = \frac{1}{R + Ls} \tag{12.8}$$

Figure 12.6 shows the Z plane root locus and the Bode plot of the closed-loop PI-based current-controlled system highlighting the bandwidth of the control loop.

In order to obtain a good dynamic response and improve grid disturbance rejection, a grid voltage feed-forward is used, as depicted in Figure 12.4. This leads in turn to stability problems related to the delay introduced by the filter usually adopted to measure the grid voltage.

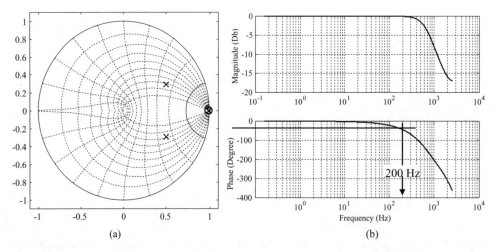

Figure 12.6 Z-plane root locus with the two poles optimally damped, i.e. $\zeta = 0.707$, and the Bode plot of the closed-loop system ($1/T_s = 5000$ Hz)

Grid Current Control

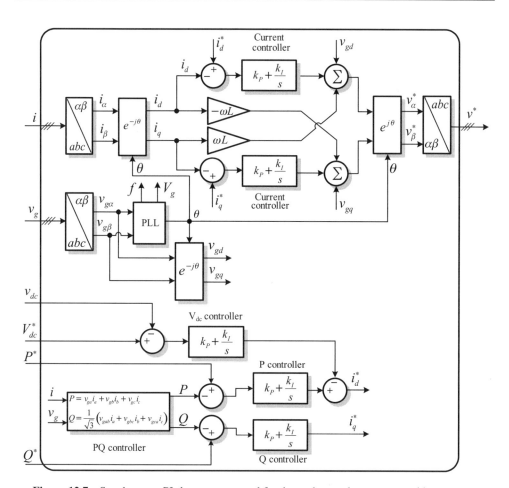

Figure 12.7 Synchronous PI dq current control for three-phase voltage-source grid converters

In order to overcome the limit of the PI in dealing with sinusoidal reference and harmonic disturbances, the PI control is implemented in a dq frame rotating with angular speed ω, where $\omega = 2\pi f$ and f is the grid frequency. The scheme of the classical control in a rotating frame is reported in Figure 12.7 and the rotating frame is defined as synchronous, as already pointed out.

If the dq frame is oriented such that the d axis is aligned on the grid voltage vector the control is called voltage oriented control (VOC). The reference current d component is made of two terms: one, i^*_{dd}, can be used to perform the DC voltage regulation while the other, i^*_d, is in charge of active power control. The reference current q component i^*_q is selected in view of reactive power control (a unity power factor is achieved if i^*_q is zero). The dq controller core is

$$G_{PI}(s)_{dq} = \begin{bmatrix} k_P + \dfrac{k_I}{s} & 0 \\ 0 & k_P + \dfrac{k_I}{s} \end{bmatrix} \tag{12.9}$$

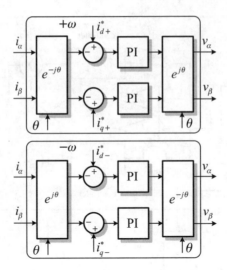

Figure 12.8 PI controllers considering both positive- and negative-sequence components (without voltage feed-forward and cross-coupling compensation)

It should be noted that under unbalanced conditions, in order to compensate the harmonics generated by the inverse sequence present in the grid voltage, both positive- and negative-sequence reference frames are required (see Figure 12.8). Obviously, when using this approach, double the computational effort must be applied.

In a single-phase system, the use of a rotating frame is not possible unless a virtual system is coupled to the real frame in order to simulate a two-axis environment [10]. The advantage is that well-proven PI controllers can be used, as DC variables need to be controlled and an independent Q control is achieved. The solution is reported in Figure 12.9. A phase delay block creates the virtual quadrature component that allows emulation of a two-phase system. Alternative systems to create such a component are discussed in Chapter 4. Starting from this point the controller behaves like a three-phase case and the v_β component of the command voltage is ignored for calculation of the duty cycle.

12.3.3 Deadbeat Control

The deadbeat controller belongs to the family of predictive controllers [8]. They are based on a common principle: to foresee the evolution of the controlled quantity (the current) and on the basis of this prediction to choose the state of the converter (ON–OFF predictive) [11] or the average voltage produced by the converter (predictive with a pulse width modulator) [12]. The starting point is to calculate its derivative to predict the effect of the control action. In other words, the controller is developed on the basis of the model of the filter and of the grid, which is used to predict the system dynamic behaviour: the controller is therefore inherently sensitive to model and parameter mismatches. Then the information on the model is used to decide the switching state of the converter with the aim to minimize the possible commutations

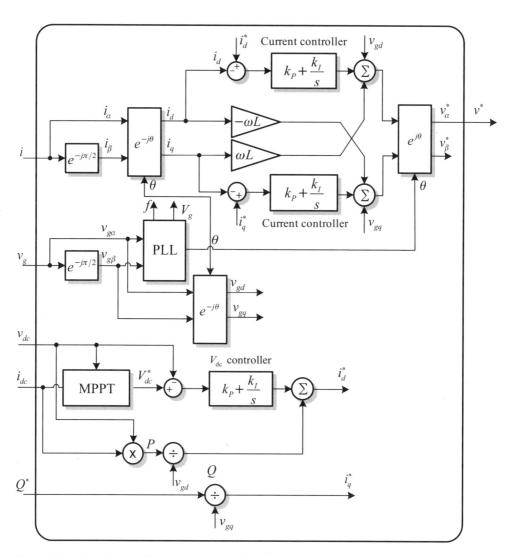

Figure 12.9 Synchronous PI *dq* current control for single-phase voltage-source grid converters (MPPT is the algorithm adopted to extract the maximum power from the renewable source)

(ON–OFF predictive) or the average voltage that the converter has to produce in order to null it. The controller is defined as 'deadbeat' where the error at the end of the next sampling period is zero. It can be demonstrated that it is the fastest current controller allowing nulling of the error after two sampling periods (see Figure 12.10).

During transient conditions typical of distributed power generation systems such as sudden variation of the power produced by the source or voltage sag due to grid faults, the deadbeat controller can be proven to have superior performances in limiting the peak current [13].

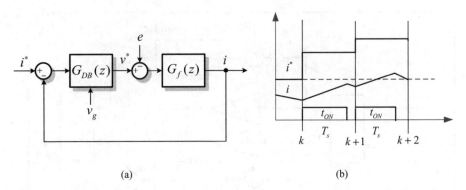

Figure 12.10 (a) The deadbeat control scheme and (b) its control action nulling the error after two sampling periods

The design of the deadbeat control will be developed on the basis of a single-phase equivalent of the system. The single-phase circuit mathematical model is

$$\frac{di(t)}{dt} = -\frac{R}{L}i(t) + \frac{1}{L}(v(t) - e(t)) \qquad (12.10)$$

where $i(t)$ is the current, $v(t)$ is the inverter voltage, $e(t)$ is the grid voltage and R and L are the total resistance and inductance downstream of the converter.

The discretized solution (considering the sample delay due to the elaboration) results in

$$i(k+1) = e^{-(R/L)T_s} i(k) - \frac{1}{R}\left(e^{-(R/L)T_s} - 1\right)(v(k) - e(k)) \qquad (12.11)$$

where T_s is the sampling period. If we denote $a = e^{-(R/L)T_s}$, $b = -\frac{1}{R}\left(e^{-(R/L)T_s} - 1\right)$ and

$$u(k) = v(k) - e(k) \qquad (12.12)$$

the following equation is derived:

$$i(k+1) = ai(k) + bu(k) \qquad (12.13)$$

From (12.13) and applying Z-transformation the following equation is derived:

$$I(z)z = aI(z) + bU(z) \qquad (12.14)$$

Therefore:

$$I(z) = \frac{b}{z-a}U(z) = \frac{bz^{-1}}{1-az^{-1}}U(z) \qquad (12.15)$$

Grid Current Control

so the plant transfer function is

$$G_f(z) = \frac{I(z)}{U(z)} = \frac{bz^{-1}}{1 - az^{-1}} \tag{12.16}$$

In the following a simplified deadbeat control design is assumed. It is supposed that there are no unstable poles and zeros in the plant transfer function.

The desired transfer function of the closed-loop system can be written as

$$G_o(z) = \frac{1}{z^k} = z^{-k} \tag{12.17}$$

where k is the delay of the closed-loop control system. The deadbeat controller aim is that the actual current equals the reference at the end of two sampling periods, and hence $k = 2$. The mathematical model of the controller results in

$$G_{DB}(z) = \frac{1}{G_f(z)} \frac{z^{-2}}{1 - z^{-2}} \tag{12.18}$$

Therefore, applying the controller design described by (12.18) at the plant transfer function of (12.16), the controller transfer function results as

$$G_{DB}(z) = \frac{1}{b} \frac{1 - az^{-1}}{z - z^{-1}} \tag{12.19}$$

and in the end results as

$$U(z)z = U(z)z^{-1} + \frac{1}{b}\Delta I(z) - \frac{a}{b}\Delta I(z)z^{-1} \tag{12.20}$$

Then the algorithm of the designed controller is

$$v(k+1) = v(k-1) + \frac{1}{b}\Delta i(k) - \frac{a}{b}\Delta i(k-1) + e(k+1) - e(k-1) \tag{12.21}$$

In case the resistance R is neglected, $a = 1$ and $b = T_s/L$ and

$$v(k+1) = -v(k) + \frac{1}{b}\Delta i(k) + e(k+1) + e(k) \tag{12.21a}$$

However, this is not an acceptable approximation, especially in the case of a distributed generation system connected to a single-phase transmission system where the resistance is not negligible, as discussed in Chapter 11. Hence to neglect the grid-side resistance an error is generated in the control action, as shown in Figure 12.11.

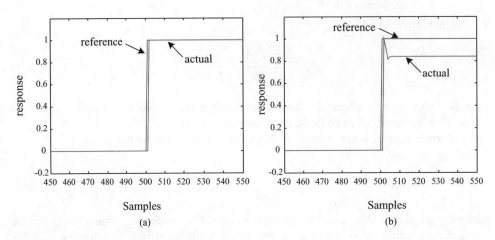

Figure 12.11 Comparison between (a) (12.21) and (b) (12.21a) in the case $(R/L)T_s = 0.1$; if the term is equal to 0.01 the two responses are equal

The deadbeat current controller has theoretically a very high bandwidth since the closed-loop transfer function given by (12.17) consists of two poles in the origin. Hence the tracking of a sinusoidal signal as shown by Figure 12.12(a) is very good. However, if the PWM and the saturation of the control action is considered (Figure 12.12(b)) the deadbeat controller exhibits a slower response. One of the main problems of the deadbeat controller is related to the parameter mismatches that generate a tracking error and also stability problems, as shown in Figure 12.13 for the controller of (12.21) for the sake of simplicity. The use of an observer may relieve these problems, making the controller more robust with respect to parameter mismatches and controller delays [14].

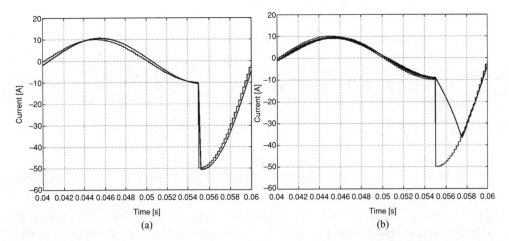

Figure 12.12 (a) Tracking of a sinusoidal signal without PWM and (b) with PWM and considering the saturation of the control action

Grid Current Control

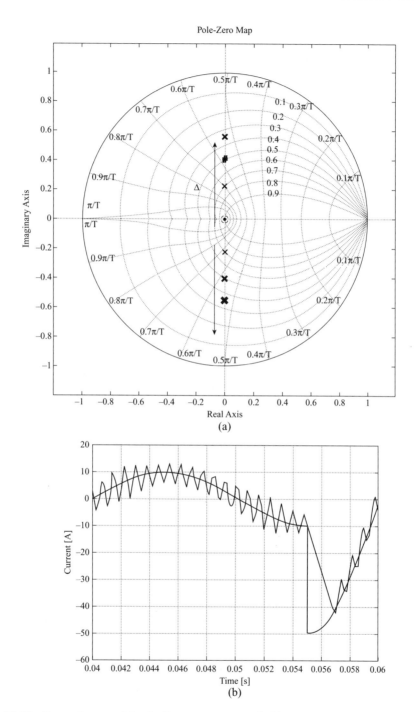

Figure 12.13 Zero-pole map of the deadbeat current-controlled loop as (a) the Δ (error between the inductance and its estimated value) increases and (b) instability due to Δ (deadbeat controller without PWM)

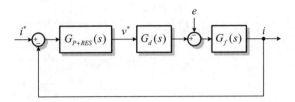

Figure 12.14 The current loop of an inverter with a P+resonant (PR) controller

12.3.4 Resonant Control

The tracking of periodical signals and the rejection of periodical disturbances are problems familiar to experts in mechanical system (such as pendulum or robots) control [15]. The inverter control has been traditionally developed as a direct extension of the control of electrical motors and of DC converters. In both cases the set-point is a constant signal as well as the main disturbance. In particular, the use of a rotating frame has allowed the control of the AC current to be treated as the control of DC virtual currents (the projections of the AC current vector in a frame that is synchronous with that vector).

However, it is possible to avoid the use of frame transformations using a controller developed on the basis of the internal model principle (the theory is reported Appendix C). This principle states that it is sufficient to include the model of the disturbance within the controller in order to ensure perfect rejection.

The more straightforward implementation of the internal model principle is the second-order generalized integrator (GI) [16]. The theory beyond this is discussed in Appendix C. The GI is a double integrator that achieves an infinite gain at a certain frequency, also called resonance frequency, and almost no gain exists outside this frequency. Thus, it can be used as a notch filter in order to compensate the harmonics in a very selective way. The current loop of the inverter with a proportional plus resonant controller is depicted in Figure 12.14. It is possible to note that the grid voltage feed-forward action is no longer needed. This solution has been successfully applied to single-phase PV inverters [17]. However, its three-phase implementation allows an interesting consideration with respect to the use of the PI controller in a synchronous frame.

The proportional plus resonant controller can be obtained through a frame transformation. The PI controller (12.9) implemented in a synchronous frame and denoted here as $G_{DC}(s)$ can be transformed into a stationary frame through a frequency modulated process that can be mathematically expressed as

$$G_{AC}(s) = G_{DC}(s - j\omega) + G_{DC}(s + j\omega) \qquad (12.22)$$

where $G_{AC}(s)$ represents the equivalent stationary frame transfer function. Therefore, for the ideal and nonideal integrators of $G_{DC}(s) = k_I/s$ and $G_{DC}(s) = k_I/(1 + (s/\omega_c))$ (k_I and $\omega_c \ll \omega$ represent the controller gain and cutoff frequency respectively), the derived generalized AC integrators $G_{AC}(s)$ are respectively expressed as follows:

$$G_{AC}(s) = \frac{Y(s)}{E(s)} = \frac{2k_I s}{s^2 + \omega^2} \qquad (12.23)$$

$$G_{AC}(s) = \frac{Y(s)}{E(s)} = \frac{2k_I \left(\omega_c s + \omega_c^2\right)}{s^2 + 2\omega_c s + (\omega_c^2 + \omega^2)} \approx \frac{2k_I \omega_c s}{s^2 + 2\omega_c s + \omega^2} \qquad (12.24)$$

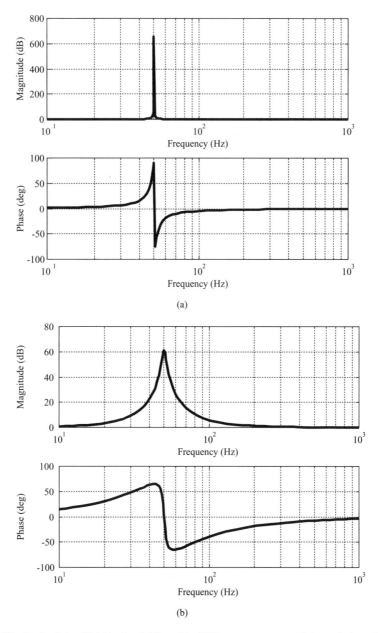

Figure 12.15 Bode plots of (a) ideal and (b) nonideal PR compensators with $k_P = 1, k_I = 20, \omega = 314$ rad/s and $\omega_c = 10$ rad/s

Equation (12.23), when grouped with a proportional term k_P, gives the ideal PR controller with an infinite gain at the AC frequency of ω (see Figure 12.15(a)) and no phase shift and gain at other frequencies. For k_P, it is tuned in the same way as for a PI controller, and it basically determines the dynamics of the system in terms of bandwidth, phase and gain margins. To avoid stability problems associated with an infinite gain, (12.24) can be used instead of (12.23)

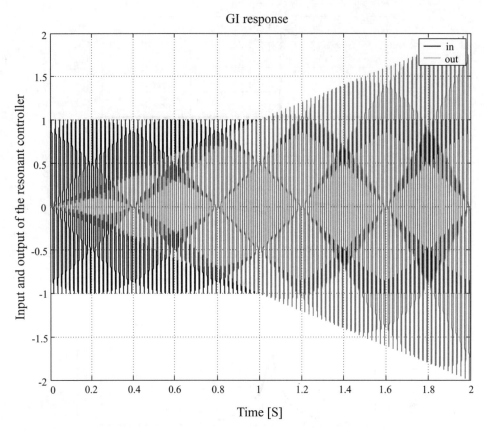

Figure 12.16 Response of the resonant controller (12.23) when excited by a sinusoidal input whose frequency matches the controller resonant frequency

to give a nonideal PR controller and, as illustrated in Figure 12.15(b), its gain is now finite, but still relatively high for enforcing a small steady-state error. Another feature of (12.24) is that, unlike (12.23), its bandwidth can be widened by setting ω_c appropriately, which can be helpful for reducing sensitivity towards (for example) slight frequency variations in a typical utility grid. (For (12.23), k_I can be tuned for shifting the magnitude response vertically, but this does not give rise to a significant variation in bandwidth.) It has been demonstrated that the same results using PI controllers in a synchronous frame can be obtained using generalized integrators (which offer infinite gain at a certain frequency) in a stationary reference frame.

Figure 12.16 shows the time-domain behaviour of a resonant controller (12.23) when excited by a sinusoidal input whose frequency matches the controller resonant frequency.

A PI controller in a synchronous frame plus a PI controller in a counter-synchronous frame (a frame that rotates in the opposite direction with respect to the VOC dq frame previously defined in Figure 12.8) is equivalent to a P+resonant controller in a stationary frame as demonstrated in reference [16] and reported in Appendix C.

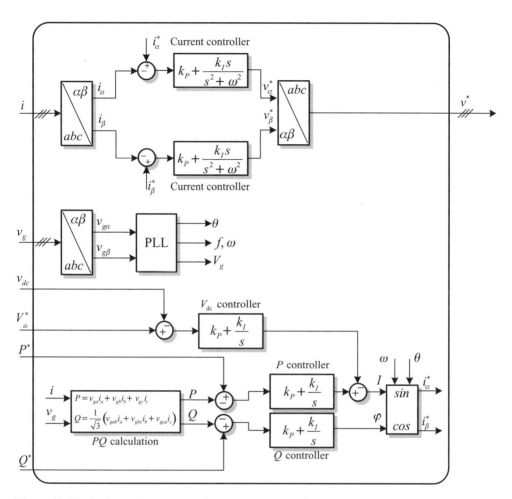

Figure 12.17 Stationary P+resonant $\alpha\beta$ current control for three-phase voltage-source grid converters

The scheme of the resonant control in a stationary frame in the case of a three-phase system is reported in Figure 12.17. It can be observed, comparing Figure 12.17 and Figure 12.7, that the complexity of the calculations has been significantly reduced as there is no more need for the grid voltage feed-forward and cross-coupling terms. Moreover, in order to have a perfect equivalence the scheme of Figure 12.7 should be used twice (one for the positive sequence and the other for the negative sequence), as reported in Figure 12.8.

12.3.5 Harmonic Compensation

The decomposition of signals into harmonics with the aim of monitoring and controlling them is a matter of interest for various electric and electronic systems. There have been many efforts to approach scientifically typical problems (e.g. faults, unbalance, low-frequency EMI) in

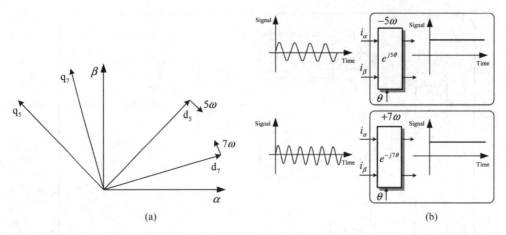

Figure 12.18 Multiple synchronous reference frames: (a) tracking the 5th (negative sequence) and 7th (positive sequence) harmonics typical of (b) three-phase power systems

power systems (power generation, conversion and transmission) through harmonic analysis. The use of multiple synchronous reference frames (MSRFs), early proposed for the study of induction machines, allows compensation for selected harmonic components in the case of two-phase motors, unbalance machines or in grid-connected systems (see Figure 12.18).

The harmonic components of power signals can be represented in stationary or synchronous frames using phasors. In the case of synchronous reference frames each harmonic component is transformed into a DC component (frequency shifting) [18]. If other harmonics are contained in the input signal, the DC output will be disturbed by a ripple that can easily be filtered out. The filtered signal can be transformed back and the result is that the harmonic whose pulsation is the frame angular speed passes through the process while the others are stopped. The result is a very selective action obtained with a first-order low-pass filter implemented in a synchronous frame; it can be demonstrated that the same result can be obtained with a higher-order digital filter in a stationary frame. The use of a synchronous frame has been very successful in controlling power electronic and electric drive systems and can be correctly interpreted as a frequency shift. Hence the detection of many harmonics can be performed using MSRFs.

Then a PLL system can be used to detect the frequency and phase of the harmonic in order to select the proper speed of the synchronous frame. Hence the reliability of the harmonic detection still relies on the PLL behaviour.

If the current controller should be immune to the grid voltage harmonic distortion (mainly the 3rd and 5th in single-phase systems and the 5th and 7th in three-phase systems), then the major possibilities are based on low-pass and high-pass filters. In the first case it is possible to compensate selectively each harmonic using a frame rotating at a speed multiple of the fundamental one or a nested multiple frame rotating at reduced speed, or to use resonant filters in the same stationary $\alpha\beta$ frame [19].

In the following the approach based on low-pass filtering, both with synchronous dq frames and stationary $\alpha\beta$ frames, is discussed, since the approach based on high-pass filtering is used more for active filters.

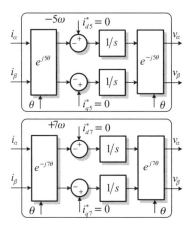

Figure 12.19 Method for compensating the positive sequence of the 5th and 7th harmonics in the *dq* control structure

12.3.5.1 Harmonic Compensation by Means of Synchronous *dq* Frames

As already pointed out, there are two possible approaches employing rotating frames and low-pass filtering. The first (shown in Figure 12.19) is to consider one frame for each harmonic; hence if the harmonic controller is designed for an active filter application that should compensate the 5th and 7th harmonics generated by a phase-controlled thyristor bridge or diode rectifier. Then two controllers should be implemented in two frames rotating at -5ω and 7ω (because the 5th harmonic generated by that load is an inverse sequence and the 7th is a direct one) (see Figure 12.20). The second possibility is to consider nested frames, i.e. to implement in the main synchronous frame two controllers in two frames rotating at 6ω and -6ω (see Figure 12.21). It should be noted that it is not possible to use only a 6ω rotating frame to compensate both the 5th inverse and 7th direct sequences, on the basis of the idea that both generate a 6th harmonic in the synchronous frame, because $-5\omega - \omega = -6\omega$ and $7\omega - \omega = 6\omega$. Hence they generate six-order harmonics of different sequences. As a conclusion both solutions are equivalent also in terms of the implementation burden because in both cases two controllers are needed.

12.3.5.2 Harmonic Compensation by Means of Stationary $\alpha\beta$ Frames

Besides single frequency compensation (obtained with the generalized integrator tuned at the grid frequency), selective harmonic compensation can also be achieved by cascading several resonant blocks tuned to resonate at the desired low-order harmonic frequencies to be compensated. As an example, the transfer functions of an ideal and a nonideal harmonic compensator (HC) designed to compensate for the 3rd, 5th and 7th harmonics (as they are the most prominent harmonics in a typical current spectrum) are given as

$$G_h(s) = \sum_{h=3,5,7} \frac{2k_{Ih}s}{s^2 + (h\omega)^2} \qquad (12.25)$$

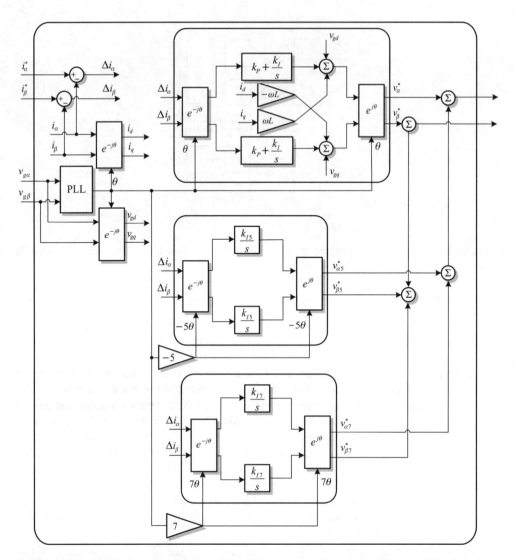

Figure 12.20 Current control structure of the three-phase grid inverter with 5th and 7th harmonic compensation using 5th and 7th *dq* reference frames

$$G_h(s) = \sum_{h=3,5,7} \frac{2k_{Ih}\omega_c s}{s^2 + 2\omega_c s + (h\omega)^2} \quad (12.26)$$

where h is the harmonic order to be compensated and k_{Ih} represents the individual resonant gain, which must be tuned relatively high (but within the stability limit) in order to minimize the steady-state error. An interesting feature of the HC is that it does not affect the dynamics of the fundamental PR controller, as it compensates only frequencies that are very close to the selected resonant frequencies.

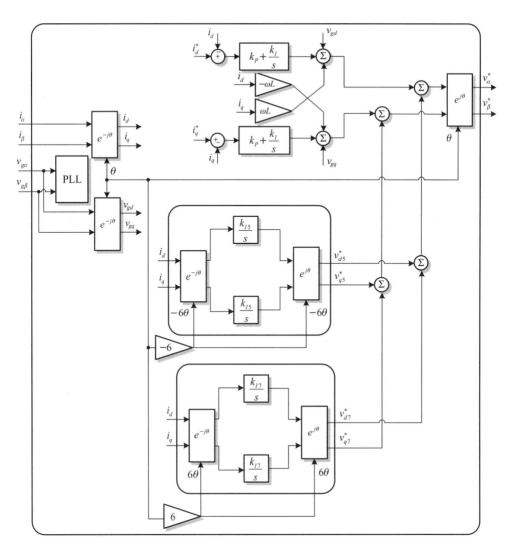

Figure 12.21 Current control structure of the three-phase grid inverter with 5th and 7th harmonic compensation using 6th and −6th dq reference frames nested in the main

Figure 12.22 shows the resonant compensators as well as the Bode plot of the closed loop. It has been shown in Figure 12.13 and commented in the text that as ω_c gets smaller, $G_h(s)$ becomes more selective (narrower resonant peaks). However, using a smaller ω_c will make the filter more sensitive to frequency variations, lead to a slower transient response and make the filter implementation on a low-cost 16-bit DSP more difficult due to coefficient quantization and round-off errors. In practice, ω_c values between 5 and 15 rad/s have been found to provide a good compromise.

Figure 12.23 shows the harmonic compensation that is possible to achieve with the resonant harmonic compensators; similar results can be achieved with compensation using the

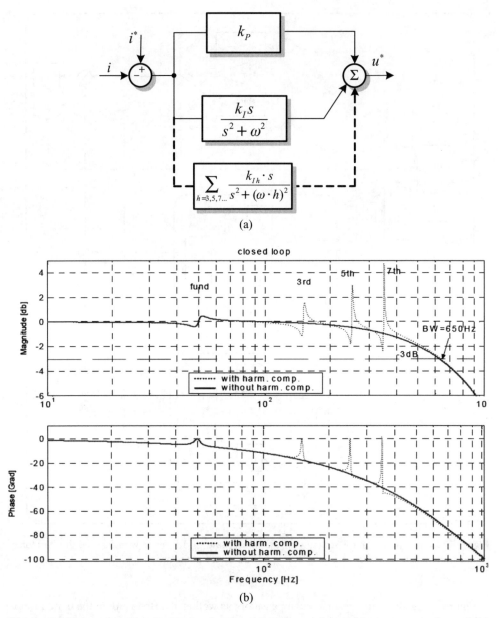

Figure 12.22 Resonant filter for filtering 3rd, 5th and 7th harmonics: (a) block representation and (b) Bode plot of the closed loop

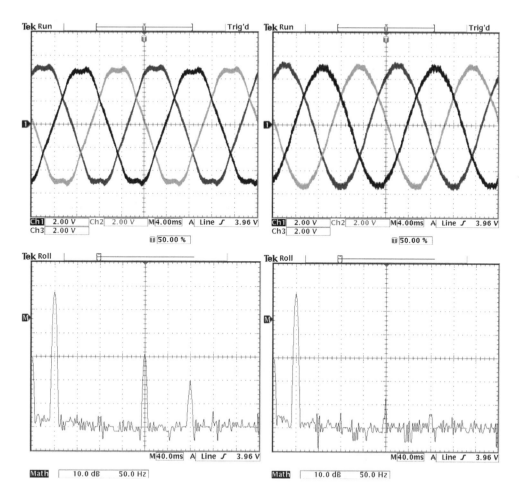

Figure 12.23 Harmonic compensation of the 5th and 7th harmonics in the grid current, showing time domain (top) and frequency domain (bottom)

synchronous dq frames. A third option that can give similar results is to use resonant compensators in dq frames [20].

12.4 Modulation Techniques

The most widely used open-loop pulse-width modulation methods are carrier-based where the pulse widths are determined by comparing a modulating waveform and a triangular carrier. Characteristic parameters of these strategies are the ratio between amplitudes of modulating and carrier waves (called the modulation index M) and the ratio between frequencies of the same signals (called the carrier index m). These techniques differ for the modulating wave chosen with the goal to obtain a lower harmonic distortion, to shape the harmonic spectrum and to guarantee a linear relation between the fundamental output voltage and modulation index in a wider range. The space vector modulations are developed on the basis of the space vector representation of the converter AC side voltage. While the carrier-based representation can be adopted for both single-phase and three-phase converters, the space vector based one can be adopted only for three-phase converters. Since a complete review of modulations, already proposed in reference [21], is beyond the scope of this section the carrier-based approach will be adopted in the following in order to offer some basic concepts useful for single-phase and three-phase grid-connected converters.

According to the adopted implementation, analogue or digital, the modulations are defined as natural sampled or regular sampled methods respectively. Regular sampling can be symmetric, when the modulating wave is sampled at a frequency equal to that of the carrier wave, or asymmetric, when the modulating wave is sampled at twice the frequency of the carrier wave.

Among these techniques we can also mention the suboptimal and flat-top modulation strategies, only for three-phase converters, where the asymmetrical sampled modulating wave is derived from a procedure of optimization of a particular performance factor, e.g. the minimization of the output current THD. This procedure results in a modulating wave composed of a sinusoidal fundamental component with a zero sequence signal that allows the linear operating range of the modulation to be extended. Thus in the overmodulation range the use of particular shaped modulating waves allows better performances to be obtained with regards to both signal quality and control effectiveness [8].

As these improvements of the PWM low-frequency range are obtained by modifying the modulating signals, improvements in the PWM carrier and sidebands harmonics (above the switching frequency) can be obtained by acting on the carrier signal. The resulting modulations are called random modulation techniques. Their aim is to modify the harmonic spectrum of the standard PWM voltage signal, characterized by some dominant harmonics grouped around multiples of the switching frequency. These harmonics produce an annoying noise that can disturb people near the PWM electric drives (e.g. the workers of a factory or the passengers on a metro train). Random modulations can guarantee a reduction of the noise but have never been employed in industrial mass-produced drives. Moreover, the spread of harmonics in the PWM spectrum makes it more difficult to design the grid filter. For these reasons the random techniques are not discussed in this chapter.

The pulse-width modulation strategy has a deep impact not only on the quality of the grid current but also on the design of the grid-connected power converter and on the grid filter. In fact the chosen PWM determines the converter current ripple and hence the design of the inductor with respect to the saturation level of the magnetic core. Moreover, depending on the produced current ripple and on the current ripple acceptable by the grid it is possible to design an LCL filter to connect the converter to the grid, as discussed in Chapter 11.

The PWM techniques can be compared on the basis of range of linear operation, switching losses and produced harmonic distortion. The first parameter is the range where the control characteristic (fundamental voltage as a function of the modulation index) of the PWM technique is a linear function (Figure 12.24). It is recommended that the PWM converter operates within this range in order not to introduce a nonlinearity in the control loop. Hence the maximum fundamental voltage that the PWM converter can produce operating with a linear behaviour is usually considered in the converter design process. If this voltage is not equal or higher than the grid voltage, it is needed to increase the chosen DC voltage level. Hence a modulation with an extended range of linear operation needs a lower DC voltage level.

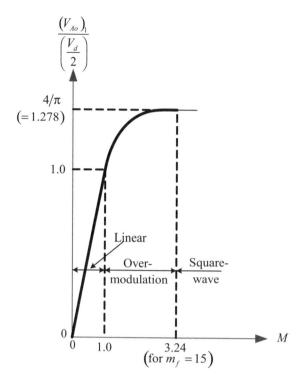

Figure 12.24 Range of linear operation

The second parameter, the switching losses, depends on the chosen modulation technique that determines the carrier and sideband harmonics. The switching losses are an important part of the overall converter losses and have an impact on the overall efficiency.

The third parameter, the produced harmonic distortion, certifies the harmonic quality of the PWM techniques. For example, it can be used for the weighted total harmonic distortion:

$$WTHD = \frac{\sqrt{\sum_{n=2}^{\infty} \frac{V_n^2}{n^2}}}{V_1} \tag{12.27}$$

The first two parameters have a strong influence on the converter design.

In the following, single-phase modulations (unipolar and bipolar types) and three-phase suboptimal modulations are briefly reviewed. Then for higher power applications or for applications characterized by several separated sources, they are typically adopted for more complex power converter configurations (e.g. neutral point clamped) or for the use of cascaded-connected converters or parallel-connected converters. The multilevel modulation, for the first two cases, and the interleaved modulation, for the last one, is briefly reviewed. The section is closed by a discussion on the operating limits of the grid converter, which are functions of the chosen AC voltage control.

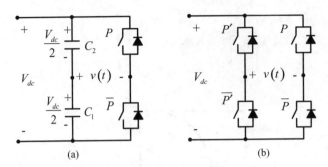

Figure 12.25 (a) Single-phase half-bridge and (b) full-bridge

12.4.1 Single-Phase

The basic single-phase inverter topologies can be half-bridge and full-bridge (Figure 12.25). The main elements for a comparison, in the case of grid-connected applications, are:

- The number of bidirectional switches (two for the half-bridge and four for the full-bridge).
- The switches voltages (double of the rated line voltage for the half-bridge and the rated line voltage for the full-bridge).
- The capacitor current, which also has a fundamental frequency component in the case of the half-bridge.
- The number of sensors (the half-bridge topology also needs one voltage sensor more to manage the balance between the two capacitor voltages).
- The algorithm complexity needed for the two converters (the half-bridge also needs a controller for the DC voltage balance).
- The modulation (the half-bridge allows only a bipolar PWM while the full-bridge allows a unipolar PWM with a better harmonic content).

In the case of asymmetrical sampling (sampling of the modulating signal on both the carrier edges), the produced bipolar voltage (shown in Figure 12.25) is

$$v(t) = \frac{4V_{dc}}{\pi} \sum_{\substack{m=0 \\ m>0}}^{\infty} \sum_{\substack{n=1 \\ n=-\infty}}^{\infty} \frac{1}{q} J_n\left(q\frac{\pi}{2}M\right) \sin\left([m+n]\frac{\pi}{2}\right) \cos(m\omega_c t + n\omega_0 t) \quad (12.28)$$

where M is the amplitude modulation coefficient, ω_0 is the pulsation of the modulating signal, ω_c is the pulsation of the carrier signal, J_n is the Bessel function of order n and $q = m + n(\omega_0/\omega_c)$. In the bipolar PWM signal reported in (12.28) the odd harmonic sideband components around the odd multiples of the carrier fundamental and even harmonic sideband components around even multiples of the carrier fundamental are completely eliminated as well as even low-ordered baseband harmonics.

In the case of the full-bridge it would be possible to modulate the two legs together to obtain a bipolar PWM signal or to adopt the unipolar modulation where, as shown in Figure 12.26,

Grid Current Control

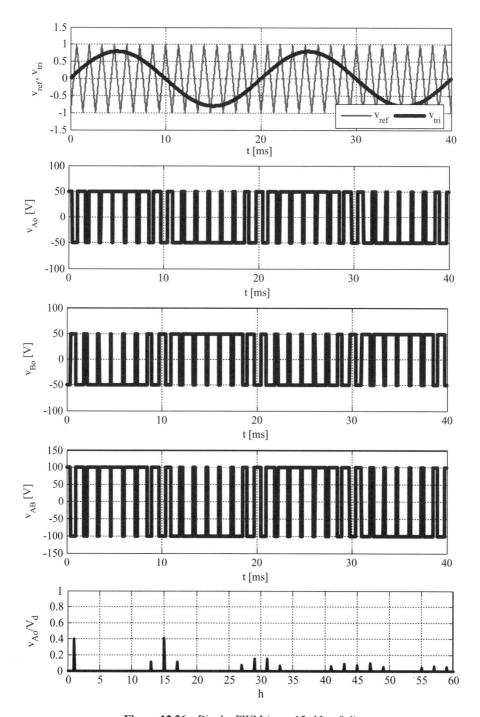

Figure 12.26 Bipolar PWM ($m = 15$, $M = 0.4$)

the two legs of the bridge are modulated with opposite modulating signals. In the case of asymmetrical sampling, the produced voltage is

$$v(t) = 2V_{dc}M\cos(\omega_0 t)$$
$$+ \frac{8V_{dc}}{\pi} \sum_{m=1}^{\infty} \sum_{n=-\infty}^{\infty} \frac{1}{2m} J_{2n-1}(m\pi M) \cos([m+n-1]\pi) \cos(2m\omega_c t + [2n-1]\omega_0 t)$$

(12.29)

Due to the unipolar PWM the odd carrier and associated sideband harmonics are completely cancelled, leaving only odd sideband harmonics $(2n-1)$ terms and even $(2m)$ carrier groups (Figure 12.27).

12.4.2 Three-Phase

The basic three-phase modulation is obtained by applying a bipolar modulation to each of the three legs of the converter. The modulating signals have to be 120° displaced. The phase-to-phase voltages are three levels of PWM signals that do not contain triple harmonics. If the carrier frequency is chosen as a multiple of three, the harmonics at the carrier frequency and at its multiples are absent. Figure 12.28 summarizes the previous considerations. Moreover, in Figure 12.28 the DC current and the transistor/diode current are also reported.

In the case of three-phase modulations it is possible to increase the range of linear operation and decrease the switching losses with respect to the single-phase case by adding a zero sequence signal to the modulating signals. This zero sequence signal has no influence on the grid due to the fact that the neutral is not connected.

Practically, depending on the form of the zero sequence voltage added to the modulating signal, there are six to seven methods of interest (Figure 12.29). The classical sinusoidal modulation, indicated with SPWM (sinusoidal PWM), has no zero sequence components. Then there are continuous and discontinuous modulations.

The continuous modulations reported in Figure 12.29 are:

- Sinusoidal PWM with the third harmonic injected (THIPWM). If the third harmonic has an amplitude of 25 % of the fundamental the minimum current harmonic content is achieved; if the third harmonic is 17 % of the fundamental the maximal linear range is obtained.
- Suboptimum modulation (subopt). A triangular signal is added to the modulating signal. In the case where the amplitude of the triangular signal is 25 % of the fundamental the modulation corresponds to the space vector modulation (SVPWM), with symmetrical placement of the zero vectors in the sampling time.

The discontinuous modulations formed by unmodulated 60° segments reported in Figure 12.29 are:

- Symmetrical flat top modulation, also called DPWM1.
- Asymmetrical shifted right flat-top modulation, also called DPWM2. It is worth noting that there is also the 'asymmetrical shifted left flat-top modulation', called DPWM0.

The use of unmodulated segments aims to obtain lower switching losses (average 33 %).

Grid Current Control

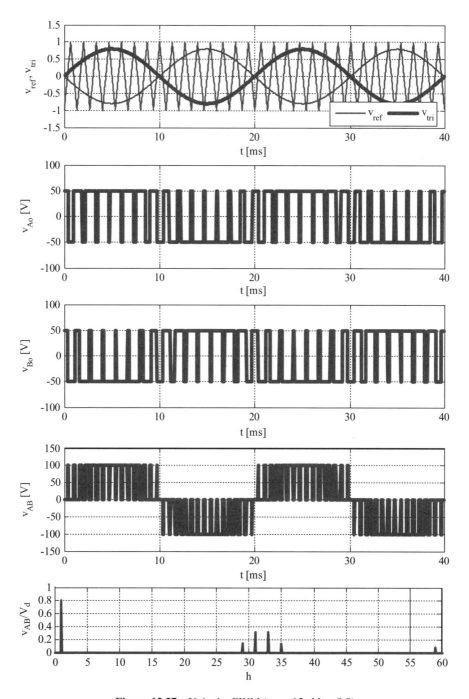

Figure 12.27 Unipolar PWM ($m = 15$, $M = 0.8$)

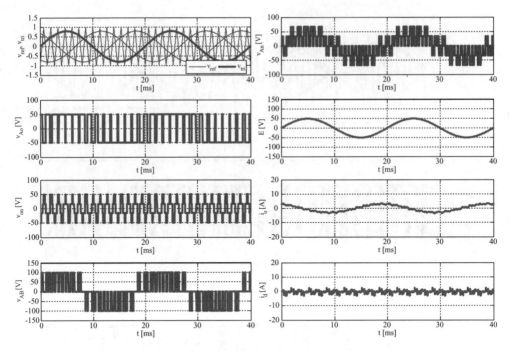

Figure 12.28 Three-phase sinusoidal PWM

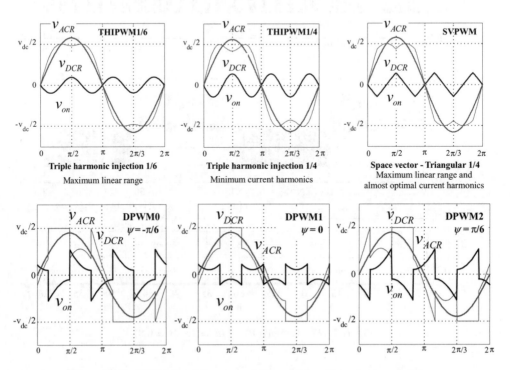

Figure 12.29 Different continuous and discontinuous three-phase modulations

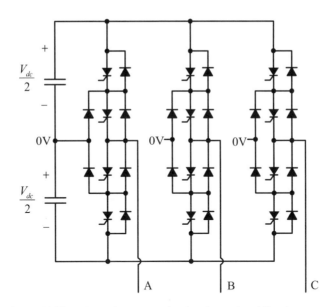

Figure 12.30 Three-phase neutral point clamped multilevel converter

12.4.3 Multilevel Modulations

Multilevel voltage source converters have been introduced for MVA applications, but their application has also been extended to very low power level cases such as audio amplifiers. In fact, the use of multilevel converters allows not only reduced stress for the semiconductor devices, greatly reducing failures and significantly extending the life of the converter, the possibility to avoid step-up/step-down transformers and/or stacks of series-connected semiconductors, but also lower harmonic pollution and hence smaller AC filters. As a consequence, even if the dominant use of multilevel converters is to deal with high-voltage conversion, the correct design of the multilevel modulation also allows the harmonic content of the voltage waveform to be reduced and hence smaller AC filters can be adopted.

The neutral point clamped multilevel converter (Figure 12.30) is one of the most successful converters in high-power wind turbine systems, as described in Chapter 6. The multilevel configuration weak point, the presence of many DC voltages, can be turned into an advantage in the case where many low voltage DC sources are available, as in the case of PV strings (as discussed in Chapter 2). Hence the use of a cascade-bridge multilevel configuration, as reported in Figure 12.31, can solve the problem of connecting many low-voltage PV strings with different irradiance/temperature conditions and leaving the possibility of managing the different power transfer without DC/DC converters. Moreover, in case electric isolation is needed (as requested by some national grid codes) it is possible to use transformers with different turn ratios to increase the number of levels that it is possible to obtain by one DC voltage source [22].

However, the solution is still limited by the high conduction losses (the same current flows through many semiconductors) as well as control and modulation problems. The control problems are generated by the fact that many DC voltages have to be controlled by the same

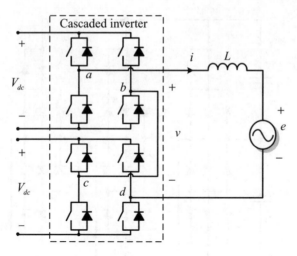

Figure 12.31 Five-level single-phase series-bridge cascaded inverter

current [24, 25] and modulation problems are related to the difficulties in minimizing the harmonic content if the DC voltages are different [23].

The first multilevel converter has been the neutral point clamped type, designed to supply an induction motor drive. Later on other topologies have been introduced, such as the flying capacitor and the cascaded H-bridge, based on the use of series-connected H-bridge cells.

A complete study of the multilevel modulations is beyond the scope of the book and hence mainly the multilevel modulation for cascaded structures (Figure 12.32) will be briefly discussed. In fact, in this case it is straightforward to demonstrate that by adopting unipolar asymmetrical modulation for each bridge (as the one discussed in previous chapters) a proper displacement among the carrier minimum WTHD is achieved. The produced PWM voltage is [21]

$$v(t) = NV_{dc}M\cos(\omega_0 t)$$
$$+ \frac{4V_{dc}}{\pi} \sum_{m=1}^{\infty} \sum_{n=-\infty}^{\infty} \frac{1}{2m} J_{2n-1}(m\pi M) \cos([m+n-1]\pi)$$
$$\sum_{i=1}^{N} \cos(2m\omega_c t + [2n-1]\omega_0 t + 2m\theta_i) \quad (12.30)$$

where N is the number of cascaded converters and θ_i is the relative phase of the carrier signal applied to leg A of each converter. If the carriers are displaced as

$$\theta_i = \frac{(i-1)\pi}{N} \quad (12.31)$$

for $\forall m \neq kN, k = 1, 2, 3, \ldots$, it is possible to achieve harmonic cancellation up to the $2N$th carrier multiple. In case the DC links are different (this can happen frequently if the PV arrays are under different irradiance conditions) it is possible to modify this technique in order to optimize the WTHD, as discussed in reference [23].

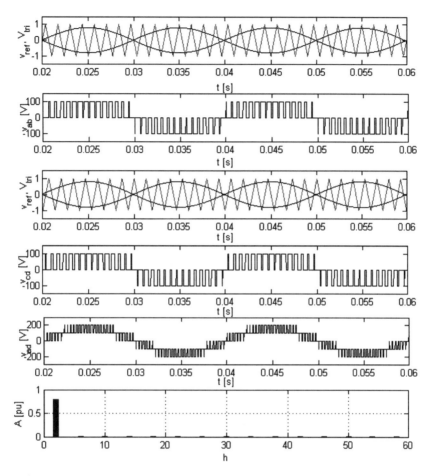

Figure 12.32 Five-level multilevel PWM generated by a single-phase series-bridge cascaded inverter ($m = 15$, $M = 0.8$)

Figure 12.33 shows a five-level single-phase neutral point clamped inverter. Basically there are three possible NPC modulations:

- Alternative phase opposition (APOD) where carriers in adjacent bands are phase shifted by 180°.
- Phase opposition disposition (POD), where the carriers above the reference zero point are out of phase with those below zero by 180°.
- Phase disposition (PD), where all the carriers are in phase across all bands.

Figure 12.34 shows the results of a single-phase bridge (Figure 12.33) where each leg is modulated using the PD strategy capable of obtaining the best WTHD.

As regards the current control, the NPC and the cascaded structures differ. In the case of the NPC there is no substantial difference with respect to what was presented in the previous

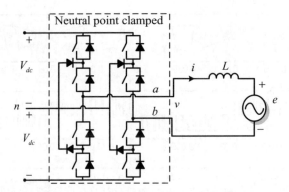

Figure 12.33 Five-level single-phase neutral point clamped inverter

sections since all the reviewed strategies are based on the use of the average model, which remains unchanged. In the case of the cascaded structure, the main difference is that several bridges are connected in series, sharing the current, and hence the current controllers should be coordinated [24,25].

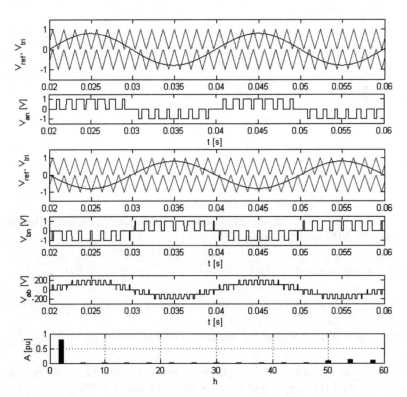

Figure 12.34 Five-level multilevel PWM generated by a single-phase neutral point clamped inverter ($m = 15$, $M = 0.8$)

12.4.4 Interleaved Modulation

In the case where two or more converters are connected in parallel in order to increase the power rating of the overall conversion stage, as described in Chapter 6 for high-power converters, or to exploit several independent sources, as described in Chapter 2 for photovoltaic systems, shifting the carrier signals (interleaving modulation) reduces the harmonic content in the current using the same principle that leads cascaded converters to have a reduced harmonic content of the voltage. In fact, in both cases the harmonic signatures of the voltages produced by the converters are equal in magnitude but have phase differences generated by the fact that the carriers are shifted (see (12.30) and (12.31)). However, while in the case of cascaded converters the voltages are summed and hence the phase opposition of some harmonics leads to their cancellation in the overall voltage, in the case of parallel-connected converters the currents are summed and hence the obtained phase opposition of the harmonics leads to a reduced harmonic content in the current. Of course, in order to achieve the desired result in the case of parallel connected converters, each of them needs to have an inductance before it is connected in parallel to the others. In fact, since the harmonic cancellation can only be achieved in the current it is needed to transform voltage sources in current sources in order to parallel them safely.

Figure 12.35 shows the effects of interleaving modulation in the case of two parallel H-bridges, as shown in Figure 12.36. The ripple of the current can be reduced to 1/4.

12.5 Operating Limits of the Current-Controlled Converter

The current control methods have been reviewed under the assumption of an ideal behaviour of the grid-connected converter, without considering its physical nature (transistor and antiparallel

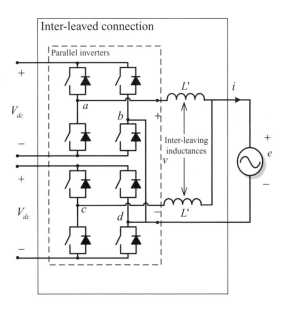

Figure 12.35 Interleaved inverters

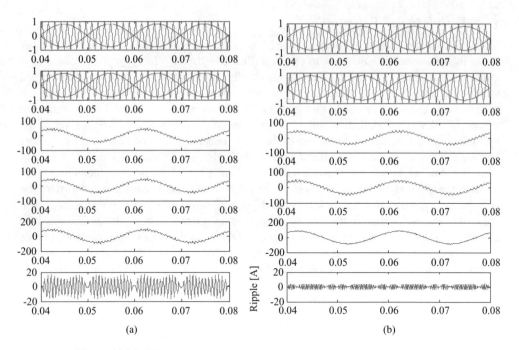

Figure 12.36 Two H-bridges (a) without and (b) with interleaved modulation

diodes) and the power rating. In the following some issues related to a real grid-connected converter will be discussed.

The grid-connected inverter has to produce an AC voltage at least equal to the grid voltage in order to control the injected grid current. In case of a fault on the grid the inverter may remain connected while not contributing with its current to the short-circuit current. Obviously this feature depends on the capability of the WT to limit or dissipate the produced power. In any case the control of the injected current to zero is indispensable in allowing an effective current control. The possibility of producing an AC voltage equal or higher with respect to the grid voltage depends on the modulation techniques and on the chosen DC voltage level. The natural DC link voltage, defined as the voltage obtainable if the transistors are not operating and their freewheeling diodes make the bridge act as a standard diode bridge, is $\sqrt{2}E$ for the single-phase case and $\sqrt{6}E$ for the three-phase case. If this condition is not fulfilled full control of the grid current is not possible.

However, to keep the switching losses down, it is desired to operate with a DC link voltage as low as possible. Typically the reference for the controlled DC link voltage is chosen as 5–10% above the natural DC link voltage (e.g. $600V_{dc}$ on a $400V_{ac}$ grid). If a unity power factor is obtained:

$$V^2 = E^2 + V_d^2 \tag{12.32}$$

The voltage drop across the inductor (V_d) depends on the reactance of the inductance at the input frequency and upon the input current. The magnitude of the switching voltage vectors depends on the DC voltage level. This means that the maximum AC voltage (V) the inverter can generate in the linear PWM region also depends on the DC link voltage. The higher the DC and thus the possible converter voltage, the higher the voltage across the inductor can be obtained. As this inductor voltage is driving the current, this means that the higher the DC voltage, the higher the input power (for a given inductance).

Once defined, the modulation index M is

$$M = 2\sqrt{2}\frac{V}{V_{dc}} \tag{12.33}$$

where V_{dc} is the average DC voltage. If a sinusoidal modulation is used the maximum modulation index in the linear region is $M_{MAX} = 1$; thus the maximum RMS fundamental voltage that is obtainable in the linear region, if the DC link is charged to 600 V, is $V_{MAX} = 212$ V. If a space vector modulation is adopted, $M_{MAX} = 1.154$ and $V_{MAX} = 245$ V.

Assuming the grid side resistance to be zero and neglecting the converter losses:

$$P = 3EI = 3E\frac{V_d}{\omega L} \tag{12.34}$$

Substituting (12.32) and (12.33) in (12.34) gives

$$P = 3E\frac{\sqrt{(M^2 V_{dc}^2/8) - E^2}}{\omega L} \tag{12.35}$$

This means that the higher the DC voltage and the smaller the inductance, the higher is the power rating of the converter.

Another important limitation is related to the sharing of the current between the transistor and the antiparallel diode. This depends on the displacement between the grid voltage and the converter voltage and on the modulation index. The cosine of the angle between the grid voltage vector and the converter voltage vector is

$$\cos\delta = \frac{E}{V} = \frac{E}{\sqrt{E^2 + V_d^2}} = \frac{1}{\sqrt{1 + (V_d/E)^2}} \tag{12.36}$$

At unity power factor the angle δ equals the angle between the converter fundamental voltage and the current. This angle influences the conduction ratio of the diodes and the transistors.

The modulation index has influence too. From (12.32) and (12.33):

$$M = \frac{2}{\sqrt{3}}\frac{\sqrt{E^2 + V_d^2}}{Ek_{boost}} = \frac{2}{\sqrt{3}}\frac{\sqrt{1 + (V_d/E)^2}}{k_{boost}} \tag{12.37}$$

where k_{boost} is the gain respect to the natural DC voltage and if the p.u. notation for L is used, $L = V_d/E$.

If the inverter is generating power the transistors will conduct the major part of the time. Assuming that $M = 1$ and $\delta = 0$, the conduction ratio of transistors to diodes would be 0:1.

From (12.36) and (12.37) result that the product of the modulation index M and the cosine of the displacment will always be constant:

$$M \cos \delta = \frac{2}{\sqrt{3} k_{boost}} \quad (12.38)$$

This factor is a good approximation to the current sharing ratio of transistors and diodes. The transistors are conducting about 93 % of the current while the load on the diodes is very low. In this situation the load on the transistors is slightly higher than for normal inverter operation on an induction motor. This is due to the small displacement angle δ of the grid inverter compared to a typical phase angle of the stator current of an induction motor.

12.6 Practical Example

It is requested that the current control, implemented on a digital system, should be designed for a three-phase converter connected to the grid.

Data

(a) LCL-filter parameters
(b) AC line-to-line grid voltage amplitude
(c) DC bus voltage
(d) Sampling frequency
(e) Switching frequency

Required steps for design/simulation

1. Choice of the modulation. The ratio of the natural DC voltage to the actual DC voltage determines the required maximum needed modulation index (neglecting possible grid overvoltage) and whether a zero sequence injection is needed in the modulating signal to increase the linear operating range of the PWM.
2. Choice of the current control PI in a synchronous dq frame.
3. Design of the current control:
   (a) Compensation of the cross-coupling terms $\omega L i_d(t)$ and $\omega L i_q(t)$ that make the two current equations not independent (see (12.3)).
   (b) Feed-forward of the grid voltage dq components $e_d(t)$ and $e_q(t)$.
   (c) Once the compensation of the grid influence and of the cross-coupling effects are completed, the controllers for both d and q axis currents can be designed on the basis of the following plant transfer function:

$$G(s) = \frac{1/R}{1 + Ts} \quad (12.39)$$

The two current loop controllers could be designed on the basis of the same time constant $T = L/R$. Two PI-based controllers can be used to perform the control action.

They have the following form in the S domain:

$$D(s) = \frac{k_P(1+T_I s)}{T_I s} \quad (12.40)$$

with k_P the proportional gain and T_I the time constant of the integrator. The delays present in the current control loop are the modulator delay and the processing delay. The modulator has a time constant that is usually settled to half of the sampling period ($T_M = T_s/2$) because this is the average time of the modulator to produce the desired voltage (chosen by the controller). Then there is one processing delay T_s due to the computational device. The two delays can be grouped together and the time constant is the sum of the two time constants.

Choosing the PI integrator time constant T_I equal to the plant time constant T, the current closed-loop transfer function in the S domain is

$$H(s) = \frac{\frac{2k_P}{3T_s L}}{s^2 + \frac{2}{3T_s}s + \frac{2k_P}{3T_s L}} \quad (12.41)$$

This means that

$$\omega_n^2 = \frac{2k_P}{3T_s L}$$
$$\zeta\omega_n = \frac{1}{3T_s} \quad (12.42)$$

Choosing to have a system optimally damped (i.e. with a 5% overshoot) leads to $\zeta = 0.707$ and thus to

$$k_P = \frac{L}{3T_s} \quad (12.43)$$

If the current control loop is adjusted to be optimally damped the following first-order approximation can be useful when calculating the bandwidth of the system:

$$H_c(s) \approx \frac{1}{1+3T_s s} \quad (12.44)$$

and the bandwidth frequency f_{bi} is

$$f_{bi} = \frac{1}{6\pi T_s} \approx \frac{1}{20 T_s} \quad (12.45)$$

(d) The digital implementation of the PI-based regulator obtained with the backward discretization rule is

$$D(z) = k_P \frac{\left(1 + \frac{T_s}{T_I}\right)z - 1}{z - 1} \qquad (12.46)$$

However, in order to limit the integral action and to apply a suitable anti-wind-up system for the integrator, the term $1/s$ is discretized separately and the controller can be written in the time domain as

$$u_P(k) = k_P \Delta i(k)$$
$$u_I(k) = \frac{T_s}{T_I} u_P(k) + u_I(k-1) \qquad (12.47)$$
$$u_{PI}(k) = u_P(k) + u_I(k)$$

Anti-wind-up can be designed to limit both u_I and u_{PI} with the same saturation values or by adding the following term to u_I:

$$u_{AW}(k) = [(\Delta i(k) > U_{I,MAX})(U_{I,MAX} - \Delta i(k)) \\ + (\Delta i(k) < -U_{I,MAX})(-U_{I,MAX} - \Delta i(k))]/T_s \qquad (12.48)$$

Moreover, (12.47) should be rewritten for the d and q axes considering the cross-coupling terms:

$$u_{Pd}(k) = k_P \Delta i_d(k)$$
$$u_{Id}(k) = \frac{T_s}{T_I} u_{Pd}(k) + u_{Id}(k-1)$$
$$u_{PId}(k) = u_{Pd}(k) + u_{Id}(k)$$
$$v_{d,av}(k) = e_d(k) - u_{PId}(k) + \omega L i_q(k)$$
$$\qquad (12.49)$$
$$u_{Pq}(k) = k_P \Delta i_q(k)$$
$$u_{Iq}(k) = \frac{T_s}{T_I} u_{Pq}(k) + u_{Iq}(k-1)$$
$$u_{PIq}(k) = u_{Pq}(k) + u_{Iq}(k)$$
$$v_{q,av}(k) = e_q(k) - u_{PIq}(k) - \omega L i_d(k)$$

Required steps for laboratory setup

1. Before starting the system:
   (a) Check the correct wiring of the system and of the coordination of sensors and PWM signals. (Are the current sensor of the phase 'a' and the voltage sensor of the phase 'a' on the same wire? Is the PWM driving signal of phase 'a' connected to the converter leg of phase 'a'?) If 'no' then modify the wiring.

(b) Check the synchronization of the inverter grid. Is the duty cycle commanded by the control in phase with the grid voltage? If 'no' go to 1.a.
(c) Is the duty cycle in saturation ? If 'no' check the anti-wind-up (12.48).
2. Starting up the system for the first time. Use a four-channel oscilloscope with two AC voltage probes (one for the grid and the other for the converter voltages), an AC current probe and a DC voltage probe (for the DC voltage) in the single acquisition mode with a trigger on the grid current. If overcurrent is found at startup:
(a) Is the AC voltage a PWM signal ? If 'no' check the PWM and the control signal (saturation) and go to 1.b.
(b) Is the DC voltage increasing ? If 'yes' check the current sensor's positive terminal connection.
3. Verify the system performance:
(a) Are the d/q currents tracking their references ? If 'no' check the integral action and the saturation of the controller.
(b) Is there any evidence of resonance? If 'yes' detune the current controller, check inductance values and consider inductor saturation or LCL filter resonance (Chapter 11).
(c) After programming a step in the d reference current, is the actual current response characterized by an overshoot in accordance with the designed system damping? If 'no' check the inductance values and consider inductor saturation or LCL filter resonance (Chapter 11).

12.7 Summary

This chapter closes the book by discussing some basic concepts on modulation and current control of grid-connected converters. In fact the modulation and the current control are the core of the grid-connected converter and the actuators of the active/reactive power control. The chapter has been structured to present some of the possible alternatives in the choice of the PWM and current control strategies.

The main modulation techniques (bipolar/unipolar, single-phase/three-phase, multilevel and interleaved) as well as the basic mathematical elements to deal with them and their harmonic content have been presented.

As for the current control: the main dq frame implementation (PI and deadbeat) and the emerging resonant controller base approach in the $\alpha\beta$ stationary frame have been considered. Finally, harmonic compensation strategies have been considered.

A practical example of how to design and implement a three-phase converter grid connected has been presented.

References

[1] Blaabjerg, F., Teodorescu, R., Liserre, M. and Timbus, A. V., 'Overview of Control and Grid Synchronization for Distributed Power Generation Systems'. *IEEE Transactions on Industrial Electronics*, **53**(5), October 2006, 1398–1408.
[2] Heier, S, *Grid Integration of Wind Energy Conversion Systems*, John Wiley & Sons, Ltd, 2006.
[3] IEEE Std 1547-2003, *IEEE Standard for Interconnecting Distributed Resources with Electric Power Systems*, 2003.
[4] IEC Standard 61727, *Characteristic of the Utility Interface for Photovoltaic (PV) Systems*, 2002.

[5] IEC Standard 61400-21, *Wind Turbine Generator Systems. Part 21: Measurements and Assessment of Power Quality Characteristics of Grid Connected Wind Turbines*, 2002.
[6] IEC Standard 61000-4-7, *Electromagnetic Compatibility, General Guide on Harmonics and Interharmonics Measurements and Instrumentation*, 1997.
[7] IEC Standard 61000-3-6, *Electromagnetic Compatibility, Assessment of Emission Limits for Distorting Loads in MV and HV Power Systems*, 1996.
[8] Kazmierkowski, M., Krishnan, R. and Blaabjerg, F., *Control in Power Electronics – Selected Problems*, Academic Press, 2002.
[9] Krein, P. T., Bentsman, J., Bass, R. M. and Lesieutre, B. L., 'On the Use of Averaging for the Analysis of Power Electronics Systems'. *IEEE Transactions on Power Electronics*, **5**, April 1990, 182–190.
[10] Zhang, R., Cardinal, M., Szczesny, P. and Dame, M., 'A Grid Simulator with Control of Single-Phase Power Converters in D–Q Rotating Frame'. In *Proceedings of PESC 2002*, Vol. **2**, pp. 1431–1436.
[11] Kolar, J. W., Ertl, H. and Zach, F. C., 'Analysis of On- and Off-line Optimized Predictive Current Controllers for PWM Converter Systems'. *IEEE Transactions on Power Electronics*, **6**(3), July 1991, 451–462.
[12] Wu, R., Dewan, S. B. and Slemon, G. R., 'Analysis of a PWM AC to DC Voltage Source Converter under the Predicted Current Control with a Fixed Switching Frequency'. *IEEE Transactions on Industry Applications*, **27**(4), July/August 1991, 756–764.
[13] Timbus, A., Liserre, M., Teodorescu, R., Rodriguez, P. and Blaabjerg, F., 'Evaluation of Current Controllers for Distributed Power Generation Systems'. *IEEE Transactions on Power Electronics*, **24**(3), March 2009, 654–664.
[14] Mattavelli, P., Spiazzi, G. and Tenti, P., 'Predictive Digital Control of Power Factor Preregulators with Input Voltage Estimation Using Disturbance Observers'. *IEEE Transactions on Power Electronics*, **20**(1), January 2005, 140–147.
[15] Hung, J. Y., 'Feedback Control with Posicast'. *IEEE Transactions on Industrial Electronics*, **50**, February 2003, 94–99.
[16] Teodorescu, R., Blaabjerg, F., Liserre, M. and Chiang Loh, P., 'A New Breed of Proportional-Resonant Controllers and Filters for Grid-Connected Voltage-Source Converters'. *IEE Proceedings on Electric Power Applications* (in press).
[17] Teodorescu, R., Blaabjerg, F., Borup, U. and Liserre, M., 'A New Control Structure for Grid-Connected LCL PV Inverters with Zero Steady-State Error and Selective Harmonic Compensation'. In *Proceedings of APEC'04*, Vol. **1**, 2004, pp. 580–586.
[18] Moreno, V. M., Liserre, M., Pigazo, A. and Dell'Aquila, A., 'A Comparative Analysis of Real-Time Algorithms for Power Signal Decomposition in Multiple Synchronous Reference Frames'. *IEEE Transactions on Power Electronics*, **22**, July 2007, 1280–1289.
[19] Limongi, L., Bojoi, R., Griva, G. and Tenconi, A., 'Digital Control Schems – Performance Comparison of DSP-Based Current Controllers for Three-Phase Active Power Filters'. *IEEE Industrial Electronics Magazine*, 2009.
[20] Liserre, M., Teodorescu, R. and Blaabjerg, F., 'Multiple Harmonics Control for Three-Phase Grid Converter Systems with the Use of PI-RES Current Controller in a Rotating Frame'. *IEEE Transactions on Power Electronics*, **21**(3), May 2006, 836–841.
[21] Holmes, D. G. and Lipo, T., *Pulse Width Modulation for Power Converters, Principles and Practice*, New York: IEEE Press, 2003.
[22] Kang, F.-S., Park, S.-J., Cho, S. E., Kim, C.-U. and Ise, T., 'Multilevel PWM Inverters Suitable for the Use of Stand-alone Photovoltaic Power Systems'. *IEEE Transactions on Energy Conversion*, **20**(4), December 2005, 906–915.
[23] Liserre, M., Pigazo, A., Monopoli, V. G., Dell'Aquila, A. and Moreno, V. M., 'Multilevel Phase-Shifting Carrier PWM Technique in Case of Non-equal DC-Link Voltages'. In *Proceedings of IECON 2006*, Paris, November 2006.
[24] Dell'Aquila, A., Liserre, M., Monopoli, V. G. and Rotondo, P., 'Overview of PI-Based Solutions for the Control of DC Buses of a Single-Phase H-Bridge Multilevel Active Rectifier'. *IEEE Transactions on Industry Applications*, **44**(3), June 2008, 857–866.
[25] Dell'Aquila, A., Liserre, M., Monopoli, V. G. and Rotondo, P., 'An Energy-Based Control for an *n*-H-Bridges Multilevel Active Rectifier'. *IEEE Transactions on Industrial Electronics*, **52**(3), June 2005, 670–678.

Appendix A

Space Vector Transformations of Three-Phase Systems

A.1 Introduction

A generic three-phase electrical system consists of a set of three voltages and three currents interacting with each other to deliver electrical power. A practical three-phase system cannot be considered as the simple addition of three independent single-phase subsystems. Actually, particular relations exist between the phase variables of a three-phase system, such as those resulting from the Kirchhoff laws or regarding phase sequences, which invite the application of certain space vector transformations to obtain a more elegant and meaningful representation of its variables. Generally, the control system of a power converter connected to a three-phase system is based on these transformed variables. This appendix reviews the most commonly used space vector transformations and highlights their applications in control grid-connected power converters.

A.2 Symmetrical Components in the Frequency Domain

In 1918, a young engineer from the Westinghouse Electric and Manufacturing Company in Pittsburgh, C. L. Fortescue, proposed a method for analysing unbalanced polyphase networks, which offered a new point of view in the analysis of three-phase systems and soon became known as the *method of symmetrical components* [1]. In simple mathematical terms, this method consists of a system of Lagrangian reference frames following the different sequences existing in a generic polyphase system, which results in a coordinate system especially suited to analyse all types of polyphase problems. The symmetrical components method allows decomposition of the steady-state phasors of an unbalanced three-phase system into a set of sequence components, namely the positive-, the negative- and the zero-sequence components. This new approach not only offered an elegant and systematic method for analysing polyphase systems under unbalanced sinusoidal conditions, but also allowed a rigorous explanation, with mathematical and physical meaning, of the phenomena that take place under such operating conditions.

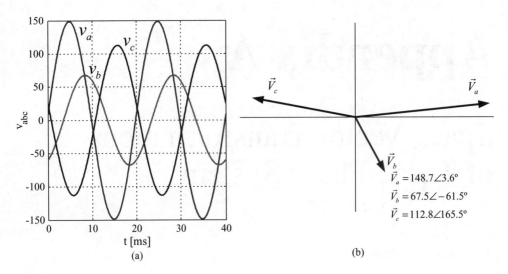

Figure A.1 Unbalanced three-phase system: (a) instantaneous voltage waveforms and (b) phase voltage phasors

The steady-state voltage waveforms of a three-phase unbalanced system together with their phasor representation on a Gauss plane are shown in Figure A.1.

Applying the symmetrical components method, these three unbalanced phasors representing the three-phase voltages can be transformed into a new set of three phasors representing the sequence components of one of the phases of the three-phase system. For example, the positive-, negative- and zero-sequence phasors of phase a ($\vec{V}_a^+$, $\vec{V}_a^-$ and $\vec{V}_a^0$) can be calculated by the following transformation matrix:

$$\mathbf{V}_{+-0(a)} = [T_{+-0}] \mathbf{V}_{abc} \tag{A.1}$$

with

$$\mathbf{V}_{abc} = \begin{bmatrix} \vec{V}_a \\ \vec{V}_b \\ \vec{V}_c \end{bmatrix} = \begin{bmatrix} V_a \angle \theta_a \\ V_b \angle \theta_b \\ V_c \angle \theta_c \end{bmatrix}, \quad \mathbf{V}_{+-0(a)} = \begin{bmatrix} \vec{V}_a^+ \\ \vec{V}_a^- \\ \vec{V}_a^0 \end{bmatrix} = \begin{bmatrix} V_a^+ \angle \theta_a^+ \\ V_a^- \angle \theta_a^- \\ V_a^0 \angle \theta_a^0 \end{bmatrix}$$

$$[T_{+-0}] = \frac{1}{3} \begin{bmatrix} 1 & \alpha & \alpha^2 \\ 1 & \alpha^2 & \alpha \\ 1 & 1 & 1 \end{bmatrix} \tag{A.2}$$

where $\alpha = e^{j2\pi/3} = 1\angle 120°$ is known as the Fortescue operator. The phasors representing the sequence components for phases b and c are given by

$$\begin{aligned} \vec{V}_b^+ &= \alpha^2 \vec{V}_a^+; & \vec{V}_b^- &= \alpha \vec{V}_a^- \\ \vec{V}_c^+ &= \alpha \vec{V}_a^+; & \vec{V}_c^- &= \alpha^2 \vec{V}_a^- \end{aligned} \tag{A.3}$$

Appendix A: Space Vector Transformations of Three-Phase Systems

The inverse transformation to pass from the phasors representing symmetrical components of phase a to the phasors representing the unbalanced phase voltages is given by

$$\mathbf{V}_{abc} = [T_{+-0}]^{-1} \mathbf{V}_{+-0(a)} \tag{A.4}$$

with

$$[T_{+-0}]^{-1} = \begin{bmatrix} 1 & 1 & 1 \\ \alpha^2 & \alpha & 1 \\ \alpha & \alpha^2 & 1 \end{bmatrix} \tag{A.5}$$

As an example, the phasors representing the sequence components of the unbalanced voltages of Figure A.1, together with their corresponding instantaneous waveforms, are shown in Figure A.2.

A.3 Symmetrical Components in the Time Domain

Lyon extended the work of Fortescue and applied the method of the symmetrical components in the time domain [2]. When the Fortescue transformation matrix of (A.2) is applied to the following set of three-phase unbalanced sinusoidal waveforms:

$$\mathbf{v}_{abc} = \begin{bmatrix} v_a \\ v_b \\ v_c \end{bmatrix} = \mathbf{v}_{abc}^+ + \mathbf{v}_{abc}^- + \mathbf{v}_{abc}^0$$

$$= V^+ \begin{bmatrix} \cos(\omega t) \\ \cos(\omega t - \frac{2\pi}{3}) \\ \cos(\omega t + \frac{2\pi}{3}) \end{bmatrix} + V^- \begin{bmatrix} \cos(\omega t) \\ \cos(\omega t + \frac{2\pi}{3}) \\ \cos(\omega t - \frac{2\pi}{3}) \end{bmatrix} + V^0 \begin{bmatrix} \cos(\omega t) \\ \cos(\omega t) \\ \cos(\omega t) \end{bmatrix} \tag{A.6}$$

the resultant instantaneous variables are given by

$$\mathbf{v}_{+-0} = [T_{+-0}] \mathbf{v}_{abc} \tag{A.7}$$

$$\mathbf{v}_{+-0} = \begin{bmatrix} \vec{v}^+ \\ \vec{v}^- \\ v^0 \end{bmatrix} = \begin{bmatrix} \frac{1}{2}V^+ e^{j\omega t} + \frac{1}{2}V^- e^{-j\omega t} \\ \frac{1}{2}V^+ e^{-j\omega t} + \frac{1}{2}V^- e^{j\omega t} \\ V^0 \cos(\omega t) \end{bmatrix} \tag{A.8}$$

It is worth mentioning that the Lyon transformation is usually defined by the following normalized matrix:

$$[T'_{+-0}] = \sqrt{3}\,[T_{+-0}] = \frac{1}{\sqrt{3}} \begin{bmatrix} 1 & \alpha & \alpha^2 \\ 1 & \alpha^2 & \alpha \\ 1 & 1 & 1 \end{bmatrix} \tag{A.9}$$

where $[T'_{+-0}]^{-1} = [T'_{+-0}]^T$.

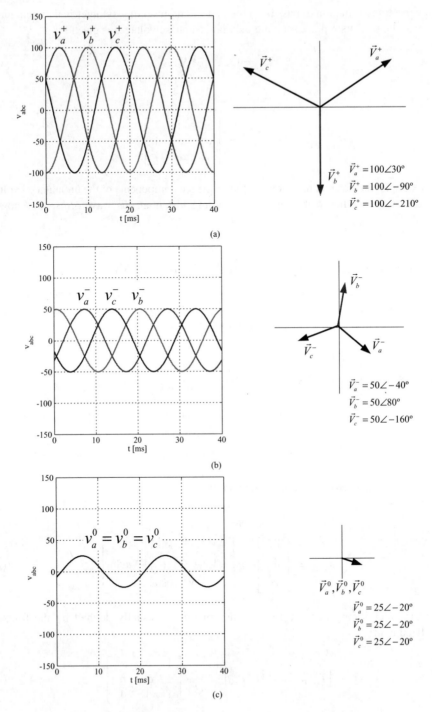

Figure A.2 Sequence components of the unbalanced three-phase system of Figure A.1: (a) positive-sequence phasors, (b) negative-sequence phasors and (c) zero-sequence phasors

Appendix A: Space Vector Transformations of Three-Phase Systems

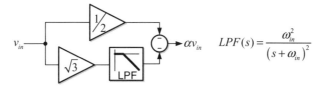

Figure A.3 Simple implementation of the *a* operator in the time domain

From (A.8) it can concluded that, independently of the scale factor used in the Lyon transformation, the resulting vector consists of two complex elements, $\vec{v}^+$ and $\vec{v}^-$, plus a real element v^0. The complex elements $\vec{v}^+$ and $\vec{v}^-$ can be represented by instantaneous space vectors, having the same amplitude and rotating in opposite directions. Therefore, $\vec{v}^+$ and $\vec{v}^-$ should not be mistaken for the positive- and negative-sequence voltage vectors v^+_{abc} and v^-_{abc}. The real element v^0 is directly related to the zero-sequence component of the original three-phase voltage vector.

To calculate the positive- and negative-sequence voltage vectors, v^+_{abc} and v^-_{abc}, from the unbalanced input vector v_{abc}, it is necessary to translate the Fortescue operator, α, from the frequency domain to the time domain. This translation can be performed by using a simple time-shifting operator, provided the frequency of the sinusoidal input is a well-known magnitude. In such a case, a $2/3T$ time-shifted sinusoidal signal, with T the signal period, can be understood as a 120° leaded version of the original sinusoidal signal. This operator in the time domain is named a in this appendix. Since $\alpha = -1/2 + j\sqrt{3}/2$, the a operator can be implemented by using a proper filter to generate the 90° phase-shifting associated to the j operator [3]. As an example, Figure A.3 shows a simple implementation of the a operator based on a second-order low-pass filter (LPF) tuned at the input frequency with a damping factor $\xi = 1$. The a^2 operator can be implemented by multiplying the output signal of the LPF by -1.

Therefore, the following expressions can be used to calculate the instantaneous positive- and negative-sequence components of v_{abc}:

$$v^+_{abc} = [T_+]\, v_{abc}; \quad \begin{bmatrix} v^+_a \\ v^+_b \\ v^+_c \end{bmatrix} = \frac{1}{3} \begin{bmatrix} 1 & a & a^2 \\ a^2 & 1 & a \\ a & a^2 & 1 \end{bmatrix} \begin{bmatrix} v_a \\ v_b \\ v_c \end{bmatrix} \quad (A.10)$$

$$v^-_{abc} = [T_-]\, v_{abc}; \quad \begin{bmatrix} v^-_a \\ v^-_b \\ v^-_c \end{bmatrix} = \frac{1}{3} \begin{bmatrix} 1 & a^2 & a \\ a & 1 & a^2 \\ a^2 & a & 1 \end{bmatrix} \begin{bmatrix} v_a \\ v_b \\ v_c \end{bmatrix} \quad (A.11)$$

A.4 Components $\alpha\beta0$ on the Stationary Reference Frame

Since the complex elements $\vec{v}^+$ and $\vec{v}^-$ in (A.8) are not independent from each other, only three independent real components can be found among the elements resulting from the transformation of (A.7). Therefore, a possible set of independent elements can be defined from (A.8) as $\{\Re(\vec{v}^+), \Im(\vec{v}^+), v^0\}$, although other combinations are also possible. At this point, the following real transformation matrix can be written:

$$\begin{bmatrix} \Re(\vec{v}^+) \\ \Im(\vec{v}^+) \\ v^0 \end{bmatrix} = \frac{1}{3} \begin{bmatrix} 1 & \Re(\alpha) & \Re(\alpha^2) \\ 0 & \Im(\alpha) & \Im(\alpha^2) \\ 1 & 1 & 1 \end{bmatrix} \begin{bmatrix} v_a \\ v_b \\ v_c \end{bmatrix} \qquad (A.12)$$

A similar reasoning caused Clarke to reformulate the Lyon transformation and to propose the following transformation matrix [4]:

$$v_{\alpha\beta 0} = [T_{\alpha\beta 0}] \, v_{abc} \qquad (A.13)$$

$$\begin{bmatrix} v_\alpha \\ v_\beta \\ v_0 \end{bmatrix} = \sqrt{\frac{2}{3}} \begin{bmatrix} 1 & -\frac{1}{2} & -\frac{1}{2} \\ 0 & \frac{\sqrt{3}}{2} & -\frac{\sqrt{3}}{2} \\ \frac{1}{\sqrt{2}} & \frac{1}{\sqrt{2}} & \frac{1}{\sqrt{2}} \end{bmatrix} \begin{bmatrix} v_a \\ v_b \\ v_c \end{bmatrix} \qquad (A.14)$$

where $[T_{\alpha\beta 0}]^{-1} = [T_{\alpha\beta 0}]^T$.

It is worth remarking here that the input and output vectors have the same norm when the normalized transformation of (A.14) is applied, i.e.

$$v_\alpha^2 + v_\beta^2 + v_0^2 = v_a^2 + v_b^2 + v_c^2 \qquad (A.15)$$

As a consequence, when the normalized transformation of (A.14) is applied to the voltages and currents of a three-phase system, power calculations will give rise to equivalent results for both the *abc* and the $\alpha\beta 0$ reference frames, i.e.

$$p = v_{\alpha\beta 0} \cdot i_{\alpha\beta 0} = v_{abc} \cdot i_{abc} \qquad (A.16)$$

The transformation $[T_{\alpha\beta 0}]$ can be rescaled as shown in the following equation when the amplitude of the sinusoidal signals on the *abc* and the $\alpha\beta 0$ reference frames should be equal, i.e. when $\hat{V}_\alpha = \hat{V}_a$:

$$\begin{bmatrix} v_\alpha \\ v_\beta \\ v_0 \end{bmatrix} = \frac{2}{3} \begin{bmatrix} 1 & -\frac{1}{2} & -\frac{1}{2} \\ 0 & \frac{\sqrt{3}}{2} & -\frac{\sqrt{3}}{2} \\ \frac{1}{\sqrt{2}} & \frac{1}{\sqrt{2}} & \frac{1}{\sqrt{2}} \end{bmatrix} \begin{bmatrix} v_a \\ v_b \\ v_c \end{bmatrix} \qquad (A.17)$$

The $\alpha\beta 0$ reference frame is graphically depicted in Figure A.4. In this figure, the $\alpha\beta$ plane holds all the symmetrical vectors, i.e. those $v_{\alpha\beta}$ vectors with no zero sequence ($v_a + v_b + v_c = 0$), whereas the 0 axis is aligned with the space diagonal of the *abc* reference frame.

Appendix A: Space Vector Transformations of Three-Phase Systems

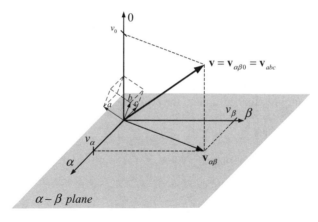

Figure A.4 Graphical representation of the $\alpha\beta 0$ reference frame

A.5 Components *dq*0 on the Synchronous Reference Frame

Any voltage vector rotating on the $\alpha\beta$ plane can be expressed on a synchronous reference frame by using the Park transformation [5]. As depicted in Figure A.5, the synchronous reference frame, also known as the *dq* reference frame, is based on two orthogonal *dq* axes, rotating at frequency ω, which are placed at the $\theta = \omega t$ angular position on the $\alpha\beta$ plane. Thanks to its rotating character, this transformation has been extensively used in the analysis of electrical machines.

The transformation matrix to translate a voltage vector from the $\alpha\beta 0$ stationary reference frame to the *dq*0 synchronous reference frame is given by

$$\boldsymbol{v}_{dq0} = \begin{bmatrix} T_{dq0} \end{bmatrix} \boldsymbol{v}_{\alpha\beta 0} \tag{A.18}$$

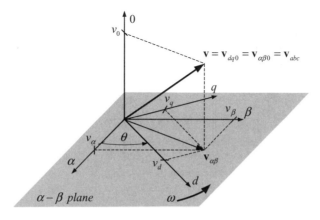

Figure A.5 Graphical representation of the *dq*0 reference frame

$$\begin{bmatrix} v_d \\ v_q \\ v_0 \end{bmatrix} = \begin{bmatrix} \cos(\theta) & \sin(\theta) & 0 \\ -\sin(\theta) & \cos(\theta) & 0 \\ 0 & 0 & 1 \end{bmatrix} \begin{bmatrix} v_\alpha \\ v_\beta \\ v_0 \end{bmatrix} \quad (A.19)$$

where $[T_{dq0}]^{-1} = [T_{dq0}]^T$. Therefore, the transformation matrix to translate a voltage vector from the abc stationary reference frame to the $dq0$ synchronous reference frame is given by

$$\boldsymbol{v}_{dq0} = [T_\theta]\, \boldsymbol{v}_{abc} \quad (A.20)$$

$$\begin{bmatrix} v_d \\ v_q \\ v_0 \end{bmatrix} = \sqrt{\frac{2}{3}} \begin{bmatrix} \cos(\theta) & \cos(\theta - \frac{2\pi}{3}) & \cos(\theta + \frac{2\pi}{3}) \\ -\sin(\theta) & -\sin(\theta - \frac{2\pi}{3}) & -\sin(\theta + \frac{2\pi}{3}) \\ \frac{1}{\sqrt{2}} & \frac{1}{\sqrt{2}} & \frac{1}{\sqrt{2}} \end{bmatrix} \begin{bmatrix} v_a \\ v_b \\ v_c \end{bmatrix} \quad (A.21)$$

where $[T_\theta]^{-1} = [T_\theta]^T$.

The normalized transformations shown in (A.14) and (A.21) allow the norm of the voltage vector to be conserved in all the reference frames; thus

$$v_d^2 + v_q^2 + v_0^2 = v_\alpha^2 + v_\beta^2 + v_0^2 = v_a^2 + v_b^2 + v_c^2 \quad (A.22)$$

Expressing voltage and currents using space vectors allows instantaneous phenomena in three-phase systems to be studied using an efficient and elegant formulation. This formulation is particularly useful to control active and reactive power components in three-phase systems.

References

[1] Fortescue, C. L. 'Method of Symmetrical Coordinates Applied to the Solution of Polyphase Networks'. *Transactions of the AIEE*, Part II, **37**, 1918, 1027–1140.
[2] Lyon, W. V., *Application of the Method of Symmetrical Components*, New York: McGraw-Hill, 1937.
[3] Iravani, M. R. and Karimi-Ghartemani, M., 'Online Estimation of Steady State and Instantaneous Symmetrical Components'. *Proceedings of the IEE on Generation, Transmission and Distribution*, **150**(5), September 2003, 616–622.
[4] Clarke, E., *Circuit Analysis of AC Power Systems*, Vol. 1, New York: John Wiley & Sons, Inc., 1950.
[5] Park, R. H., 'Two Reaction Theory of Synchronous Machines. Generalized Method of Analysis – Part I'. In *Proceedings of the Winter Convention of the AIEE*, 1929, pp. 716–730.

Appendix B

Instantaneous Power Theories

B.1 Introduction

Accurate knowledge of the currents to be injected into the grid for delivering a given power under generic grid conditions is a critical matter in the design of grid-connected power converters. The interaction between the power converter and the grid depends both on the power exchanged between both systems and on the strategy applied to determine the waveform of the injected currents.

Conventional definitions of active, reactive and apparent powers are based on theories developed and adopted in the 1930s. Such definitions have been applied successfully by electrical engineers in regular three-phase systems where voltages and currents were sinusoidal and balanced. However, changes that occurred in power systems throughout the twentieth century gave rise to a deep debate about the calculation of power components in such systems. Some factors causing such discussion are:

1. Currently, there is widespread usage of power electronics in power systems and it is expected that there will be a massive integration of this technology in future electricity networks. Currents and voltages associated with these equipments are not always sinusoidal and power flows should be properly calculated.
2. Intensive studies about power definitions in three-phase systems have been carried out in the last 30 years [1] and higher attention is currently paid to effects related to currents flowing through the neutral conductor, which has originated in an extension of the conventional instantaneous power theories in three-phase systems. New power measurement standards reflect this trend by including new definitions for such cases in which currents and voltages are not balanced sinusoidally [2, 3].
3. Traditional instrumentation is designed to work with sinusoidal waveforms at 50/60 Hz frequency and is prone to mistakes when waveforms are distorted and/or unbalanced. Moreover, advances in microprocessors allow new measuring equipments to be more flexible and precise, which are able to calculate electrical magnitudes based on complex mathematical models.

Grid Converters for Photovoltaic and Wind Power Systems Remus Teodorescu, Marco Liserre, and Pedro Rodríguez
© 2011 John Wiley & Sons, Ltd

4. Power quality has been an issue of increasing interest over the last few years. Correct quantification of power flows is necessary in order to know accurately the costs of maintaining high-power quality rates.
5. New grid codes regulating the connection of renewable energy sources to the grid, mainly wind turbines and photovoltaics systems, pay especial attention to the active and reactive power to be delivered to the grid for supporting the grid frequency and voltage in the steady state and for improving the stability of the system during transient faults [4, 5].

Instantaneous power theories are a relevant issue in the design of power converter controllers for integrating renewable energies into the electrical grid. Although the first power definitions in the time domain came from the 1930s [6], it was not until the 1980s that Akagi stated an instantaneous power theory (p-q) that was well adopted by researchers and engineers working with power converters connected to three-phase three-wire systems [7, 8]. Later, in the 1990s, new theories appeared, dealing with three-phase four-wire systems [9–12]. In spite of some criticism, all these theories awoke a big interest among engineers and offered them a new tool for controlling grid-connected converters and analysing their interaction with the electrical grid.

The growing interest aroused in this area during the last few years should not induce one to think that the study of the active and nonactive components of the current in multiphase systems is a recent issue. In the early 1930s, Fryze set the basis of instantaneous active and nonactive current calculations in the time domain for single-phase systems. It was in the 1950s when Buchholz made a very transcendent contribution about the decomposition of currents in a generic multiphase system [13]. Later, in the early 1960s, Depenbrock extended Buchholz's work and proposed the FBD method – in honour of Fryze, Buchholz and Depenbrock – to calculate instantaneously the active currents in a generic multiphase system [14].

In the studies conducted by Buchholz and Depenbrock, they considered that the power was transferred from a source to a load by using a generic multiphase system, in which all of its conductors have the same capability for transporting electrical energy. When the FDB method is applied to a three-phase three-wire system, the resultant active currents match those calculated using the p-q theory – proposed by Akagi later. Actually, the expression determining the active currents in the p-q theory is a particular case of the FDB method. However, the conceptual contribution performed by Akagi goes further than a simple current decomposition. Akagi introduced the *imaginary power* concept, which explains, clearly and with physical meaning, that the instantaneous exchange of energy occurred between the phases of a three-phase three-wire system as a consequence of the nonactive currents flowing through its conductors.

When working with three-phase four-wire systems, the active currents resulting from the FBD method differ from those obtained applying the instantaneous power theories proposed in the 1990s [15]. This is because both formulations stem from different approaches. While Depenbrock proposed the FBD method from a formal analysis based on the Kirchhoff laws of a generic multiphase system, most of the other theories come up from the mathematical updating of the p-q theory originally proposed by Akagi.

The aim of this appendix is to make a short review of the most relevant theories on both instantaneous power calculation and current decomposition in order to acquire the necessary knowledge to calculate the reference currents for implementing advanced functionalities such as instantaneous active/reactive power control or harmonics compensation in grid-connected

Appendix B: Instantaneous Power Theories

power converters under generic grid conditions. Taking into account that there are excellent books specifically focused on the instantaneous power theories [16], the number of power theories reviewed in this appendix has been limited to:

- 1932 – *Fryze*. Single-phase power definitions in the time domain.
- 1950 – *Buchholz*. Active currents in multiphase systems.
- 1962 – FBD method by *Depenbrock*. Calculation of power currents (instantaneous) for generic multiphase systems.
- 1983 – *p-q* theory by *Akagi* and others. Introduction of the instantaneous imaginary power.
- Generalization and modification of the *p-q* theory.
    - 1992 – *Willems*. Calculation of the current active component in multiphase systems (with a neutral conductor).
    - 1994 – *Nabae* and others. Reformulation of the *p-q* theory for three-phase four-wire systems.
    - 1996 – *Peng* and *Lai*. Vector formulation of the instantaneous imaginary power theory for three-phase four-wire systems.

B.2 Origin of Power Definitions at the Time Domain for Single-Phase Systems

Fryze can be considered as the main precursor of the modern power definitions in the time domain [17]. Power formulation proposed by Fryze was not based on the Fourier analysis, which made its practical application simpler since equipment like modern real-time spectrum analysers did not exist at that time.

From two generic single-phase voltage and current signals, v and i, Fryze dealt with the following magnitudes in his power calculations:

- Rms voltage:

$$V = \sqrt{\frac{1}{T}\int_0^T v^2 dt} \quad (B.1)$$

- Rms current:

$$I = \sqrt{\frac{1}{T}\int_0^T i^2 dt} \quad (B.2)$$

- Active power:

$$P = \frac{1}{T}\int_0^T p\, dt = \frac{1}{T}\int_0^T v\, i\, dt \quad (B.3)$$

- Apparent power:

$$S = VI \quad (B.4)$$

- Reactive power:
$$Q = \sqrt{S^2 - P^2} \tag{B.5}$$

- Active power factor:
$$\lambda_a = \frac{P}{S} \tag{B.6}$$

- Nonactive power factor:
$$\lambda_n = \sqrt{1 - \lambda_a^2} = \frac{Q_F}{S} \tag{B.7}$$

Fryze verified that λ_a reaches its maximum ($\lambda_a = 1$) if and only if the instantaneous current i is proportional to the instantaneous voltage v. He laid the foundations of the method to divide the instantaneous current i into two instantaneous orthogonal components, namely the active current i_a and the nonactive current i_n.

The instantaneous active component of the current, i_a, must develop the same active power as the original current i, which means that

$$P = \frac{1}{T}\int_0^T v\, i\, dt = \frac{1}{T}\int_0^T v\, i_a\, dt = \frac{1}{T}\int_0^T v\,(Gv)\, dt = G\frac{1}{T}\int_0^T v^2\, dt = GV^2, \tag{B.8}$$

where G can be understood as the equivalent conductance of the system averaged over the grid period T. Therefore, the instantaneous active current can be calculated by

$$i_a = G\, v; \quad G = \frac{P}{V^2} \tag{B.9}$$

and the instantaneous nonactive current is given by

$$i_n = i - i_a \tag{B.10}$$

The orthogonal relationship between both current components implies that

$$\frac{1}{T}\int_0^T i_a\, i_n\, dt = \frac{G}{T}\int_0^T v\, i_n\, dt = \frac{G}{T}\int_0^T v\,(i - i_a)\, dt = 0 \quad \Rightarrow \quad I^2 = I_a^2 + I_n^2 \tag{B.11}$$

where I_a and I_n are the rms values of the active and nonactive components of the original current respectively.

B.3 Origin of Active Currents in Multiphase Systems

In 1950, Buchholz extended Fryze's work to systems with multiple phase and conductors [13]. According to the Buchholz's approach, this kind of system can be represented by a

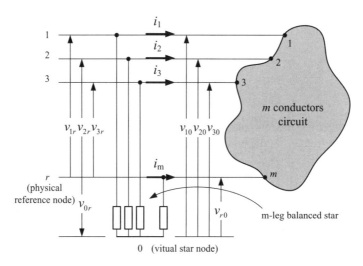

Figure B.1 Multiphase system

homogeneously structured circuit in which none of the conductors is treated as an especial conductor. In this homogeneous circuit, the voltages of the m conductors are referenced to a virtual node '0':

$$v_{k0} = v_{kr} - v_{0r}; \quad v_{0r} = \frac{1}{m}\sum_{k=1}^{m} v_{kr}; \quad k \in \{1, \ldots, m\} \tag{B.12}$$

where v_{kr} is the voltage of the kth terminal measured with respect to the physical reference node 'r', which can be arbitrarily chosen. Hence, the v_{0r} voltage is the floating voltage of the virtual node '0' with respect to the arbitrary physical reference node 'r'.

Applying the Kichhkoff laws to the m-terminal circuit of Figure B.1, it is found that

$$\sum_{k=1}^{m} v_{k0} = 0; \quad \sum_{k=1}^{m} i_k = 0 \tag{B.13}$$

Expressions of (B.13) are always true, independently of the values of the phase voltages and currents and the node 'r' chosen as a physical reference.

The instantaneous power collectively developed by the m phases of the system is independent of the node chosen as a reference for measuring the phase voltages, i.e.

$$p_\Sigma = \sum_{k=1}^{m} v_{kr}i_k = \sum_{k=1}^{m} (v_{k0} + v_{0r})\, i_k = \sum_{k=1}^{m} v_{k0}i_k + v_{0r}\underbrace{\sum_{k=1}^{m} i_k}_{0} = \sum_{k=1}^{m} v_{k0}i_k \tag{B.14}$$

As the instantaneous active power is a single value describing the energy consumption rate of a multiphase system, the *collective value* concept was introduced by Buchholz to represent collectively the voltages and currents in a multiphase system. These voltage, v_Σ, and current, i_Σ, instantaneous collective values are respectively defined as

$$v_\Sigma = \sqrt{\sum_{k=1}^{m} v_{k0}^2}; \quad i_\Sigma = \sqrt{\sum_{k=1}^{m} i_k^2} \qquad (B.15)$$

Buchholz also defined the rms collective value of voltage and current in multiphase systems, which are suitable for calculating powers under steady-state sinusoidal operating conditions:

$$V_\Sigma = \sqrt{\frac{1}{T} \int_0^T v_\Sigma^2 \, dt}; \quad I_\Sigma = \sqrt{\frac{1}{T} \int_0^T i_\Sigma^2 \, dt} \qquad (B.16)$$

Buchholz studied multiphase systems in detail much earlier than he determined their instantaneous active currents. Actually, it was in 1922 [18] when he proposed the following expression for calculating the apparent power in a multiphase system:

$$S_\Sigma = V_\Sigma I_\Sigma \qquad (B.17)$$

The modern standard IEEE Std 1459-2010 [3], which pretends to arrive at generalized definitions for the measurement of electric power quantities under sinusoidal, nonsinusoidal, balanced and unbalanced conditions, adopts Buchholz's definition for the apparent power, renaming it as the *effective apparent power*, S_e, and textually states in its introduction that:

> For sinusoidal unbalanced or for nonsinusoidal balanced or unbalanced situations, S_e allows rational and correct computation of the power factor. This quantity was proposed in 1922 by the German engineer F. Buchholz and in 1933 was explained by the American engineer W. M. Goodhue.

Taking into account the fact that the collective active power of the multiphase system can be calculated by

$$P_\Sigma = \frac{1}{T} \int_0^T p_\Sigma \, dt \qquad (B.18)$$

there exists a nonactive power, N_Σ, that is defined in the IEEE Std 1459-2000 as N, which verifies that

$$S_\Sigma^2 = P_\Sigma^2 + N_\Sigma^2 \qquad (B.19)$$

From these definitions, Buchholz stated that the instantaneous current in each conductor of a multiphase system, i_k, can be divided into an active component, i_{ak}, and a nonactive component, i_{nk}, according to

$$i_k = i_{ak} + i_{nk} \begin{cases} i_{ak} = G_a v_{k0}; \quad G_a = \dfrac{P_\Sigma}{V_\Sigma^2} \\ i_{nk} = i_k - i_{ak} = i_k - G_a v_{k0} \end{cases} \qquad (B.20)$$

Appendix B: Instantaneous Power Theories

In (B.20), G_a can be considered as an equivalent active conductance of the multiphase system. The nonactive current of (B.20) does not contribute to the averaged collective active power P_Σ, i.e.

$$p_{\Sigma a} = \sum_{k=1}^{m} v_{k0} i_{ak} = G_a \sum_{k=1}^{m} v_{k0}^2 \Rightarrow \frac{1}{T}\int_0^T p_{\Sigma a}\, dt = G_a \frac{1}{T}\int_0^T \sum_{k=1}^{m} v_{0k}^2\, dt = G_a V_\Sigma^2 = P_\Sigma \tag{B.21}$$

Therefore, in terms of the net energy transfer, the nonactive current i_{nk} can be cancelled out by a suitable compensator, which allows the collective rms value of the total current flowing through the conductors to be reduced and the generation and transmission capacity of the power system to be increased.

Buchholz proved that, based on Cauchy–Schwarz inequalities:

- $p_{\Sigma a}$ is a function of time.
- $p_{\Sigma a}$ is a constant if and only if v_Σ is a constant as well.
- The set of instantaneous active currents, i_{ak}, has permanently the smallest possible instantaneous current collective value, i_Σ, to supply the instantaneous collective active power $p_{\Sigma a}$.
- For a given collective rms value of the multiphase voltage, V_Σ, the set of active currents, i_{ak}, leads to the smallest possible current collective rms value, I_Σ, to supply the collective active power P_Σ.

B.4 Instantaneous Calculation of Power Currents in Multiphase Systems

In the formulation proposed by Buchholz, the equivalent active conductance G_a cannot be calculated without introducing a certain delay, since the values of p_Σ and v_Σ cannot be properly determined until a certain time interval necessary to averaging them has passed. Therefore, the instantaneous active currents defined by (B.20) can only be properly identified in real time if steady-state conditions are assumed, i.e. if it is possible to predict the values adopted by the voltages and currents in each period. This assumption cannot be applied in practice and the conductance G_a should be permanently recalculated after varying the system operating conditions. In most cases, this calculation takes at least one grid period. Thus, a really instantaneous calculation of the active and nonactive current components is not possible using the expressions of (B.20).

In his PhD dissertation published in 1962 [14], Depenbrock introduced an instantaneous formulation to calculate a new set of active currents, defined as *power currents*, in generic multiphase systems. This formulation, named the FBD method in honour of Fryze, Buchholz and Depenbrock, was not presented at a German scientific event of wide resonance until 1980 [19], and the first complete and detailed presentation of this method was not written in English until 1993 [20]. Maybe for this reason the FDB method did not have the same repercussions as other theories formulated later.

According to Depenbrock's approach, the instantaneous current in each phase of a multi-phase system, i_k, can be divided into two instantaneous components, namely the *power* current, i_{pk}, and the *powerless* current, i_{zk}, which can by calculated by

$$i_k = i_{pk} + i_{zk} \begin{cases} i_{pk} = g_p v_{k0}; & g_p = \dfrac{p_\Sigma}{v_\Sigma^2} \\ i_{zk} = i_k - i_{pk} = i_k - g_p v_k \end{cases} \quad (B.22)$$

Currents of (B.22) can be calculated instantaneously, without any delay, even under nonperiodic conditions. The powerless currents, i_{zk}, do not contribute to the instantaneous active power, p_Σ, collectively delivered by the phases of the system. This means that the instantaneous active power supplied by the multiphase system is the same when the currents flowing by the conductors are either the power currents, i_{pk}, or the original currents, i_k, i.e.

$$\sum_{k=1}^{m} v_{k0} i_{pk} = \sum_{k=1}^{m} v_{k0} g_p v_{k0} = g_p \sum_{k=1}^{m} v_{k0}^2 = g_p v_\Sigma^2 = p_\Sigma \quad (B.23)$$

Depenbrock demonstrated that, for a given collective voltage, v_Σ, the power currents, i_{pk}, permanently present the lowest collective value, $i_{p\Sigma}$, to supply the instantaneous active power, p_Σ.

Depenbrock introduced a new set of currents, defined as the *variation* currents, i_{vk}, which relates the power currents, i_{pk}, with the active currents, i_{ak}, previously defined by Buchholz. The variation current in each conductor of the multiphase system can be calculated by

$$i_{vk} = i_{pk} - i_{ak} = i_{nk} - i_{zk} \quad (B.24)$$

The variation currents are zero only if $g_a = G_a$. In the rest of the cases

$$\sum_{k=1}^{m} v_{0k} i_{vk} = \sum_{k=1}^{m} v_{0k} i_{pk} - \sum_{k=1}^{m} v_{0k} i_{ak} = p_\Sigma - p_{\Sigma a} \neq 0 \quad (B.25)$$

However, it is always true that

$$\frac{1}{T} \int_0^T \sum_{k=1}^{m} v_{0k} i_{vk} \, dt = \frac{1}{T} \int_0^T p_\Sigma \, dt - \frac{1}{T} \int_0^T p_{\Sigma a} \, dt = P_\Sigma - P_\Sigma = 0 \quad (B.26)$$

The expression of (B.25) implies that the variation currents, i_{vk}, generate instantaneous exchanges of energy between the source and the load. However, as the expression (B.26) shows, these currents do not give rise to any net transfer of energy over a grid period.

As demonstrated by Buchholz, the active currents, i_{ak}, present the lowest rms collective value, $I_{\Sigma a}$, supplying the active power demanded by a load, P_Σ. Therefore, it is always true that

$$\sum_{k=1}^{m} i_{pk}^2 \geq \sum_{k=1}^{m} i_{ak}^2 \Leftrightarrow I_{p\Sigma} \geq I_{a\Sigma} \quad (B.27)$$

B.5 The *p-q* Theory

In 1983, Akagi and others proposed an instantaneous power theory for three-phase three-wire systems [7, 8], which was based on expressing voltages and currents as space vectors by using the Clarke transformation, which is defined by (A.14) in Appendix A. This theory is commonly known as the *p-q theory*. Later, Akagi extended the use of this theory to three-phase four-wire systems [21]. Therefore, in the p-q theory, the phase-to-neutral voltages and currents can be writen as

$$\boldsymbol{v}_{\alpha\beta 0} = \begin{bmatrix} T_{\alpha\beta 0} \end{bmatrix} \boldsymbol{v}_{abc}, \quad \boldsymbol{i}_{\alpha\beta 0} = \begin{bmatrix} T_{\alpha\beta 0} \end{bmatrix} \boldsymbol{i}_{abc}, \tag{B.28}$$

where $\boldsymbol{v}_{\alpha\beta 0} = [v_\alpha, v_\beta, v_0]^T$ and $\boldsymbol{i}_{\alpha\beta 0} = [i_\alpha, i_\beta, i_0]^T$. Using these variables on the $\alpha\beta 0$ reference frame, Akagi defined the following instantaneous powers:

$$\begin{bmatrix} p_{\alpha\beta} \\ q_{\alpha\beta} \\ p_0 \end{bmatrix} = \begin{bmatrix} M_{\alpha\beta 0} \end{bmatrix} \boldsymbol{i}_{\alpha\beta 0}, \quad \begin{bmatrix} M_{\alpha\beta 0} \end{bmatrix} = \begin{bmatrix} v_\alpha & v_\beta & 0 \\ -v_\beta & v_\alpha & 0 \\ 0 & 0 & v_0 \end{bmatrix} \tag{B.29}$$

The power $p_{\alpha\beta}$ term (named p in the original formulation) was defined as the instantaneous real power and the power term p_0 was defined as the instantaneous zero-sequence power. The addition of both powers gives rise to the instantaneous active power delivered collectively by the three phases of the three-phase system, $p_{3\phi}$, i.e.

$$p_{3\phi} = p_{\alpha\beta} + p_0 = v_\alpha i_\alpha + v_\beta i_\beta + v_0 i_0 = v_a i_a + v_b i_b + v_c i_c. \tag{B.30}$$

Logically, the unit for $p_{\alpha\beta}$, p_0 and $p_{3\phi}$ is the watt (W).

The real contribution in Akagi's power theory was the introduction of the imaginary power, $q_{\alpha\beta}$ (named q in the original formulation). The imaginary power results from the product of voltage and currents from different phases of the $\alpha\beta 0$ three-phase system ($q_{\alpha\beta} = v_\alpha i_\beta - v_\beta i_\alpha$). Consequently, the unit for the imaginary power should not have to be either W or VA or var. Akagi proposed to use the volt-ampere imaginary (vai) as a valid unit for this power [16]. Over time, however, this power has been commonly known as the instantaneous reactive power and its unit has been the volt-ampere reactive (var) in many publications. Likewise, the instantaneous real power, p, has been eventually known as the instantaneous active power.

Aredes presented in 1996 a clear interpretation of the physical meaning of the power terms resulting from Akagi's power theory when it is applied to three-phase four-wire systems [22]. Aredes concluded that the instantaneous real power, $p_{\alpha\beta}$, and the instantaneous imaginary power, $q_{\alpha\beta}$, resulted from the interaction of voltages and currents with positive and negative sequences; i.e. the zero-sequence components of voltages and currents do not contribute to $p_{\alpha\beta}$ and $q_{\alpha\beta}$. Moreover, these powers can consist of constant and oscillatory terms, namely

$$p_{\alpha\beta} = \overline{p}_{\alpha\beta} + \tilde{p}_{\alpha\beta} \tag{B.31}$$

$$q_{\alpha\beta} = \overline{q}_{\alpha\beta} + \tilde{q}_{\alpha\beta} \tag{B.32}$$

$$p_0 = \overline{p}_0 + \tilde{p}_0 \tag{B.33}$$

The constant power terms $\overline{p}_{\alpha\beta}$ and $\overline{q}_{\alpha\beta}$ result from the interaction between positive- and negative-sequence voltages and currents with the same frequency and sequence, whereas the oscillatory power terms $\tilde{p}_{\alpha\beta}$ and $\tilde{q}_{\alpha\beta}$ result from the interaction between positive- and negative-sequence voltages and currents with either a different frequency or sequence. On the other hand, v_0 and i_0 represent AC single-phase voltages and currents respectively, and their interaction always gives rise to power oscillations, $\tilde{p}_0$, which can be accompanied by a certain mean different to zero, $\overline{p}_0$. Aredes stated that the total energy flow per time unit, $p_{3\phi}$, resulting from the interaction of voltages and currents in a three-phase four-wire system, is always equal to the sum of the real power, $p_{\alpha\beta}$, and the zero-sequence power, p_0, as indicated in (B.30). He also stated that the imaginary power, $q_{\alpha\beta}$, represents an energy quantity that is being exchanged among the phases of the system. This means that $q_{\alpha\beta}$ does not contribute to the energy transfer between the two systems at any time. The term 'energy transfer' should be understood in a general manner, referring not only to the net transfer of energy from one system to another but also to the energy oscillation between them. Akagi, in a paper published in 1999 [21], ratified Aredes' interpretation of the p-q theory power terms in three-phase four-wire systems.

By inverting the matrix $[M_{\alpha\beta 0}]$ in (B.29), it is possible to find the currents to be injected into the grid, under generic voltage conditions, in order to deliver given values of instantaneous real and imaginary (active and reactive) powers set as references, i.e.

$$i^*_{\alpha\beta 0} = [M_{\alpha\beta 0}]^{-1} \begin{bmatrix} p^*_{\alpha\beta} \\ q^*_{\alpha\beta} \\ p^*_0 \end{bmatrix} \; ; \; [M_{\alpha\beta 0}]^{-1} = \frac{1}{v_\alpha^2 + v_\beta^2} \begin{bmatrix} v_\alpha & -v_\beta & 0 \\ v_\beta & v_\alpha & 0 \\ 0 & 0 & 1/v_0 \end{bmatrix} \quad (B.34)$$

The voltage matrix of (B.34) shows a singularity in the formulation of the p-q theory when applied to three-phase four-wire systems, since the calculation of i_0^* presents discontinuities if $v_0 = 0$. By removing the third row and column of $[M_{\alpha\beta 0}]^{-1}$, the zero-sequence current to be injected into the grid is forced to be equal to zero. In such a case, the positive- and negative-sequence currents to be injected into the grid to deliver a given active and reactive (real and imaginary) power setpoint, $[p^*_{\alpha\beta}, q^*_{\alpha\beta}]$, are given by

$$i^*_{\alpha\beta} = \begin{bmatrix} i^*_\alpha \\ i^*_\beta \end{bmatrix} = \frac{1}{|v|^2_{\alpha\beta}} \begin{bmatrix} v_\alpha & -v_\beta \\ v_\beta & v_\alpha \end{bmatrix} \begin{bmatrix} p^*_{\alpha\beta} \\ q^*_{\alpha\beta} \end{bmatrix} ; \; |v|^2_{\alpha\beta} = v_\alpha^2 + v_\beta^2 \quad (B.35)$$

The active currents calculated by using (B.35), i.e. those currents obtained by making $q^*_{\alpha\beta} = 0$ in (B.35), match the ones calculated by using the expression (B.22) of the FDB method for a three-phase three-wire system.

When it is assumed that $v_0 = 0$, the currents calculated by (B.35) satisfy the instantaneous power conservation principle, i.e.

$$\left(v_\alpha^2 + v_\beta^2\right)\left(i_\alpha^2 + i_\beta^2\right) = |v|^2_{\alpha\beta} |i|^2_{\alpha\beta} = s^2 = p^2_{3\phi} + q^2_{\alpha\beta} \quad (B.36)$$

In (B.36), $|v|_{\alpha\beta}$ and $|i|_{\alpha\beta}$ can be understood as instantaneous collective values for voltage and current in the $\alpha\beta$ domain. According to Buchholz's power definitions, the product of both instantaneous collective values gives rise to the instantaneous apparent power, s. As in the conventional power theory, any increment in the instantaneous imaginary power $q_{\alpha\beta}$ entails an increment in the instantaneous apparent power s as well, but it does not result in any instantaneous energy transfer between the two systems linked by $v_{\alpha\beta}$ and $i_{\alpha\beta}$.

In the moment $v_0 \neq 0$, the instantaneous collective value of the voltage will increase and hence the value of the instantaneous apparent power s will increase as well. In such a case, the p-q theory gives rise to the following inequality:

$$(v_\alpha^2 + v_\beta^2 + v_0^2)(i_\alpha^2 + i_\beta^2) = |v|_{\alpha\beta 0}^2 |i|_{\alpha\beta}^2 = s^2 \neq p_{3\phi}^2 + q_{\alpha\beta}^2 \qquad (B.37)$$

This incoherence in fulfilling the power conservation principle under generic voltage conditions gave way to new formulations of the p-q theory. In this appendix, just three examples of such modified formulations will be presented in the following. However, instantaneous power definitions in three-phase four-wire systems are still nowadays a source of controversy and new formulations are constantly proposed [23, 24].

B.6 Generalization of the p-q Theory to Arbitrary Multiphase Systems

In 1992, Willems extended the p-q theory to generic multiphase systems [9]. According to Willems' approach, the voltages and currents in an m-phase system could be represented by using m-dimensional instantaneous vectors, v and i. Thus, the instantaneous active power delivered by the systems might be calculated by means of the following dot product:

$$p_{m\phi} = v \cdot i \qquad (B.38)$$

In his formulation, the instantaneous active current vector, i_p, can be calculated by projecting i over v. He arrived at this point by setting the condition that the instantaneous current vectors i and i_p should deliver the same active power, i.e.

$$p_{m\phi} = v \cdot i = v \cdot i_p \qquad (B.39)$$

which explicitly implies that

$$i_p = \frac{v \cdot i}{|v|^2} \cdot v = \frac{p_{m\phi}}{|v|^2} \cdot v = \frac{p_{m\phi}}{|v|} \cdot \frac{v}{|v|} = |i| \cdot \vec{v} \qquad (B.40)$$

where $|v|$ and $\vec{v}$ represent the length and the unit vector of v respectively. The current and voltage vectors of a three-phase four-wire systems are graphically depicted in Figure B.2.

Therefore, the instantaneous active current vector i_p is codirectional with i. On the other hand, the instantaneous nonactive current vector could be calculated by

$$i_q = i - i_p \qquad (B.41)$$

This reactive current vector v is orthogonal, such that

$$v \cdot i_q = 0 \qquad (B.42)$$

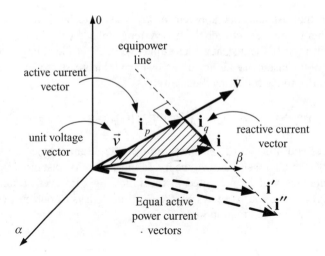

Figure B.2 Multiphase system

Therefore, the instantaneous reactive power could be associated with

$$q = |v| \cdot |i_q| \tag{B.43}$$

According to Willems' formulation, as opposed to Akagi's *p-q* theory, the zero-sequence currents affect the calculation of both the active and the reactive powers. The currents calculated using Willems' formulation to deliver a given active/reactive power set-point only match the ones calculated by Akagi's *p-q* theory in the case of three-phase three-wire systems. Otherwise, the currents resulting from both methods are different. Likewise, and considering a general case, the active currents calculated from Willems' formulation do not match the ones calculated from the FDB method. This discrepancy stems from the different conceptions of the role of the neutral conductor to power delivery in both methods.

B.7 The Modified *p-q* Theory

In 1995, Nabae and others, by using an appropriate algebraic formulation, modified the original *p-q* theory to make it suitable to be applied to three-phase four-wire systems [25]. Starting from the voltage and current vectors shown in (B.28), this modified theory proposed the following expression to calculate instantaneous real and imaginary powers:

$$\begin{bmatrix} p_{3\phi} \\ q_\alpha \\ q_\beta \\ q_0 \end{bmatrix} = [M'_{\alpha\beta 0}] \begin{bmatrix} i_\alpha \\ i_\beta \\ i_0 \end{bmatrix} \; ; \; [M'_{\alpha\beta\gamma}] = \begin{bmatrix} v_\alpha & v_\beta & v_0 \\ 0 & -v_0 & v_\beta \\ v_0 & 0 & -v_\alpha \\ -v_\beta & v_\alpha & 0 \end{bmatrix} \tag{B.44}$$

Appendix B: Instantaneous Power Theories

In (B.44), the instantaneous imaginary powers q_α, q_β and q_0 result from calculating the cross-product of $\boldsymbol{v}_{\alpha\beta 0}$ and $\boldsymbol{i}_{\alpha\beta 0}$, i.e.

$$\boldsymbol{q} = \begin{bmatrix} q_\alpha \\ q_\beta \\ q_0 \end{bmatrix} = \boldsymbol{v}_{\alpha\beta 0} \times \boldsymbol{i}_{\alpha\beta 0} \tag{B.45}$$

In this theory, the systematic algebraic formulation to study the interaction between the voltage and current vectors is not accompanied by a circuital analysis that allows the physical meaning to be revealed of each imaginary power term, as it is very difficult to identify the energy exchanges associated with each of these power terms in the actual three-phase system. Moreover, these three imaginary power terms are not linearly independent, given that

$$\begin{vmatrix} 0 & -v_\gamma & v_\beta \\ v_\gamma & 0 & -v_\alpha \\ -v_\beta & v_\alpha & 0 \end{vmatrix} = 0 \tag{B.46}$$

Therefore, there are only two characteristic terms in the set of instantaneous imaginary powers formed by q_α, q_β and q_0. The explanation about the physical relationship between these imaginary power terms is not a trivial issue, however, as their interdependence turns into evidence when the expression (B.44) is analysed. For given grid voltage conditions, this expression reveals that there are only three degrees of freedom for the currents to be injected into the grid, i.e. i_α, i_β and i_0. Therefore, only three power terms can be independently controlled by using these three independent currents. As the instantaneous real power, $p_{\alpha\beta}$, is the power of interest, only two additional independent power terms can be defined, either to deal with real or imaginary powers.

Even though q_α, q_β and q_0 are not independent, the set of currents to be injected into the grid to deliver a given real and imaginary power set-point under generic grid voltage conditions can be calculated by

$$\boldsymbol{i}^* = \begin{bmatrix} i_\alpha^* \\ i_\beta^* \\ i_0^* \end{bmatrix} = \frac{1}{|\boldsymbol{v}|_{\alpha\beta 0}^2} \begin{bmatrix} v_\alpha & 0 & v_0 & -v_\beta \\ v_\beta & -v_0 & 0 & -v_\gamma \\ v_0 & v_\beta & -v_\alpha & 0 \end{bmatrix} \begin{bmatrix} p_{3\phi}^* \\ q_\alpha^* \\ q_\beta^* \\ q_0^* \end{bmatrix} ; \quad |\boldsymbol{v}|_{\alpha\beta 0} = \sqrt{v_\alpha^2 + v_\beta^2 + v_0^2} \tag{B.47}$$

If it is set as a condition that $q_\alpha^* = q_\beta^* = q_0^* = 0$ in (B.47), the instantaneous active currents to be injected into the grid match the ones calculated by using the expression (B.40) from Willems' formulation.

Finally, it is worth pointing out that the formulation proposed by Nabae and others always accomplishes the instantaneous power conservation principle, i.e.

$$s^2 = |\boldsymbol{v}|_{\alpha\beta 0}^2 |\boldsymbol{i}|_{\alpha\beta 0}^2 = p_{3\phi}^2 + q_\alpha^2 + q_\beta^2 + q_0^2 \tag{B.48}$$

B.8 Generalized Instantaneous Reactive Power Theory for Three-Phase Power Systems

In 1996, Peng and Lai proposed a generalization of the instantaneous reactive power theory by using a very elegant vector formulation [26], in which they conferred a vector connotation to the instantaneous reactive power of a generic three-phase system. Akagi had already treated the imaginary power as a vector in his first studies and stated that the imaginary power vector was always orthogonal to the $\alpha\beta$ plane. Peng and Lai extended this approach to generic systems and defined the instantaneous reactive power vector q by means of the following cross-product:

$$q = v \times i \tag{B.49}$$

This definition for the reactive power vector is independent of the reference frame used to express v and i, provided that it is an orthogonal reference frame. In (B.49), Peng and Lai used the natural abc reference frame, so the instantaneous reactive power vector was defined as

$$q = \begin{bmatrix} q_a \\ q_b \\ q_c \end{bmatrix} = \begin{bmatrix} \begin{vmatrix} v_b & v_c \\ i_b & i_c \end{vmatrix} & \begin{vmatrix} v_c & v_a \\ i_c & i_a \end{vmatrix} & \begin{vmatrix} v_a & v_b \\ i_a & i_b \end{vmatrix} \end{bmatrix}^{\mathrm{T}} \tag{B.50}$$

This vector could also have been written in the following more compact form:

$$q = \begin{bmatrix} q_a \\ q_b \\ q_c \end{bmatrix} = \begin{bmatrix} 0 & -v_c & v_b \\ v_c & 0 & -v_a \\ -v_b & v_a & 0 \end{bmatrix} \begin{bmatrix} i_a \\ i_b \\ i_c \end{bmatrix} \tag{B.51}$$

Analysing (B.51), it is observed that

$$\begin{vmatrix} 0 & -v_c & v_b \\ v_c & 0 & -v_a \\ -v_b & v_a & 0 \end{vmatrix} = 0 \tag{B.52}$$

which leads to the same conclusions as in the presentation of the modified p-q theory proposed by Nabae and others, namely that this algebraic formulation results in a set of three instantaneous reactive powers that are not independent from each other, being difficult to explain their relationship and physical meaning.

From the following real magnitudes:

$$|v| = \sqrt{v \cdot v} = \sqrt{v_a^2 + v_b^2 + v_c^2}, \quad |i| = \sqrt{i \cdot i} = \sqrt{i_a^2 + i_b^2 + i_c^2}$$

$$p_{3\phi} = v \cdot i, \quad q = \sqrt{q_a^2 + q_b^2 + q_c^2} \tag{B.53}$$

Peng and Lai proposed the following expressions, where i_p and i_q are the instantaneous active and reactive current vectors respectively, s is the instantaneous apparent power and λ is the

Appendix B: Instantaneous Power Theories

instantaneous power factor:

$$i_p = \frac{p_{3\phi}}{|v|^2}v; \quad i_q = \frac{q \times v}{|v|^2}; \quad s = |v| \, |i|; \quad \lambda = \frac{p_{3\phi}}{s} \quad \text{(B.54)}$$

Even though some of these expressions had already been proposed in previous formulations, the definition of i_q is actually a novelty in this generalized approach of the reactive power theory.

Peng and Lai stated and justified four theorems that collect the main properties of the instantaneous components of the current and the power in three-phase systems. These theorems are adapted to the nomenclature used in this appendix as follows:

Theorem 1. The three-phase current vector, i, is always equal to the sum of the instantaneous active current vector, i_p, and the instantaneous reactive current vector, i_q, i.e., $i = i_p + i_q$.

Theorem 2. The instantaneous reactive current vector, i_q, is orthogonal to v and the instantaneous active current vector, i_p, is parallel to v, namely, $v \cdot i_q = 0$ and $v \times i_p = 0$, with $\times$ and $\cdot$ as the cross and dot products respectively.

Theorem 3. $|i|^2 = |i_p|^2 + |i_q|^2$, $s^2 = p_{3\phi}^2 + q^2$ and $|i|^2 = (p_{3\phi}^2 + q^2)/|v|^2$, where $|i_p|$ and $|i_q|$ represent the norm, or collective values or 'length', of the instantaneous active and reactive current vectors.

Theorem 4. If $i_q = 0$, then the norm of i, namely $|i|$, becomes minimal for transmitting the same instantaneous active power and the maximal instantaneous power factor is achieved, namely $\lambda = 1$.

B.9 Summary

This appendix has presented a short review of some of the instantaneous power theories used to determine the reference currents for power converters connected to three-phase systems. The appendix did not aim to provide an exhaustive study on the instantaneous current components and power definitions. For this reason, many of the theories and methods used in applications such as power conditioners or power analysers have not been mentioned for the sake of not extending this appendix unnecessarily.

The current analysis proposed by Buchholz, and its subsequent extension proposed by Depenbrock through the FBD method, leads to the maximum efficiency in delivering instantaneous active power in a generic multiphase system in which all the electrical conductors are treated as equals, including the neutral conductor. It is worth remarking here that this is not the case in most actual electrical networks, where the neutral conductor is considered as an auxiliary conductor with a lower section than the phase conductors and devoted to reduce voltage transients and allow the flow of current during asymmetrical faults. Medium voltage distribution systems and high-voltage transmission systems do not have any neutral conductor but their phase voltages are referenced to earth potential. Therefore, asymmetrical faults will result in currents flowing through the earth circuit, which does not present the same impedance as the phase conductors.

The p-q theory proposed by Akagi is suitable to be used in three-phase three-wire systems and in four-wire systems in which, as a condition, the neutral current is set equal to zero. The concept of the instantaneous imaginary power meant a significant contribution to the study of three-phase systems in the time domain. The extrapolation of this theory to three-phase four-wire systems resulted in some singularities that gave way to the proposal of new theories and methods.

References

[1] Filipski, P. S., Baghzouz, Y. and Cox, M. D., 'Discussion of Power Definitions Contained in the IEEE Dictionary'. *IEEE Transactions on Power Delivery*, **9**, July 1994, 1237–1244.
[2] DIN 40110 Teil 2, *Mehrleiter-Stromkreise*, Berlin: Beuth-Verlag, November 2002.
[3] IEEE Std 1450-2010, *IEEE Standard Definitions for Power Measurement of Electric Power Quantities under Sinusoidal, Nonsinusoidal, Balanced or Unbalanced Conditions*, IEEE, 19 March 2010. ISBN 978-0-7381-6059-7.
[4] E.ON Netz Gmbh, 'Grid Code for High and Extra High Voltage', April 2006.
[5] E.ON Netz GmbH, 'Requirements for Offshore Grid Connections in the E.ON Netz Network', April 2008.
[6] Fryze, S., 'Wirk-, Blind- und Scheinleistung in elektrischen Stromkreisen mit nicht-sinusförmigem Verlauf von Strom und Spannung'. *ETZ-Arch. Elektrotech*, **53**, 1932, 596–599, 625–627, 700–702.
[7] Akagi, H., Kanazawa, Y. and Nabae, A., 'Generalized Theory of the Instantaneous Reactive Power in the Tree-Phase Circuits'. In *Proceedings of the IEEE International Power Electronic Conference (IPEC'83)*, 1983, pp. 1375–1386.
[8] Akagi, H., Kanazawa, Y. and Nabae, A., 'Instantaneous Reactive Power Compensator Comprising Switching Devices without Energy Storage Components'. *IEEE Transactions on Industry Applications*, **20**, 1984, 625–630.
[9] Willems, J. L., 'A New Interpretation on the Akagi–Nabae Power Components for Nonsinuoidal Three-Phase Situations'. *IEEE Transactions on Instruments and Measurements*, **41**, August 1992, 523–527.
[10] Peng, F. Z., Ott Jr, G. W. and Adams, D. J., 'Harmonic and Reactive Power Compensation Based on the Generalized Reactive Power Theory for Three-Phase Four-Wire Systems'. *IEEE Transactions on Instruments and Measurements*, **45**, February 1996, 293–297.
[11] Nabae, A., Cao, L. and Tanaka, A., 'A Universal Theory of Instantaneous Active–Reactive Current and Power Including Zero-Sequence Component'. In *Proceedings of the IEEE ICHQP'96*, October 1996, pp. 90–95.
[12] Kim, H. S. and Akagi, H., 'The Instantaneous Power Theory on the Rotating p-q-r Reference Frames'. In *Proceedings of the IEEE PEDS'99 Conference*, July 1999, pp. 422–427.
[13] Buchholz, F., *Das Begriffsystem Rechtleistung. Wirkleistung, totale Blindleistung*, Munich, Germany: Selbstverlag, 1950 (in German).
[14] Depenbrock, M., 'Untersuchungen über die Spannungs und Leistungsverhältnisse bei Umrichtern ohne Energiespeicher', Dr. Ing. Dissertation, Technisch Universität Hannover, Germany, 1962 (in German).
[15] Depenbrock, M., Staudt, V. and Wreder, H., 'A Concise Assessment of Original and Modified Instantaneous Power Theory Applied to Four-Wire Systems'. In *Proceedings of the Power Conversion Conference (PCC'02)*, 2002, pp. 60–67.
[16] Akagi, H., Watanabe, E. H. and Aredes, M., *Instantaneous Power Theory and Applications to Power Conditioning*, Wiley–IEEE Press, March 2007. ISBN 10: 0470107618.
[17] Fryze, S., 'Active, Reactive and Apparent Power in Non-sinusoidal Systems'. *Przeglad Elektrot.*, **7**, 1931, 193–203 (in Polish).
[18] Buchholz, F., 'Die Drehstrom-Scheinliestung bei Ungleichmassiger Belastung Der Drei Zweige'. *Licht und Kraft*, **2**, 1922, 9–11 (in German).
[19] Depenbrock, M., 'Wirk- und Blindleistungen periodscher Ströme in Ein- u. Mehrphasensystemen mit periodischen Spannungen beleibiger Kurvenform'. In *ETG-Fachberichte 6 - Blindleistung*, 1980, pp. 17–59.
[20] Depenbrock, M., 'The FBD-Method, a Generally Applicable Tool for Analysing Power Relations'. *IEEE Transactions on Power Systems*, **8**, May 1993, 381–386.

[21] Akagi, H., Ogasawara, S. and Kim, H., 'The Theory of Instantaneous Power in Three-Phase Four-Wire Systems: A Comprehensive Approach'. In *Conference Records of the IEEE–IAS Annual Meeting*, 1999, pp. 431–439.

[22] Aredes, M., 'Active Power Line Conditioners', PhD Dissertation, Technisch Universität Berlin, Germany, 1996.

[23] Depenbrock, M., Staudt, V. and Wreder, H., 'A Theoretical Investigation of Original and Modified Instantaneous Power Theory Applied to Four-Wire Systems'. *IEEE Transactions on Industrial Applications*, **39**, July–August 2003, 1160–1167.

[24] Staudt, V. and Wreder, H., 'Compensation Strategies for Four-Conductor Systems'. In *Proceedings of the International Conference on Work, Power Definitions and Measurements*, October 2003, pp. 139–146.

[25] Nabae, A., Nokano, H. and Togasawa, S., 'An Instantaneous Distortion Current Compensator without any Coordinate Transformation'. In *Proceedings of the International Power Electronics Conference*, 1995, pp. 1651–1655.

[26] Peng, F. and Lai, J., 'Generalized Instantaneous Reactive Power Theory for Three-Phase Power Systems'. *IEEE Transactions on Instrumentation and Measurement*, **45**, February 1996, 293–297.

Appendix C

Resonant Controller

C.1 Introduction

The resonant controller can be obtained either by applying the internal model principle or by using the coordinate transformation. The two methods are briefly described in the following.

C.2 Internal Model Principle

The internal model principle states that if the models of the reference and of the disturbance are included in the feedback control loop a good reference tracking and a good disturbance rejection capability is ensured. If the goal is to track and reject periodic signals that can be decomposed into sinusoidal components (harmonics), this procedure results in the design of controllers that have a pair of poles on the imaginary axis at the frequencies of the harmonics to track and/or reject. In fact, the Laplace transform of a sinusoidal signal as the normalized grid voltage, i.e. a disturbance in the current control loop as shown in Figure C.1, is

$$E(s) = \frac{\omega}{s^2 + \omega^2} \tag{C.1}$$

The transfer function (C.1) will be used to demonstrate the effectiveness of a resonant controller tuned on the fundamental frequency to null the error due to the grid voltage.

Considering the block diagram as depicted in Figure C.1, the current, the consequence of the disturbance, will be defined as

$$I(s) = -\frac{G_f(s)}{1 + G_f(s)G_d(s)G_c(s)} E(s) \tag{C.2}$$

Considering that

$$G_f(s) = \frac{1}{Ls + R} \tag{C.3}$$

Grid Converters for Photovoltaic and Wind Power Systems   Remus Teodorescu, Marco Liserre, and Pedro Rodríguez
© 2011 John Wiley & Sons, Ltd

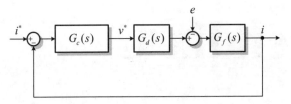

Figure C.1 The current control loop of an inverter: $G_c(s)$ is the controller, $G_d(s)$ is the delay of PWM and of the computational device and $G_f(s)$ models the filter and the grid

$$E(s) = \frac{\omega}{s^2 + \omega^2} \tag{C.4}$$

$$G_c(s) = \frac{k_P s^2 + k_I s + k_P \omega^2}{s^2 + \omega^2} \tag{C.5}$$

$$G_d(s) = \frac{1}{(1.5T_s)s + 1} \tag{C.6}$$

where L and R are respectively the total inductance and resistance of the grid and the filter, T is the sample time and k_P and k_I are the proportional and resonant gain. k_P is tuned in order to ensure that the overall system has a damping factor of 0.707. A high k_I is usually adopted in order to obtain sufficient attenuation of the tracking error in case the frequency of the grid is subjected to changes. However, a high k_I can cause too high an overshoot. The optimum gain k_I can be adopted considering that the grid frequency is stiff and it is only allowed to vary in a narrow range, typically $\pm 1\%$
(C.2) results into:

$$\frac{\Delta I(s)}{E(s)} = \frac{(s^2 + \omega^2) \cdot (0.5T_s s + 1)}{(Ls + R) \cdot (s^2 + \omega^2) \cdot (0.5T_s s + 1) + (k_P s^2 + k_I s + k_P \omega^2)} \tag{C.7}$$

It can be proven that (C.7) is zero for $s = j\omega$.

C.3 Equivalence of the PI Controller in the dq Frame and the P+Resonant Controller in the $\alpha\beta$ Frame

The process can be derived by inverse transforming the synchronous controller back to the stationary $\alpha\beta$ frame $G_{dq}(s) \rightarrow G_{\alpha\beta}(s)$. The inverse transformation can be performed by using the following 2×2 matrix:

$$G_{\alpha\beta}(s) = \frac{1}{2}\begin{bmatrix} G_{dq1} + G_{dq2} & jG_{dq1} - jG_{dq2} \\ -jG_{dq1} + jG_{dq2} & G_{dq1} + G_{dq2} \end{bmatrix} \tag{C.8}$$

$$G_{dq1} = G_{dq}(s + j\omega) \tag{C.9}$$

$$G_{dq2} = G_{dq}(s - j\omega) \tag{C.10}$$

Appendix C: Resonant Controller

Given that $G_{dq}(s) = k_I/s$ and $G_{dq}(s) = k_I/(1+(s/\omega_c))$, the equivalent controllers in the stationary frame for compensating positive-sequence feedback error are therefore expressed respectively as

$$G_{\alpha\beta}^{+}(s) = \frac{1}{2} \begin{bmatrix} \dfrac{2k_I s}{s^2+\omega^2} & \dfrac{2k_I \omega}{s^2+\omega^2} \\ -\dfrac{2k_I \omega}{s^2+\omega^2} & \dfrac{2k_I s}{s^2+\omega^2} \end{bmatrix} \quad \text{(C.11)}$$

$$G_{\alpha\beta}^{+}(s) \approx \frac{1}{2} \begin{bmatrix} \dfrac{2k_I \omega_c s}{s^2+2\omega_c s+\omega^2} & \dfrac{2k_I \omega_c \omega}{s^2+2\omega_c s+\omega^2} \\ -\dfrac{2k_I \omega_c \omega}{s^2+2\omega_c s+\omega^2} & \dfrac{2k_I \omega_c s}{s^2+2\omega_c s+\omega^2} \end{bmatrix} \quad \text{(C.12)}$$

Similarly, for compensating negative sequence feedback error, the required transfer functions are expressed as

$$G_{\alpha\beta}^{-}(s) = \frac{1}{2} \begin{bmatrix} \dfrac{2k_I s}{s^2+\omega^2} & -\dfrac{2k_I \omega}{s^2+\omega^2} \\ \dfrac{2k_I \omega}{s^2+\omega^2} & \dfrac{2k_I s}{s^2+\omega^2} \end{bmatrix} \quad \text{(C.13)}$$

$$G_{\alpha\beta}^{-}(s) \approx \frac{1}{2} \begin{bmatrix} \dfrac{2k_I \omega_c s}{s^2+2\omega_c s+\omega^2} & -\dfrac{2k_I \omega_c \omega}{s^2+2\omega_c s+\omega^2} \\ \dfrac{2k_I \omega_c \omega}{s^2+2\omega_c s+\omega^2} & \dfrac{2k_I \omega_c s}{s^2+2\omega_c s+\omega^2} \end{bmatrix} \quad \text{(C.14)}$$

Comparing (C.11) and (C.12) with (C.13) and (C.14), it is noted that the diagonal terms of $G_{\alpha\beta}^{+}(s)$ and $G_{\alpha\beta}^{-}(s)$ are identical, but their nondiagonal terms are opposite in polarity. This inversion of polarity can be viewed as equivalent to the reversal of rotating direction between the positive- and negative-sequence synchronous frames.

Combining the above equations, the resulting controllers for compensating both positive- and negative-sequence feedback errors are expressed as

$$G_{\alpha\beta}(s) = \frac{1}{2} \begin{bmatrix} \dfrac{2k_I s}{s^2+\omega^2} & 0 \\ 0 & \dfrac{2k_I s}{s^2+\omega^2} \end{bmatrix} \quad \text{(C.15)}$$

$$G_{\alpha\beta}(s) \approx \frac{1}{2} \begin{bmatrix} \dfrac{2k_I \omega_c s}{s^2+2\omega_c s+\omega^2} & 0 \\ 0 & \dfrac{2k_I \omega_c s}{s^2+2\omega_c s+\omega^2} \end{bmatrix} \quad \text{(C.16)}$$

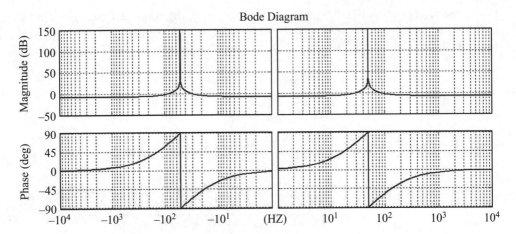

Figure C.2 Positive- and negative-sequence Bode diagrams of a PR controller

Bode plots representing (C.15) are drawn in Figure C.2, where their error-eliminating ability is clearly reflected by the presence of two resonant peaks at the positive frequency ω and negative frequency $-\omega$. Note that if (C.11) or (C.12) ((C.13) or (C.14)) is used instead, only the resonant peak at ω ($-\omega$) is present since those equations represent PI control only in the positive-sequence (negative-sequence) synchronous frame. Another feature of (C.15) and (C.16) is that they have no cross-coupling nondiagonal terms, implying that each of the α and β stationary axes can be treated as a single-phase system. Therefore, the theoretical knowledge described earlier for single-phase PR control is equally applicable to the three-phase functions expressed in (C.15) and (C.16).

Index

Page numbers in *italics* denotes a table/diagram

abnormal grid conditions (PV) 35–7
   frequency deviations 36, *36*
   reconnection after trip 36, *36*
   voltage deviations 35–6, *35*
AC current control 315–17
AC voltage control (WTS) 205, 210–19
active currents in multiphase systems 364–7, *365*
active damping 210, 300
active frequency drift (AFD) 104–5, *105*, 108, *120*
active frequency drift with positive feedback (AFDPF) 106
active frequency drift with pulsating chopping factor (AFDPCF) 107–8, *107*, *120*
active islanding detection methods 98, 104–20, *120*, 121
   frequency drift 104–10
      active frequency drift (AFD) 104–5, *105*, 108, *120*
      active frequency drift with pulsating chopping factor (AFDPCF) 107–8, *107*, *120*
      General Electric (GE) frequency shift (GEFS) 108, *108*, *120*
      reactive power variation (RPV) 108, 110, *120*
      Sandia frequency shift (SFS) 106, 107, *120*
      slip-mode frequency shift (SMS) 105–6, *105*, 108, *120*
   grid impedance estimation (GIE) 110–14
      by active reactive power variation (GIE-ARPV) 113–14, *113*
      harmonic injection (HI) 111–12, *112*, *113*, *120*
      PLL-based 114–19, *115*, *116–18*, *119*
   voltage drift 110
      Sandia voltage shift (SVS) 110, *120*
active power control in normal operation (WTS) 152–5
   frequency control 154–5, *154*
   power curtailment 153–4, *153*, *154*
active power set point, estimation of the maximum 277–9
active reactive power variation
   grid impedance estimation by (GIE-ARPV) 113–14, *113*
active rectifiers
   studies on control of under unbalanced grid voltage conditions 239–43, *239*
adaptive band-pass filter (ABPF) 74
adaptive filtering, PLLs based on 68–80
   enhanced PLL (EPLL) 70–2, *71*, 80
   second order adaptive filter 72–4, *73*
   second-order generalized integrator (SOGI) 74–8, *76*, *79*
adaptive noise cancelling (ANC) concept 68, *69*, 72
adaptive notch filter (ANF) 68, 70, 74

anti-islanding (AI) requirements (PV systems) 28, 38–41
   defined by IEC 62116 40
   defined by IEEE 1547/UL 1741 39–40, *39*
   defined by VDE 0126-1-1 40–1, *41*
arbitrary multiphase systems and p-q theory 371–2
automatic voltage regulation (AVR) 155, 162
average active-reactive control (AARC) 262–3, 264–7, *265, 266*
average power factor 38
averaging
   use of and current control 315–17
axial-flux generators 128

back-to-back converter
   full-power 127, *127*
   reduced power 124, 126–7, *126*
balanced positive-sequence control (BPSC) 263–4, 264–7, *265, 266*, 267
bi-polar modulation (BP) 7–8, *7*
boost converters 6

cage induction generators 146
capacitance impedance 299
cascade-bridge multilevel configuration 343, 344–6, *344, 345*, 347
cascaded control of the DC voltage through the AC current 213–16, *213, 215, 216*
China 1
   Grid Codes (GCs) *147*, 150–1
   power curtailment requirements (WTS) 154, *154*
chopping factor
active frequency drift with pulsating (AFDPCF) 107–8, *107*, 120
Clarke transformation 171, 172, 369
closed-loop error transfer function 55
closed-loop phase transfer function 55
collective value concept 365–6
Conergy Refuso 27
control of grid converter *see* grid converter control (for WTS)

control structures
   and PV inverters 27–8, *27*
   for unbalanced current injection 243, 244–56
crowbar system 139
CSC (current-stiff) 129, *130*, 131–2, 135
current control *see* grid current control
current harmonic requirements 313–15, *315*
   for PV 37–8, *37, 38*
current injection, control structures for unbalanced 243, 244–56
   decoupled double synchronous reference frame (DDSRF) 245–51, *245, 247, 248, 249, 250, 252*
   resonant controllers for unbalanced 251–6, *253, 255*
current limitation, flexible power control with 269–85
current vector
   locus of under unbalanced grid conditions 270–2, *271*
current-stiff (CSC) 129, *130*, 131–2, *135*

damping 296, 299–304
   active 210, 300
   instability of the undamped current control loop 300–1, *301*
   passive damping of the current control loop 300, 301–4, *302, 303, 304*
Danfoss Solar 23, 27
DC bypass (DCBP), full-bridge inverter with 17–19, *17, 18*, 21
DC current injection
   grid requirements for PV 37, *37*
DC voltage control (WTS) 205, 210–19
   cascaded control of through AC current 213–15, *213, 215, 216*
   management of 211–12, *211*
   PI-based voltage control design example 217–19, *218, 219*
   tuning procedure of the PI controller 216–17
deadbeat control 320–4, *322, 324, 325*
decoupled double synchronous reference frame (DDSRF)

current controller for unbalanced current
    injection 245–51, *245*, *247*, *248*,
    249, *250*, *252*
decoupled double synchronous reference
    frame PLL (DDSRF-PLL) 186–95,
    202
  analysis of 189–93
  decoupling network 187–9, *188*, *189*, 193
  double synchronous reference frame
      (DSRF) 186–7, *186*
  relationship between DSOGI-FLL and
      198–200
  structure and response of 193–5, *193*,
      194
Denmark
  Grid Codes (GCs) *147*, 148
  power curtailment requirements (WTS)
      153, *153*
  reactive power control (WTS) *156*, 157
  wind energy penetration 2, 3
diode bridge 138
direct power control (DPC) 205, 221, 226–7,
    227, *228*
direct torque control (DTC) 137, 226
disconnection times
  for frequency variations 36, *36*
  for voltage variations 35–6, *35*
discrete Fourier transform (DFT) 44, 48–50,
    49
distributed power generation systems
    (DPGSs) 93
Distribution System Operators (DSOs) 145
DKE 31
DOE (US Department of Energy) 2
double second-order generalized integrator
    FLL *see* DSOGI-FLL
double synchronous reference frame (DSRF)
    186–7, *186*
  current controller for unbalanced current
      injection 245–50, *245*, *248*, *249*
doubly fed induction generator (DFIG) 124,
    126, *126*, 138–9, *139*, 141, 146
$dq0$ on the synchronous reference frame
    359–60, *359*
droop control 212, 231–3, *232*, *233*, *234*

DSOGI-FLL (double second-order
    generalized integrator FLL) 195–202
  relationship between DDSRF-PLL and
      198–200
  response 200–2, *201*
  structure of 197, *197*
  use of FLL 200, *200*
dynamic hysteresis 52

effective apparent power 366
*Electromagnetic Compatibility (EMC – low
    frequency)* (IEC 61000) 34, *38*, 38
EMI 290
EN 50160: *Public Distribution Voltage
    Quality* 34–5, *35*
Enercon 146
enhanced PLL (EPLL) 70–2, *71*, 80
ENS safety device 33
Euler formula 47, 66, 67
European Wind Energy Association
    (EWEA) 2, 165, 166
external switched capacitor (ESC) detection
    97

fast Fourier transform (FFT) 49–50
fault ride-through (FRT) 146, 165
FBD method 362, 367–8, 375
Federal Energy Regulatory Commission
    (FERC) 150
field oriented control (FOC) 137, 138
filters 289–308
  AC and DC passive elements *291*, 291
  adaptive *see* adaptive filtering
  conventional 68
  damping 296, 299–304
    active 210, 300
    instability of the undamped current
        control loop 300–1, *301*
    passive damping of the current control
        loop 300, 301–4, *302*, *303*, *304*
  design considerations 291–6
  harmonic attenuation 295
  influence of capacitance impedance 299
  installed reactive power of 295–6
  low-pass LCL filter 289, 290–1, *291*

filters (Cont.)
  nonlinear behavior of 305–8, *305*
  practical examples of grid interactions
    and LCL 296–9, *297*, *298*, *299*
  and resonance frequency 295, 296, 297,
    298–9, *298*
  ripple analysis and converter-side
    inductor choice 294
  role of 289
  rules for capacitors and inductors 292,
    293, 294
  topologies 290–1
  tuned LC filter 289, 290–1, *291*
finite impulse response (FIR) filter 65
flexible positive-and negative-sequence
    control *see* FPNSC
flexible power control with current limitation
    269–85
FLL (frequency-locked loop 45, 80–9
  DSOGI *see* DSOGI-FLL
  SOGI *see* SOGI-FLL
Fortescue transformation 195, 354, 355
Fourier analysis 45–51, 64, 363
Fourier series 44, 45–7, *46*
Fourier transform 307
  discrete (DFT) 44, 48–50, *49*
  fast (FFT) 49–50
  inverse discrete (IDFT) 49
  recursive discrete (RDFT) 44, 50–1, *51*
FPNSC (flexible positive-and
    negative-sequence control) strategy
    267–85
  estimation of the maximum active and
    reactive power set-point 277–9
  estimation of the maximum current in
    each phase 274–7, *274*, *276*
  instantaneous value of the three-phase
    currents 272–4, *273*
  locus of the current vector under
    unbalanced grid conditions
    270–2
  performance of 279–85, *279*
    injection of negative-sequence reactive
      power (Case B) 282–3, *283*
    injection of positive-sequence reactive
      power (Case A) 281–2, *281*
    injection of simultaneous-and
      negative- sequence reactive
      power (Case C) 283–4, *284*
frequency control (WTS) 154–5, *154*
frequency deviation
  and grid requirements (WTS) 151–2, *152*
  and grid requirements (PV) 36, *36*
frequency deviation detector 108
frequency domain
  symmetrical components in the 353–5,
    354
frequency-domain detection methods 44,
    305–7
  discrete Fourier transform (DFT) 44,
    48–50, *49*
  Fourier series 44, 45–7, *46*
  recursive discrete Fourier transform
    (RDFT) 44, 50–1, *51*
frequency drift methods 104–10
  active frequency drift (AFD) 104–5, *105*,
    108, *120*
  active frequency drift with pulsating
    chopping factor (AFDPCF) 107–8,
    107, *120*
  General Electric (GE) frequency shift
    (GEFS) 108
  reactive power variation (RPV) 108, 110,
    120
  Sandia frequency shift (SFS) 106, 107,
    120
  slip-mode frequency shift (SMS) 105–6,
    105, 108, *120*
frequency-locked loop *see* FLL
frequency monitoring 98–9
frequency/phase-angle generator (FPG) 61
full-bridge (FB) inverter 7–11
  bipolar modulation (BP) 7–8, *7*
  with DC bypass (FB-DCBP) (Ingeteam)
    17–19, *17*, *18*, 21
  hybrid modulation 9–11, *10*
  unipolar (UP) modulation 8–9, *9*
full power converter 141–2, *142*
  back-to-back 127, *127*
full-bridge zero voltage rectifier (FB-ZVR)
    19–20, *19*, *20*, 21
full-scale converter (FSC) 146

Gaia project 229–30, *230*
gearboxes, reduced 128
General Electric (GE) frequency shift (GEFS) 108, *108*, *120*
generalized integrator (GI) 75–7, 326
generator-side control (WTS) 136–9
   doubly fed induction generator (DFIG) 124, 126, *126*, 138–9, *139*, 141, 146
   squirrel cage induction generator 126–7, *126*, 136–7, *137*
   synchronous 137–8, *137*
Germany
   frequency control requirements (WTS) 154
   frequency and voltage deviation (WTS) 151–2
   Grid Codes (GCs) *147*, 148–9
   grid disturbances 158–60, *159*
   power curtailment requirements (WTS) 153
   reactive power control (WTS) 155, *156*, 157
   wind energy penetration 2
Grid Codes (GCs) 3, 43–4, 93, 145–51, 158
   China *147*, 150–1
   Denmark *147*, 148
   development of 145–6
   evolution and factors influencing 146–51
   and filter design 289
   general message of 146
   Germany *147*, 148–9
   harmonization of 164–5
   importance of 145
   and inertia emulation (IE) 149, 165–6, *166*
   Ireland *147*, 150
   and local voltage control 165
   and power oscillation dumping (POD) 166
   requirements expected to be included in future 165–6
   Spain *147*, 149, 165–6
   UK *147*, 149–50
   United States *147*, 150
grid converter control (for WTS) 205–34
   AC and DC voltage control 205, 210–19

   cascaded control of the DC voltage through the AC current 213–15, *213*, *215*, *216*
   management of the DC link voltage 211–12, *211*
   PI-based voltage control design example 217–19, *218*, *219*
   tuning procedure of the PI controller 216–17
   direct power control (DPC) 205, 221, 226–7, *227*, *228*
   and droop control 212, 231–3, *232*, *233*, *234*
   mathematical model of L-filter inverter 207–9, *207*
   mathematical model of LCL-filter inverter 209–10, *210*
   and stand-alone or micro-grid operation 228–31, *229*, *230*
   voltage oriented control (VOC) 219–26, *220*, 319
      stationary frame
         PQ closed-loop control 224, *225*
         PQ open-loop control 222, *224*
      synchronous frame
         PQ closed-loop control 222, *223*
         PQ open-loop control 221, *222*
      virtual-flux-based approach 234–6, *236*
grid converter control (under grid faults) 237–85
   control of active rectifiers 239–43, *239*
   control structures for unbalanced current injection 243, 244–56
      decoupled double synchronous reference frame (DDSRF) 245–51, *245*, *247*, *248*, *249*, *250*, *252*
      resonant controllers 251–6, *253*, *255*
   flexible power control with current limitation 269–85
      estimation of the maximum active and reactive power set-point 277–9
      estimation of the maximum current in each phase 274–7, *274*, *276*

grid converter control (*Cont.*)
    instantaneous value of the three-phase currents 272–4, *273*
    locus of the current vector under unbalanced grid conditions 270–2, *271*, *272*
    performance of the FPNSC 279–85, *279*
  and grid synchronization system 244
  overview of for unbalanced grid voltage conditions 238–44
  power control under unbalanced grid conditions 256–69
    average active-reactive control (AARC) 262–3, 264–7, *265*, *266*
    balanced positive-sequence control (BPSC) 263–4, 264–7, *265*, *266*, 267
    flexible positive-and negative-sequence control (FPNSC) 267– 9
    instantaneous active-reactive control (IARC) 258–9, 262, 264–7, *265*, *266*
    positive-and negative-sequence control (PNSC) 260–2, 264–7, *265*, *266*
grid current control 27, 313–51
  classification of methods *316*
  deadbeat control 320–24, *322*, *324*, *325*
  harmonic compensation 329–35
  harmonic requirements 313–15, *315*
  linear current control with separated modulation 315–35
  modulation techniques 335–47
    interleaved 347, *347*, *348*
    multilevel 343–6, *343*, *344*, *345*, *346*
    single-phase 338–40, *338*, *339*, *341*
    three-phase 340, *342*
  operating limits of current-controlled converter 347–8
  PI-based control 317–20, *317*, *318*, *319*, *320*, *321*, 328
  practical example 350–51
  resonant control 326–9, *326*, *327*, *328*, *329*
  use of averaging 315–17

grid disturbances 99
grid fault(s)
  and Grid Codes 158–64
  grid converter control under *see* grid converter control (under grid faults)
  phase-to-phase 173–4, *174*, 175–6, *175*, *176*, 177
  three-phase voltage vector under 169–70, 171–80, *1670*
  unbalanced grid voltages during a 175–7, *175*
  voltage sags 158, 177–80, *178*, *179*, *180*
grid filters *see* filters
grid impedance estimation (GIE) 110–14
  by active reactive power variation (GIE-ARPV) 113–14, *113*
  by harmonic injection (HI) 111–12, *112*, *113*, *120*
grid monitoring 28, 44
grid requirements (for PV) 31–41
  abnormal grid conditions 35–7
    frequency deviations 36, *36*
    reconnection after trip 36, *36*
    voltage deviations 35–6, *35*
  anti-islanding 28, 38–41
    defined by IEC 62116 40
    defined by IEEE 1547/UL 1741 39–40, *39*
    defined by VDE 0126-1-1 40–1, *41*
  harmonization of 41
  international regulations 32–5
  power quality 37–8
    average power factor 38
    current harmonics 37–8, *37*, *38*
    DC current injection 37, *37*
grid requirements (WTS) 145–66
  active power control in normal operation 152–5
    frequency control 154–5, *154*
    power curtailment 153–4, *153*, *154*
  behaviour under grid disturbances 158–64
  frequency and voltage deviation under normal operation 151–2, *152*
  future trends 165–6
  Grid Codes *see* Grid Codes

Index 391

harmonization of Grid Codes 164–5
reactive power control in normal operation 155–8
grid-resident detection 96–7, *97*
grid synchronization (single-phase systems) 43–89, 244
   frequency-domain detection methods 44
      discrete Fourier transform (DFT) 44, 48–50, *49*
      Fourier series 44, 45–7, *46*
      recursive discrete Fourier transform (RDFT) 44, 50–1, *51*
   PLLs based on adaptive filtering 68–80
   PLLs based on in-quadrature signals 59–68, *60*
   and SOGI-FLL 80–9, *82*, *86*
grid synchronization (three-phase power converters) 169–202, 244
   decoupled double synchronous reference frame PLL (DDSRF-PLL) 186–95, 202
   double second-order generalized integrator FLL (DSOGI-FLL) 195–202
   synchronous reference frame PLL (SRF-PLL) 180–6, *183*, *185*, 202
   three-phase voltage vector under grid faults 169–70, *170*, 171–80
   usage of advanced detection systems 170

H-bridge based boosting PV inverter
   with high-frequency transformer 25, *26*
   with low-frequency transformer 25, *26*
H-bridge converters 6–21, *7*
   full-bridge (FB) inverter 7–11, *7*, *9*, *10*
   full-bridge inverter with DC bypass (FB-DCBP) (Ingeteam) 17–19, *17*, *18*, 21
   full-bridge zero voltage rectifier (FB-ZVR) 19–20, *19*, *20*, 21
   H5 inverter (SMA) 11–12, *11*, *12*, 14, 21
   HERIC inverter (Sunways) 13–14, *13*, *14*, 15, 20, 21
   REFU inverter 15–17, *15*, *16*, 21
   summary 21

H5 inverter (SMA) 11–12, *11*, *12*, 14, 21
half-bridge inverter, NPC 21–3, *21*, *22*, 25
harmonic attenuation, of LCL filter 295
harmonic compensation 329–35
   and stationary frames 330, 331–35, *334*, *335*
   and synchronous frames 330, 331, *331*, *332*
harmonic detection (HD) 99–103, *102*, *103*
harmonic injection (HI) 111–12, *112*, *113*, *120*
HERIC inverter (Sunways) 13–14, *13*, *14*, 15, 20, 21
Hilbert transform 64–5, *65*
hold range 58
hybrid modulation 9–11, *10*
hysteresis 244

IEC 61000: *Electromagnetic Compatibility (EMC – low frequency)* 34, 38, *38*
IEC 61727: *Characteristics of Utility Interface* 33, 35, 36, *36*, 37, *37*, 38
IEC 62116: *Testing Procedure of Islanding Prevention Measures* 33, 40, 41
IEC (International Electrotechnical Commission) 31, 166
IEEE 929-2000: *Recommended Practice for Utility Interface of Photovoltaic (PV) Systems* 32
IEEE 1547: *Interconnection of Distributed Generation* 32–3, *35*, *36*, 39–40, *390*
IEEE 1547.1: *Conformance Test Procedures for Equipment Interconnecting Distributed Resources with Electric Power Systems* 32, 33
IEEE 1574 37, *37*, 38, 41
imaginary power 362, 369–70, 372–3, 375–6
impedance emulation approach 233
in-quadrature signal generation
   phase detection based on 59–63
   some PLLs based on 63–8
inertia emulation (IE)
   and Grid Codes (GCs) 149, 165–6, *166*
Ingecon® Sun TL series 19

instantaneous active-reactive control (IARC) 258–9, 262, 264–7, *265*, *266*
instantaneous power theories 205, *206*, 219, 240, 256, 361–76
instantaneous value of three-phase currents 272–4, *273*
intelligent power modules (IPM) 29
Interconnection of Distributed Generation (IEEE 1547) 32–3, *35*, *36*, 39–40, *39*
interleaved modulation 347, *347*, *348*
internal model principle 326, 379–80
international regulations
 and grid requirements for PV 32–5
 see also Grid Codes
inverse discrete Fourier transform (IDFT) 49
inverse Park transform 65–8, *66*, *69*
inverter-resident detection 98
Inverters, Converters, and Controllers for use in Independent Power Systems (UL 1741) 32, 33, 39–40, 41
IPG series 25
Ireland 165
 frequency control requirements (WTS) *154*, 155
 frequency and voltage deviation (WTS) 151
 Grid Codes (GCs) *147*, 150
 power curtailment requirements (WTS) 154
 reactive power control (WTS) *157*, 158
 wind energy penetration 2
islanding
 definition 38
islanding detection 93–121
 active methods 98, 104–20, *120*, 121
  frequency drift 104–10
  grid impedance estimation 110–14
  PLL-based 114–19, *115*, *116–18*, *119*
  voltage drift 110
 external switched capacitor (ESC) detection 97
 grid-resident detection 96–7
 inverter-resident detection 98
 nondetection zone (NDZ) 94–6, *95*, 99, 100

passive methods 98–103, 121
 evaluation 103, *103*
 harmonic detection (HD) 99–103, *102*, *103*
 OUF-OUV detection 98–9, *103*
 phase jump detection (PJD) 99, *103*
 see also anti-islanding requirements
islanding detection algorithm 119, *119*

L-filter inverter 313, *314*
 mathematical model of 207–9, *207*
Laplace transform 54–5, 66, 73, 75, 254
LC trap filters 289, 290, 291, *291*
LCL filter 290, 291–304, 313, *314*, *see also* filters
LCL-filter inverter
 mathematical model of 209–10, *210*
least-mean-squares (LMS) algorithm 70, *70*
linear current control with separated modulation 315–35
local voltage control
 and Grid Codes (GCs) 165
lock range 58, 60
lock time 60
loop filter (LF) 53, 55
low-pass filter (LPF) 46–7, 66 *see also* LCL filter
low-voltage ride-through (LVRT) 239
Lyon transformation 195, 355, 357, 358

matrix converter topology 128
maximum active and reactive power set-point, estimation of 277–9
maximum power point tracking (MPPT) 28
medium-power converter (WTS) 131–2, *132*
micro-grid 228, 229, 231, *231*
mini central inverters 5, 26
Mitsubishi 29
modulation techniques 335–45
 interleaved 347, *347*, *348*
 multilevel 343–6, *343*, *344*, *345*, *346*
 random 336
 single-phase 338–40, *338*, *339*, *341*
 symmetric and asymmetric sampling 336
 three-phase 340, *342*
module-integrated inverters 5, 6

MOSFET technology 5
multicell (interleaved or cascaded) 133–4, *133*, *134*
multilevel modulations 343–6, *343*, *344*, *345*, *346*
multiphase systems
　active currents in 364–7, *365*
　instantaneous calculation of power currents in (FBD method) 367–8
　and *p-q* theory 371–2, *371*
multiple synchronous generator *127*, 128
multiple synchronous reference frames (MSRFs) 330, *330*
multistring inverters 5

National Electric Code (NEC) 32
NDZ (nondetection zone) 94–6, *95*, 99, 100
negative-sequence reactive power
　injection of in FPNSC 282–3, *283*
neutral point converters *see* NPC
neutral point multilevel converter 343, *343*, 344, 345–6, *346*
nondetection zone *see* NDZ
NORDEL 2
notch filter (NF) 248, *249*
NPC (neutral point clamped) converters 4, 21–5
　conergy NPC inverter 23–5, *23*, *24*
　half-bridge inverter 21–3 *21*, *22*, 25
　summary 25

open-loop phase transfer function 44, 217
OUF-OUV detection 98–9, *103*
over/under frequency (OUF) 98–9
over/under voltage (OUV) 98–9

P and *Q* limitation during faults and recovery 158, 160, 162
p-q theory 363, 369–71, 375–6
　and arbitrary multiphase systems 371–2, *371*
　modified 372–3
Park transform 59, 61, 62, 65–6, 80, 115, 172, 180, 181, 243, 246, 254
　inverse 65–8, *66*, *69*
passive damping 300, 301–4

passive islanding detection methods 98–103, 121
　evaluation 103, *103*
　harmonic detection (HD) 99–103, *102*, *103*
　OUF-OUV detection 98–9, *103*
　phase jump detection (PJD) 99, *103*
peak currents
　estimation of in each phase 274–7, *274*, *276*
permanent magnet machines 128
phase detector (PD) 52–3, 54, 55
　in-quadrature 59–63
phase jump detection (PJD) 99, *103*
phase-locked loop *see* PLL
phase-to-phase grid fault 173–4, *174*, 175–6, *175*, *176*, 177
photovoltaic *see* PV
PI controller
　tuning procedure 216–17
PI-based control 317–20, *317*, *318*, *319*, *320*, *321*, 328
PI-based voltage control design example 217–19, *218*, *219*
plant monitoring 28
PLL (phase-locked loop) 45, 51–80, 99
　based on adaptive filtering 68–80
　　enhanced PLL (EPLL) 70–2, *71*, 80
　　second-order adaptive filter 72–4, *73*
　　second-order generalized integrator (SOGI) 74–80, *76*, *79*
　based on in-quadrature signals 59–68, *60*
　　Hilbert transform 64–5, *65*
　　inverse Park transform 65–8, *66*
　　T/4 transport delay techniques 63–4, *64*
　basic equations for 53–4
　basic structure of 52–3, *52*
　block diagram elementary 53, *53*
　decoupled double synchronous reference frame (DDSRF-PLL) 186–95, 202
　and islanding detection 114–19, *115*, *116–18*, *119*
　key parameters of 58, 59
　linearized small signal model of 54–6, *55*
　and phase jump detection 99

PLL (phase-locked loop) (*Cont.*)
  pulling process of 59, *59*
  response 56–8, *57*
  synchronous reference frame (SRF-PLL) 180–6, *183*, *185*, 202
  and zero-crossing detection method 51–2
point of common coupling (PCC) 151, 176–7
polynomial regression analysis 49
positive-and negative-sequence control (PNSC) 260–2, 264–7, *265*, *266*
positive-sequence reactive power
  injection of in FPNSC 281–2, *281*
power currents
  instantaneous calculation of in multiphase systems 367–8
power curtailment (WTS) 153–4, *153*, *154*
power factor
  and grid requirements for PV 38
power oscillation damping (POD) 149, 166
power quality requirements
  PV 37–8
  WTS 313–14
powerless currents 368
PQ closed-loop control
  stationary frame VOC 224, *225*
  synchronous frame VOC 222, *223*
PQ open-loop control
  stationary frame VOC 222, *224*
  synchronous frame VOC 221, *222*
predictive digital filtering algorithms 52
proportional resonant (PR) controller 76, 251, 253–5, *255*
Public Distribution Voltage Quality (EN 50160) 34–5, *35*
Public Utilities Regulatory Policy Act (1978) 32
pull-in range 58
pull-in time 60
pull-out range 58
pulse-width modulation (PWM) 336–7, *337*
PV (photovoltaic) inverter structures 5–29
  categorization 5
  control structures 27–8, *27*
  functions 5
  future trends 28–9
  grid requirements for *see* grid requirements
  H-bridge 6–21, *7*
    full-bridge (FB) inverter 7–11, *7*, *9*, *10*
    full-bridge inverter with DC bypass (FB-DCBP) (Ingeteam) 17–19, *17*, *18*, 21
    full-bridge zero voltage rectifier (FB-ZVR) 19–20, *19*, *20*, 21
    H5 inverter (SMA) 11–12, *11*, *12*, 14, 21
    HERIC inverter (Sunways) 13–14, *13*, *14*, 15, 20, 21
    REFU inverter 15–17, *15*, *16*, 21
    summary 21
  H-bridge based boosting PV inverters 25, *26*
  NPC (neutral point clamped) 4, 21–5
    conergy NPC inverter 23–5, *23*, *24*
    half-bridge inverter 21–3 *21*, *22*, 25
    summary 25
    three-phase 26–7, 31
PV panels
  capitance to earth 6
  cost of 4
  lifetime of 6
  structure 6
PV power
  development 3–4
  worldwide cumulative and annual installed 3, *3*

Q requirements 155–8
quadrature signal generator (QSG) 45, 60, 62, 62–3, *62*, 63, *63*, 74, 197
  SOGI- 77–8, *78*, 79, 80, 81, 82, 84, 88, 89, 197, 200

reactive current ellipse 273–4, *273*
reactive current injection (RCI) 158, 160, 162–3, *162*
reactive power control in normal operation 155–8
reactive power set-point
  estimation of the maximum 277–9

reactive power variation (RPV) 108, 110, 120
reconnection after trip 36, *36*
recursive discrete Fourier transform (RDFT) 44, 50–1, *51*
reduced power back-to-back converter 124, 126–7, *126*
reduced power converter 141–2, *142*
reference currents, determining 256, 258–64
REFU inverter 15–17, *15*, *16*
RefuSol® 17
resonance frequency
  and filters 295, 296, 297, 298–9, *298*
resonance problems, damping solutions to *see* damping
resonant control/controller 246, 326–9, *326*, *327*, *328*, 329, 381–84
  proportional (PR) 76, 251, 253–5, *255*
  for unbalanced current injection 251–6, *253*, *255*
resuming active power 158, 160, 163
ripple analysis 294
rotating frame 208, *208*

Safety (VDE 0126-1-1) 4, 33–4, *35*, 36, *36*, 37, *37*, 40–1, *41*, 110
Sandia frequency shift (SFS) 1–7, 106, *120*
Sandia voltage shift (SVS) 110, *120*
SCADA (supervisory control and data acquisition) system 97
Scherbius drive 138
second-order adaptive filter 72–4, *73*
second-order generalized integrator *see* SOGI
selected harmonic elimination (SHE) 132, *133*
Semikron 29
short term interruptions (STI) 160
SiC diodes 28
SiC MosFet 28
Siemens 146
simultaneous active and reactive power delivery 278
simultaneous-and negative-sequence reactive power
  injection of in FPNSC 283–4, *284*

sine/cosine basic functions 46–7
single cell 129–33, *130*
  current-stiff (CSC) 129, *130*, 131–2, 135
  high-power converter 132–3, *133*
  medium-power converter 131–2, *132*
  voltage-stiff (VSC) 129–30, *130*
single-phase modulation 338–40, *338*, *339*
single-phase power definitions
  in the time domain 363–4
single-phase systems, grid synchronization *see* grid synchronization (single- phase systems)
sinusoidal integrator 75
sinusoidal signals
  and second-order generalized integrator 74–7
slip control 138
slip-mode frequency shift (SMFS) 105–6, *105*, 108, 120
SMA 11, 26, 29
smart micro-grid (SMG) 93
SOGI (second-order generalized integrator) 74–8
SOGI-FLL 80–9, *82*, 86
  double (DSOGI-FLL) 195–202
  equations for 82–5
  and FLL gain normalization 86–7, *86*,k *88*
  response of *83*
SOGI-PLL 78–80, *79*
  local dynamics 84–5
SOGI-QSG 77–8, *78*, *79*, 80, 81, *82*, 84, 88, 89, 197, 200
space vector transformations (of three-phase systems) 353–60
  β0 on stationary reference frame 357–9, *359* [add symbol]
  $dq0$ on the synchronous reference frame 359–60, *359*
  symmetrical components in the frequency domain 353–4, *354*
  symmetrical components in the time domain 355–7, *356*
Spain
  frequency and voltage deviation (WTS) 151

Spain (*Cont.*)
  Grid Codes (WTS) *147*, 149, 165–6
  grid disturbances 160–3, *161*
  power curtailment requirements (WTS) 153
  reactive power control (WTS) *156*, 157
  wind energy penetration 2
Spanish Wind Association (AEE) 149
squirrel cage induction generator 126–7, *126*, 136–7, *137*
stand-alone system 228–31, *229*, *230*
  main technical difficulties with 228
  reasons for 228
STATCOM
  performance of reference current determining strategies 264–7, *265*, *266*
stationary frame VOC *see* voltage oriented control (VOC)
stationary frames 208, *208*
  and harmonic compensation 330, 331–35, *334*, *335*
stationary reference frame
string inverters 5
SunnyBoy 4000/5000 TL 12
Sunways 13
switching frequency 4
symmetrical components
  in the frequency domain 353–5, *354*
  in the time domain 355–7, *356*
symmetrical optimum, method of 216
synchronous control systems 52
synchronous frame VOC *see* voltage oriented control (VOC)
synchronous frames
  and harmonic compensation 330, 331, *331*, *332*
synchronous generators 146
  multiple *127*, 128
synchronous generator control 137–8, *137*
synchronous reference frame

$dq0$ on 359–60, *359*
  multiple (MSRF) 330, *330*
synchronous reference frame PLL (SRF-PLL) 180–6, *183*, *185*, 202

T/4 transport delay technique 63–4, *64*
TC-82 Committee on Solar Photovoltaic Energy Systems 33
Testing Procedure of Islanding Prevention Measures (IEC 62116) 33, 40, 41
three-level voltage source converter 132, *133*
three-phase currents
  estimation of maximum current in each phase 274–7, *274*, *276*
  instantaneous value of 272–4, *273*
three-phase modulation 340, *342*
three-phase power converters, grid synchronization *see* grid synchronization (three-phase power converters)
three-phase PV inverters 26–7, 31, 40, 41
three-phase systems
  space vector transformations of 353–60
three-phase voltage vector
  under grid faults 169–70, *170*, 171–80
time domain
  simple components in 355–7, *356*
  single-phase power definitions 363–4
time-domain analysis (ripple evaluation) 294
time-domain detection methods 44–5 *see also* FLL; PLL
total harmonic distortion (THD) 314
transformerless PV inverters 31, 37
Transmission System Operators (TSOs) 145, 147
TripleLynx series 23
tuned LC filter 289, 290–1, *291*

UCTE 2
UK
  frequency control requirements (WTS) 154–5

frequency and voltage deviation (WTS) 151
Grid Codes (GCs) *147*, 149–50
reactive power control (WTS) 157, *157*
UL 1741 *Inverters, Converters, and Controllers for use in Independent Power Systems* 32, 33, 39–40, 41
unbalanced current injection *see* current injection
Underwriters Laboratories Inc. 32, 41
unipolar (UP) modulation *9*, 89
United States 1
Grid Codes (GCs) *147*, 150
reactive power control (WT system) 158
wind energy penetration 2
unsynchronized synchronous generator 127, *127*

variation currents 368
VDE 0126-1-1: *Safety* 33–4, 35, 36, *36*, 37, *37*, 40–1, *41*, 110
Vincotech 29
virtual-flux-based control 224–6, *226*
volt-ampere imaginary (vai) 369
volt-ampere reactive (var) 369
voltage-controlled oscillator (VCO) 53, 54, 55, 61, 70
voltage deviation
and grid requirements (for PV) 35–6, *35*
and grid requirements (for WTS) 151–2, *152*
voltage drift methods 110
Sandia voltage shift (SVS) 110, *120*
voltage monitoring 98–9
voltage oriented control (VOC) 219–26, *220*, 319
stationary frame
PQ closed-loop control 224, *225*
PQ open-loop control 222, *224*
synchronous frame
PQ closed-loop control 222, *223*
PQ open-loop control 221, *221*
virtual-flux-based approach 234–6, *236*
voltage quality
and EN (50160) 34–5, *35*

voltage ride-through (VRT) 158, *159*, 160, *161*, 162
voltage sags 177–80, *178*, *179*, 238
propagation of 179–80, *180*, *181*, *182*
voltage-stiff (VSC) 129–30, *130*, 131, 140
Voterra series expansion 305, 306–7, *307*

WECC LVRT standard 164
Western Electricity Coordinating Council (WECC) 150
wind energy penetration 1–3
wind farms 146, 155
embedded and nonembedded systems 150
wind forecasting 123
Wind Generation Task Force (WGTF) 150
wind power
development 1–3
worldwide cumulative and annual installed 1, *2*
WTS (wind turbine system)
grid converter control *see* grid converter control (for WTS)
grid requirements of *see* grid requirements (for WTS)
power configurations 124–8, *124*
power conversion structures *125*
WTS control 135–42, *136*
generator-side control 136–9
doubly fed induction generator (DFIG) 124, 126, *126*, 138–9, *139*, 141, 146
squirrel cage induction generator 126–7, *126*, 136–7, *137*
synchronous 137–8, *137*
grid control 140–2, *140*, *141*, *142*
WTS grid converter structures 123–35
demands of 128
direct and indirect conversion 128
multicell (interleaved or cascaded) 133–4, *133*, *134*
single cell 129–33, *130*
current-stiff (CSC) 129, *129*, 131–2, 135

high-power converter 132–3, *133*
medium-power converter 131–2, *132*
voltage-stiff (VSC) 129–30, *130*

zero voltage rectifier (ZVR), full bridge 19–20, *19*, *20*, 21
zero-cross detection methods 51–2, 99
zero-sequence voltage component 171